AF794803

Smart Chips for Smart Devices

Scrivener Publishing
100 Cummings Center, Suite 541J
Beverly, MA 01915-6106

Publishers at Scrivener
Martin Scrivener (martin@scrivenerpublishing.com)
Phillip Carmical (pcarmical@scrivenerpublishing.com)

Smart Chips for Smart Devices

VLSI Design for Next-Generation IoT Solutions

Edited by

Abhishek Kumar

and

Suman Lata Tripathi

Scrivener Publishing

WILEY

This edition first published 2026 by John Wiley & Sons, Inc., 111 River Street, Hoboken, NJ 07030, USA and Scrivener Publishing LLC, 100 Cummings Center, Suite 541J, Beverly, MA 01915, USA
© 2026 Scrivener Publishing LLC
For more information about Scrivener publications please visit www.scrivenerpublishing.com.

All rights reserved. No part of this publication may be reproduced, stored in a retrieval system, or transmitted, in any form or by any means, electronic, mechanical, photocopying, recording, or otherwise, except as permitted by law. Advice on how to obtain permission to reuse material from this title is available at http://www.wiley.com/go/permissions.

Wiley Global Headquarters
111 River Street, Hoboken, NJ 07030, USA

For details of our global editorial offices, customer services, and more information about Wiley products visit us at www.wiley.com.

The manufacturer's authorized representative according to the EU General Product Safety Regulation is Wiley-VCH GmbH, Boschstr. 12, 69469 Weinheim, Germany, e-mail: Product_Safety@wiley.com.

Limit of Liability/Disclaimer of Warranty
While the publisher and authors have used their best efforts in preparing this work, they make no representations or warranties with respect to the accuracy or completeness of the contents of this work and specifically disclaim all warranties, including without limitation any implied warranties of merchantability or fitness for a particular purpose. No warranty may be created or extended by sales representatives, written sales materials, or promotional statements for this work. The fact that an organization, website, or product is referred to in this work as a citation and/or potential source of further information does not mean that the publisher and authors endorse the information or services the organization, website, or product may provide or recommendations it may make. This work is sold with the understanding that the publisher is not engaged in rendering professional services. The advice and strategies contained herein may not be suitable for your situation. You should consult with a specialist where appropriate. Neither the publisher nor authors shall be liable for any loss of profit or any other commercial damages, including but not limited to special, incidental, consequential, or other damages. Further, readers should be aware that websites listed in this work may have changed or disappeared between when this work was written and when it is read.

Library of Congress Cataloging-in-Publication Data

ISBN 9781394335800

Front cover images courtesy of Wikimedia Commons
Chip image: Generated with AI using Adobe Firefly
Cover design by Russell Richardson

Set in size of 11pt and Minion Pro by Manila Typesetting Company, Makati, Philippines

Printed in the USA

10 9 8 7 6 5 4 3 2 1

Contents

Preface

The IoT revolution has created a demand for ever-more-sophisticated, ever-more-compact devices, making VLSI design an extremely significant concern. *Smart Chips for Smart Devices* focuses on the relationship of technology, giving researchers, engineers, and designers the know-how and resources they need to be successful in their fields and drive innovation. Also, the demand for low power, efficient, portable and reliable semiconductor chips force the research towards developing new designs and techniques with addition of advanced applications. The book meets the demands of professionals looking to advance their knowledge and contribute to the development of the next generation of smart, connected devices by acting as both a scholarly resource and a practical guide.

The book *Smart Chips for Smart Devices: VLSI Design for Next-Generation IoT Solutions* explores the state-of-the-art field of VLSI (Very-Large-Scale Integration) design, emphasizing on how it is used in the Internet of Things (IoT). This book is designed for engineers, designers, and researchers who want to learn and use VLSI principles to make more creative, scalable, and effective Internet of Things solutions. It covers both theoretical and practical elements.

The book takes a strategy that combines VLSI aspect in IoT application. The book covers a wide range of subjects that are crucial for creating integrated circuits that are dependable, powerful, and high-performing. It explores advanced circuit design, system-on-chip (SoC) architectures, and basic VLSI design ideas. Additionally, it addresses important aspects of IoT applications, like sensor integration, low power consumption, and secure communication protocols. The book also contains case studies and practical applications that show how theoretical ideas are put to use in real-world settings to address the unique requirements of IoT systems.

Authors
Abhishek Kumar
Suman Lata Tripathi

1

Integration of Nanotechnology and Machine Learning in Healthcare: Fundamentals, Applications, and Future Scope

Bindu S.[1]*, Prashanth M. K.[2] and Mohan Kishore D.[3]

[1]*Dept. of Electronics & Communication Engineering, BNM Institute of Technology, Visvesvaraya Technological University, Bengaluru, Karnataka, India*
[2]*Dept. of Chemistry, Engineering, BNM Institute of Technology, Visvesvaraya Technological University, Bengaluru, Karnataka, India*
[3]*Swami Vivekananda Yoga Anusandhana Samsthana (SVYASA) Deemed to be University, Bengaluru, Karnataka, India*

Abstract

Nanotechnology and machine learning (ML) are two transformative fields that are driving significant advancements in healthcare. Nanotechnology enables the precise manipulation of matter at the nanoscale, facilitating innovative solutions for diagnostics, drug delivery, and therapeutics. Machine learning, on the other hand, empowers systems to analyse complex datasets, optimize processes, and predict outcomes, making it a valuable tool for managing the high-dimensional data generated by nanotechnology applications. This interaction has led to groundbreaking innovations, including personalized treatment approaches, real-time monitoring devices, and predictive models for nanomaterial behaviour. Fundamental ML techniques, such as supervised learning for predicting nanoparticle interactions, unsupervised learning for clustering nanoscale properties, and reinforcement learning for optimizing synthesis processes, have been instrumental in advancing nanotechnology. Applications in healthcare include ML-optimized smart drug delivery systems, nano biosensors for early disease detection, and diagnostics tailored to individual patient profiles. Case studies illustrate the integration of AI-driven cancer therapies and nanomaterial optimization for enhanced treatment efficacy, while

**Corresponding author*: bindus@bnmit.in

Abhishek Kumar and Suman Lata Tripathi (eds.) Smart Chips for Smart Devices: VLSI Design for Next-Generation IoT Solutions, (1–30) © 2026 Scrivener Publishing LLC

the fusion of ML and nanotechnology offers immense potential. Future trends point to the continued evolution of personalized medicine, nano biosensor integration with wearable technologies, and AI-driven design of advanced nanomaterials. Together, these innovations promise a new era of precision healthcare, improving outcomes and transforming clinical practices.

Keywords: Nanotechnology in healthcare, machine learning in healthcare, healthcare innovation, smart nanomaterials, nanomedicine, AI-driven healthcare

1.1 Introduction

Nanotechnology and machine learning are the two transformative fields that have reshaped numerous industries, particularly healthcare. Nanotechnology deals with manipulating materials at the atomic or molecular scale and has enabled groundbreaking innovations in diagnostics, treatment, and personalized medicine. On the other hand, Machine Learning (ML) offers computational models that can analyse vast datasets and make predictions with less human involvement [1]. The intersection of these fields provides unique opportunities to enhance healthcare outcomes. ML algorithms can optimize the design and application of nanomaterials, enabling targeted drug delivery systems, efficient diagnostics,

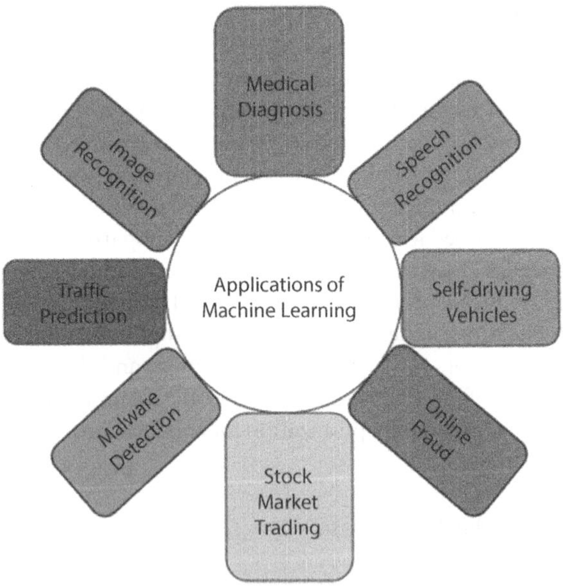

Figure 1.1 Machine learning used in different areas.

and reduced side effects [2]. This combined effect also nurtures the rapid development of novel solutions for global health challenges, from cancer treatment to infectious disease management, etc., as shown in Figure 1.1.

1.1.1 Overview of Machine Learning (ML) and Its Impact on Modern Technology

Machine learning (ML) involves developing algorithms to learn from datasets and make decisions. Artificial Intelligence is beneficial in analysing complex datasets which are generated in nanotechnology applications [2]. Machine Learning could be a supervised learning, where the models are trained on labelled dataset to predict outcomes, where the input-output relationship is explicitly defined, such as distinguishing between malignant and normal cells [3].

Secondly, it could be unsupervised learning where the algorithms identify the patterns in unlabelled datasets or identify inherent structures or patterns such as grouping nanoparticles according to their properties based on shape, size, etc. In reinforcement learning the algorithms learn the best action by using trial-and-error methods such as optimizing nanomaterial synthesis. The important characteristics of Machine Learning are in feature engineering which involves selecting and transforming data variables to improve model performance. In nanotechnology, features might include nanoparticle size, surface charge, or hydrophobicity. The other important aspect is training, the process of teaching an ML algorithm to make predictions based on a dataset. Validation ensures the model generalizes well by testing it on unseen data. In healthcare applications, this ensures models trained on experimental data can reliably predict outcomes in clinical settings.

Deep learning is a multi-layered neural network used to analyse complex and unstructured data. For example, analysing nanoscale imaging data to detect structural abnormalities, and predicting molecular interactions between nanoparticles and biological systems. Machine learning is best suited to be integrated in nanotechnology as it is reliable in the field of high-dimensional and complex datasets. For example, ML can process data from molecular simulations, nanoscale imaging, and high-throughput screening. It can generate models and optimize algorithms to accelerate the development of nanostructures with desired properties. It can predict the behaviour of nanomaterials' interactions with biological systems and helps to ensure safety and efficacy in medical applications. It has the ability to automate, analysis and provide actionable insights that makes it an invaluable tool for advancing nanotechnology in healthcare.

1.2 Nanotechnology and Its Impact on Healthcare

The human body, being a system dealing with molecular interactions, requires a technology which can understand and repair it. Such complex interactions can be possible by nanotechnology. Nanotechnology deals with manipulation at the near-atomic level to produce new materials for making devices [4]. This technology has made advancements in medicine and other essential needs including material, manufacturing methods, etc. Nanotechnology in healthcare is one of its most transformative domains. By working at smaller dimensions comparable to biological molecules like proteins and DNA, nanotechnology enables unprecedented precision and functionality. Nanotechnology can address problems that traditional technologies cannot, offering precision, targeted therapies, and minimizing side effects. This capability opens up possibilities for groundbreaking medical treatments, advanced diagnostic tools, and highly effective drug delivery systems [5]. In drug delivery, for instance, nanomaterials can ensure that medications are delivered to specific cells or tissues, optimizing therapeutic outcomes. Conventional methods in drug delivery have drawbacks which can result in suboptimal therapeutic effects and side effects. Nanomaterials, such as liposomes, dendrimers, and polymeric nanoparticles, address these issues by encapsulating drugs and then delivery to target cells or tissues [6]. For example, nanoparticles could be configured to identify and attach to malignant cells and release their therapeutic payload precisely at the tumour site while sparing healthy cells. This method has enhanced the efficacy of treatment by significantly reducing adverse side effects. Nanotechnology also plays a key role in diagnostic imaging, creating highly sensitive sensors and biomarkers that help detect diseases in their earliest stages.

In diagnostics, nanotechnology enables the design of precision sensors and biomarkers capable of detecting diseases at their initial stages [7]. Nanomaterials are used to improve the sensitivity of diagnostic imaging techniques including MRI, CT scans, and optical imaging, etc. These advancements allow for earlier detection of conditions such as cancer and other disorders, improving the likelihood of successful treatment. Furthermore, point-of-care diagnostic tools leveraging nanotechnology are making rapid and accurate testing more accessible, even in resource-limited settings. Nanotechnology is promising in the area of regenerative medicine and tissue engineering. Nanostructured materials are being used to create frameworks that mimic the cells and tissues. These scaffolds can be customized to regenerate specific tissues, such as bone, cartilage, or skin. Additionally, nanoparticles are being investigated for their ability to accelerating the healing process and improving outcomes in conditions such as chronic wounds and organ failure [8].

The integration of nanotechnology into personalized medicine is another promising area. Nanoscale devices and materials can analyse individual genetic and molecular profiles, enabling the design of tailored treatments. Theragnostic, a combination of therapy and diagnostics, leverages nanotechnology to deliver personalized treatment while simultaneously monitoring therapeutic efficacy. For example, nanoparticles used in drug release in response to specific biomarkers and provide real-time feedback on treatment progress through imaging techniques. The potential of nanotechnology in healthcare continues to expand, with ongoing research focusing on areas like artificial organs, gene therapy, and nanoscale robotic systems for precision surgeries. However, scalability and safety need to be addressed to fully realize its benefits. Ethical considerations and the cost of developing nanotechnology-based solutions also require careful attention to ensure equitable access to these innovations. Nanotechnology is shaping the future of healthcare by enabling solutions that were previously unattainable by offering opportunities for improving disease prevention, diagnosis, and treatment, promising a healthier future for all.

1.3 The Role of Machine Learning in Advancing Nanotechnology

Machine Learning (ML) is rapidly being used as an essential tool in the field of nanotechnology, especially in healthcare. The complexity of nanoscale systems and the vast amounts of data generated during nanomaterial research make traditional analytical techniques challenging. ML offers powerful methods for analyzing large datasets, modeling complex systems, and predicting material properties, which is essential for the design of more efficient and effective nanomaterials [9]. In nanomedicine, ML algorithms can assist in designing nanostructures with optimized properties, identifying suitable targets for drug delivery, and improving diagnostic accuracy. Furthermore, ML can be used to predict how the biological system interacts with nanomaterials for ensuring safety and efficacy in medical applications.

1.3.1 Objectives of Integrating ML in Healthcare through Nanotechnology

The integration of ML in healthcare *via* nanotechnology aims to revolutionize medical diagnostics, treatment methods, and personalized medicine as shown in Figure 1.2. Some primary objectives are described below.

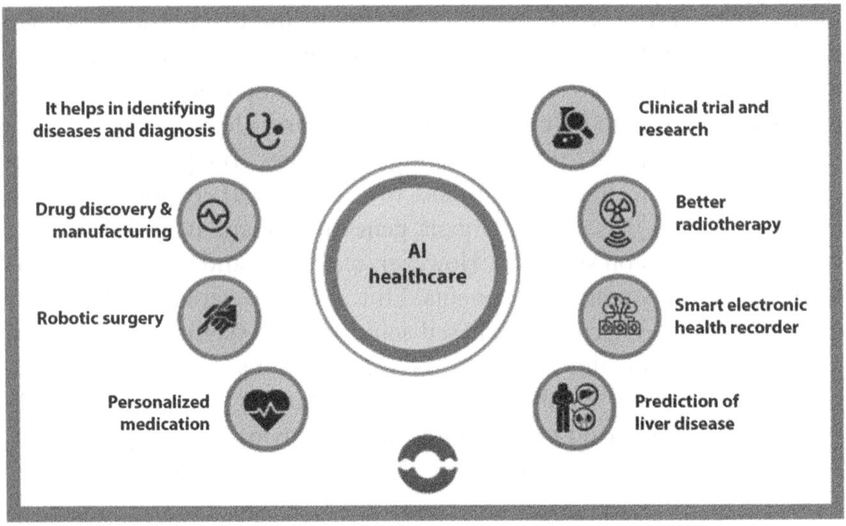

Figure 1.2 The role of AIML in healthcare [10].

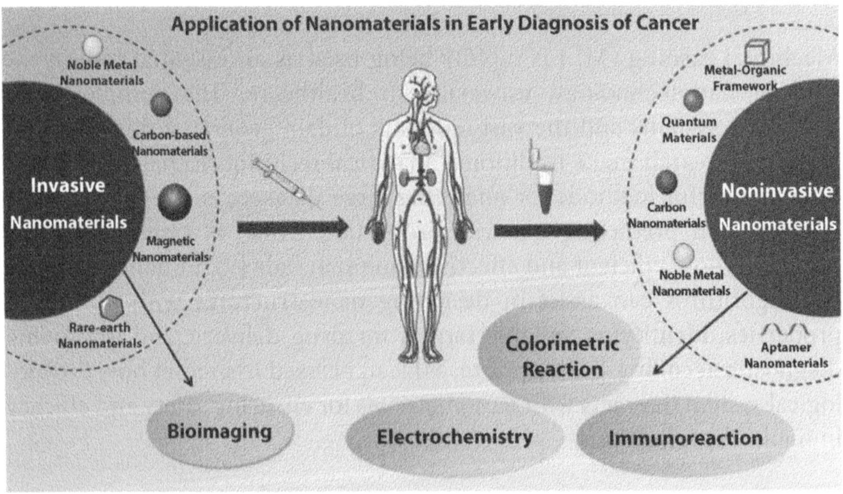

Figure 1.3 Machine learning technique used in early diagnosis using nanomaterials [12].

- ***Enhanced Diagnostics***: Machine learning can analyse data from nanotechnology-based sensors and diagnostic tools to detect diseases with greater accuracy and at earlier stages [11], potentially leading to better patient outcomes as shown in Figure 1.3.
- ***Targeted Drug Delivery***: ML can optimize nanomaterials and ensure precise medication in targeted drug delivery, where medications reach the right cells or tissues in the body, which could reduce side effects and improve treatment efficacy [13] as shown in Figures 1.4(a) and 1.4(b).
- ***Personalized Medicine***: The integration of ML with nanotechnology allows for the creation of personalized medication to an individual by enhancing therapeutic effectiveness [15].
- ***Safety and Risk Prediction***: ML models can predict potential risks or side effects of nanomaterials in the human body, ensuring that new technologies are safe before they are used in clinical applications.
- ***Accelerating Research and Development***: ML can speed up the discovery of new nanomaterials and their applications in healthcare by identifying promising candidates for drug delivery, diagnostics, and therapies from large datasets.

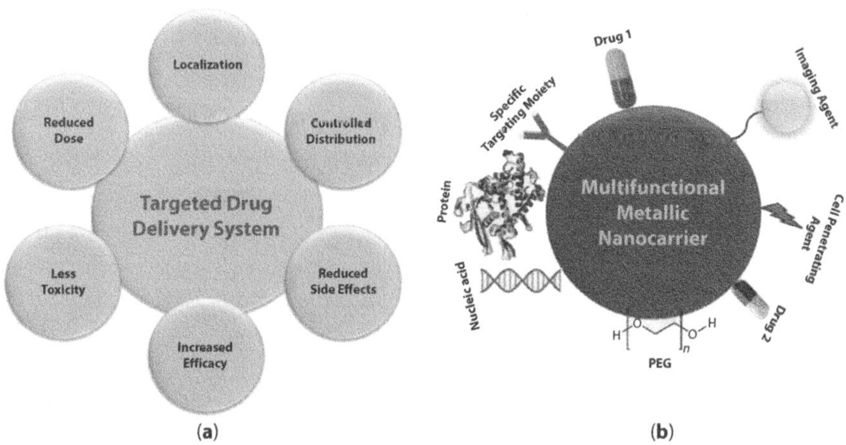

Figure 1.4 (a and b) Targeted delivery and metabolism of liposome encapsulated drug [14].

The intersection of machine learning and nanotechnology for advancing healthcare, offering the possibility to provide faster, accurate, and more personalized treatments, thus improving the overall quality of medical care is promising. With the increase in collection of healthcare data the role of ML will increase to provide accurate diagnoses and build efficient healthcare systems.

1.4 Fundamentals of Nanotechnology in Healthcare

Nanotechnology deals with materials and structures having structural size up to 100 nanometers as shown in Figure 1.5. At the nanoscale the materials exhibit beneficial properties in their physical, chemical, and other biological aspects that is unique in macroscopic counterparts.

These properties are pivotal in addressing healthcare challenges, particularly in precision medicine. The convergence of nanotechnology and healthcare is providing unprecedented opportunities for advanced diagnostics. Several types of nanomaterials are commonly used in healthcare applications, each with distinct properties suited for different medical purposes. One of the critical aspects of nanomaterial design is surface modification. The surface properties of nanomaterials significantly influence their interactions with biological tissues, immune cells, and other biomolecules. Surface modification can enhance biocompatibility and ensure targeting specific cells or tissues. This can be achieved by attaching targeting ligands, antibodies, peptides, or small molecules to the nanoparticle surface.

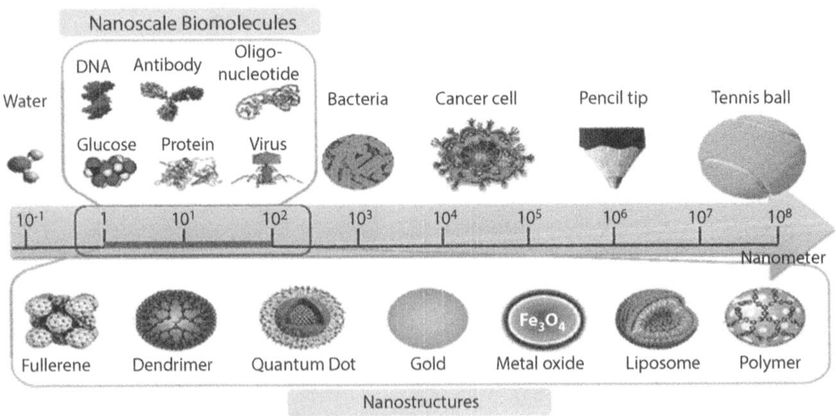

Figure 1.5 Scale of nanoparticles with some examples [16].

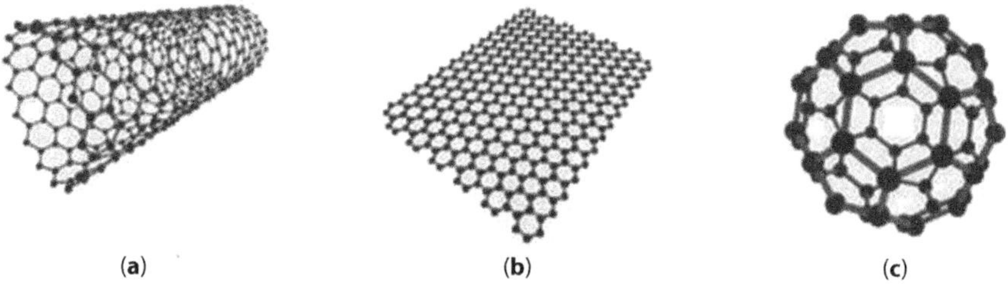

Figure 1.6 Atom structure of (a) carbon nanotubes, (b) graphene nanoplatelets, and (c) fullerene-C60 particles [17].

1.4.1 Key Nanomaterials in Healthcare

Carbon-based Nanomaterials (CBNs) are composed primarily of carbon atoms arranged in unique structures at the nanoscale as shown in Figure 1.6. These materials exhibit remarkable mechanical, electrical, thermal, and optical properties that are best suited for a range of applications in science, engineering, and medicine. Carbon nanotubes and graphene are used for drug delivery and biosensing.

1.4.1.1 Nanoparticles

Nanoparticles have sizes ranging up to100 nm and can be composed of materials such as lipids, polymers and ceramics, etc. Nanoparticles can be engineered for drug delivery, where they encapsulate drugs and release them in targeted areas of the body, minimizing side effects and improving therapeutic efficacy. Metallic nanoparticles such as gold and silver are used in imaging and in cancer therapy. Polymeric nanoparticles which are biodegradable polymers are used for controlled drug release. Lipid-based nanoparticles are used in vaccine delivery for example in COVID-19 mRNA vaccines.

1.4.1.2 Nanotubes

Nanotubes are tubular-like structures that are made up of carbon or other materials, typically ranging from 1 nm to several micrometres. Carbon nanotubes are particularly used for drug delivery, biosensing, and cancer therapy, for their excellent mechanical properties, large surface area, and ability to penetrate cells.

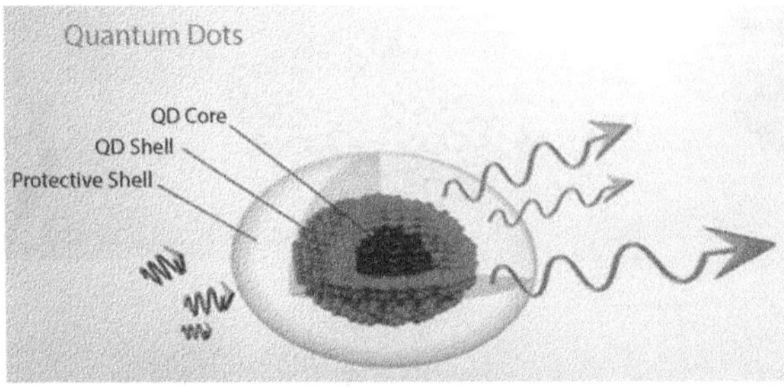

Figure 1.7 Structure of quantum dot [18].

1.4.1.3 Nanostructures

These include nanorods, nanoplates, and nanowires, which have applications in biosensing, imaging, and catalysis. The aspect ratio and geometry of nanostructures can influence their interaction with biological systems.

1.4.1.4 Quantum Dots

Quantum dots are semiconductor nanoparticles that possess unique optical properties due to quantum mechanical effects. Quantum dots are widely used in imaging, as their fluorescence properties allow for highly sensitive detection of biomolecules in medical diagnostics. Figure 1.7 shows the structure of quantum dots.

1.4.1.5 Dendrimers

Dendrimers are highly branched, tree-like structures that can be designed for specific tasks. Their surface can be modified to attach drugs, genes, or imaging agents, making them valuable for targeted drug delivery and gene therapy.

1.4.2 Machine Learning: Overview and Key Concepts

Machine learning involves developing algorithms so that the systems can learn from data and take decisions. The adaptability of ML algorithms is particularly beneficial in analyzing the complex datasets generated in nanotechnology applications [19]. Different types of machine learning are as follows:

- ***Supervised Learning***: A method based on trained the labelled datasets, where the input and output dependencies is explicitly defined. Algorithms trained on labeled data to predict outcomes (e.g., classification of cancerous vs. non-cancerous cells). Predicting the behavior of nanoparticles in drug delivery systems by training on datasets of nanoparticle compositions and their biological outcomes.
- ***Unsupervised Learning***: a method where algorithms identify patterns in unlabelled data such as in clustering nanoparticles based on properties such as size, shape, or functionalization for various medical applications.
- ***Reinforcement Learning***: a method where algorithms learn by trial-and-error method. In reinforcement learning the training is done by agents to make decisions interacting with environment or by receiving feedback in the form of rewards or penalties, for example, optimizing the synthesis processes of nanoparticles to achieve desired properties.

The Vital Concepts in Machine Learning include the Feature Engineering that involves choosing and transforming data variables to improve model performance. In nanotechnology, features might include nanoparticle size, surface charge, or hydrophobicity. The second key concept is Model Training and Validation. The process of teaching an ML algorithm to make predictions based on a dataset is training and validation, ensuring the model generalizes well by testing it on unseen data. In healthcare applications, this ensures models trained on experimental data can reliably predict outcomes in clinical settings. Deep learning is a multi-layered neural network used to analyse complex and unstructured data. For example, analysing nanoscale imaging data to detect structural abnormalities and predicting molecular interactions between nanoparticles and biological systems. Regarding relevance to analyzing complex nanotechnological data, machine learning is particularly suited to nanotechnology due to the field's reliance on high-dimensional, complex datasets [20]. For example, ML can process data from molecular simulations, nanoscale imaging, and high-throughput screening. The generative models and optimization algorithms accelerate the development of nanostructures with desired properties. Predicting how nanomaterials will interact with biological systems helps ensure safety and efficacy in medical applications. Machine learning's ability to automate analysis and provide actionable insights makes it an invaluable tool for advancing nanotechnology in healthcare.

1.5 Intersection of Machine Learning and Nanotechnology

The fusion of machine learning and nanotechnology is driven by the need to handle and interpret vast and complex datasets. While nanotechnology provides powerful tools for manipulating matter at the microscopic level, machine learning enhances the ability to predict and optimize how these tools interact with biological systems [21]. Nanotechnology research generates a plethora of experimental and simulation data, from molecular dynamics simulations to nanoscale imaging outputs. Machine learning enables researchers to extract actionable insights and accelerate innovation. The synergy between these two fields is driving numerous innovations across healthcare.

The combination of nanotechnology's precision and machine learning's ability to process and interpret vast datasets creates a feedback loop that continuously improves medical devices, treatments, and diagnostics. The result is a new wave of healthcare solutions that are smarter, more effective, and highly personalized. The combination of machine learning (ML) and nanotechnology offers groundbreaking opportunities in research and development, particularly in healthcare. The integration is driven by ML's ability to process complex datasets, optimize processes, and predict outcomes, which aligns with the experimental and computational challenges in nanotechnology.

1.5.1 ML Techniques in Design of Nanotechnology Complementing Each Other

- ***Optimizing Nanomaterials and Nanodevices***: The design of nanomaterials, such as nanoparticles or nanostructures, requires a deep understanding of how different parameters—size, shape, charge, and surface properties—affect their performance in biological environments. ML algorithms can analyse large datasets from experiments to uncover patterns and optimize the design of these nanomaterials for specific medical applications.
- ***Personalized Treatment***: ML models can analyse patient data, including genetic, proteomic, and metabolic information, to predict how nanoparticles will behave in individual patients. This allows for the customization of drug delivery

systems or diagnostic tools based on each patient's unique biological profile, leading to more effective and personalized therapies.

- ***Real-Time Monitoring and Control***: Nanotechnology-enabled devices, such as sensors or implants, can collect real-time data from patients. ML models can analyse this data and make predictions or adjustments in real time, enabling continuous monitoring and adaptive treatment without human intervention.
- ***Accelerating Drug Discovery***: ML can help analyse the large amounts of data generated in drug discovery and development, including data from nano-enabled screening technologies. ML models can predict which nanoformulations will deliver drugs most effectively to target cells, streamlining the discovery of novel drug delivery systems.

1.5.2 Predictive Modeling for Nanomaterial Behavior

The design and synthesis of nanomaterials involve numerous variables that include size, shape, composition, and surface properties. These factors significantly influence the performance of nanomaterials in medical applications like drug delivery, imaging, and diagnostics [22]. ML plays a critical role in optimizing these designs, as described below.

1.5.2.1 Predictive Modeling

ML algorithms predict how specific design parameters will affect a nanomaterial's behavior in biological systems, for example, using decision trees or support vector machines to predict the stability of nanoparticles in different pH levels for drug delivery. Understanding how nanomaterials interact with biological systems is critical for healthcare applications. Machine learning enables predictive modeling for the following:

1.5.2.2 Nanoparticle-Cell Interactions

Predicting cellular uptake of nanoparticles based on surface charge, size, and functionalization, for example, neural networks that classify nanoparticles likely to penetrate the blood-brain barrier for neurological therapies.

1.5.2.3 Nanotoxicology

Ensuring the safety of nanomaterials by predicting their potential toxic effects, for example, decision tree models trained on datasets of nanoparticle toxicity in various cell types.

1.5.2.4 Drug Release Dynamics

Modeling how nanoparticles release drugs under different physiological conditions, for example, reinforcement learning to optimize release profiles in response to specific biomarker levels.

1.6 Generative Modeling for Nanomaterial Behavior

Algorithms like Generative Adversarial Networks (GANs) can create entirely new nanomaterial configurations based on desired properties. Designing nanoparticles with optimized hydrophobic and hydrophilic regions for targeted drug delivery. Accelerating Discovery: traditional nanomaterial synthesis involves trial and error. ML models streamline this process by identifying promising candidates faster. Bayesian optimization techniques for narrowing down the best polymeric nanoparticles for vaccine delivery.

a) Nanotechnology experiments generate massive datasets from simulations, imaging techniques, and biosensor outputs. Machine learning helps analyze and interpret this data efficiently.
b) High-resolution imaging techniques such as atomic force microscopy (AFM) and electron microscopy (EM) produce complex images.
c) ML algorithms, particularly convolutional neural networks (CNNs), automate image analysis to detect structural defects, measure particle sizes, and classify material types.
d) Nanobiosensors produce real-time data streams for detecting biomarkers.
e) ML enhances signal processing to distinguish noise from meaningful patterns, improving sensor sensitivity and accuracy.

f) Nanotechnology relies on experiments involving hundreds of material configurations.
g) ML models help sift through this data to identify correlations between material properties and biological outcomes.

1.7 Applications in Healthcare

The healthcare sector is witnessing rapid advancements fuelled by innovative technologies that aim to improve patient outcomes, enhance diagnostics, and streamline therapeutic interventions. For example, use of smart drug delivery systems, nanobiosensors, machine learning (ML)-driven diagnostics, and predictive models for nanotoxicology.

Nanobiosensors are highly sensitive devices designed to detect specific biomolecules, such as proteins or DNA, in biological samples. ML algorithms enhance their functionality by improving accuracy, sensitivity, and data analysis. For example, nanobiosensors combined with ML can detect diseases like cancer or diabetes at early stages by identifying subtle biomarker changes. Quantum dot-based sensors are analyzed by ML to detect prostate-specific antigen (PSA) for early prostate cancer detection. Wearable nanobiosensors integrated with ML algorithms enable continuous tracking of vital parameters, such as glucose levels in diabetic patients. Nanomaterial-coated electrodes paired with ML to analyze glucose fluctuations in real time. ML models filter out noise and enhance the interpretation of weak signals from nanosensors, improving diagnostic accuracy. Signal processing algorithms distinguish true biomarker signals from background interference in blood samples.

Nanotechnology-enabled diagnostics generate complex data, such as nanoscale imaging or genomic profiles, that can be challenging to interpret. Machine learning excels in extracting meaningful patterns from such datasets, enabling precision diagnostics and personalized treatment plans. ML algorithms trained on nanoscale imaging datasets can identify diseases like cancer, Alzheimer's, or cardiovascular disorders earlier than traditional methods. Convolutional neural networks analyze nanoscale MRI scans to detect tumors. Integrating patient data (e.g., genetics, proteomics, and environmental factors) with ML and nanotechnology enables treatments tailored to individual needs. Nanoparticles designed for drug delivery tailored to a patient's unique cancer cell receptor profile. ML models can analyze nanobiosensor data during therapy to adjust treatments dynamically. For example, adjusting insulin doses in diabetic patients based on continuous glucose monitoring *via* nanosensors.

These technologies leverage cutting-edge science to address complex medical challenges, paving the way for personalized and precision medicine. Collectively, these innovations signify a tremendous change in healthcare by using advanced technologies for seamless integration into clinical practice by offering unprecedented potential to enhance human health and well-being. The convergence of artificial intelligence (AI) and nanotechnology is revolutionizing cancer therapy, offering unprecedented precision, efficiency, and personalization. By harnessing AI's computational power and nanotechnology's capabilities, researchers and clinicians can design and implement highly effective cancer treatments.

1.8 Application of AI with Aid of Nanotechnology

The integration of artificial intelligence (AI) into nanotechnology-based cancer therapy has catalysed significant advancements, enabling more precise, efficient, and personalized treatment strategies [23]. Nanoparticles serve as versatile tools for targeted drug delivery, ensuring therapeutic agents are directed towards cancer cells by minimizing damage to healthy tissues as shown in Figure 1.8. AI amplifies the effectiveness of these therapies by optimizing design, delivery, and treatment outcomes.

Cancer treatment has seen significant advancements through the use of nanotechnology as shown in Figure 1.8. Nanoparticles can be engineered to absorb light or heat and release it in a controlled manner. This property is used in photothermal and photodynamic therapy to destroy cancer cells selectively by exposing them to specific wavelengths of light or heat. Nanoparticles, such as liposomes or dendrimers, can deliver therapeutic genes to cancer cells, replacing defective genes or introducing new genetic material to combat cancer.

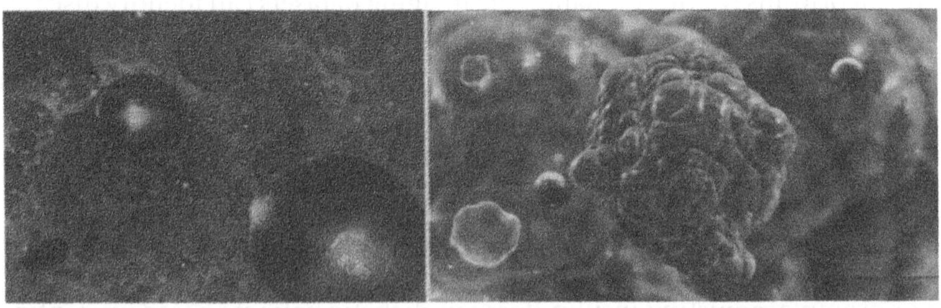

Figure 1.8 Enhanced targeted cancer therapy [24].

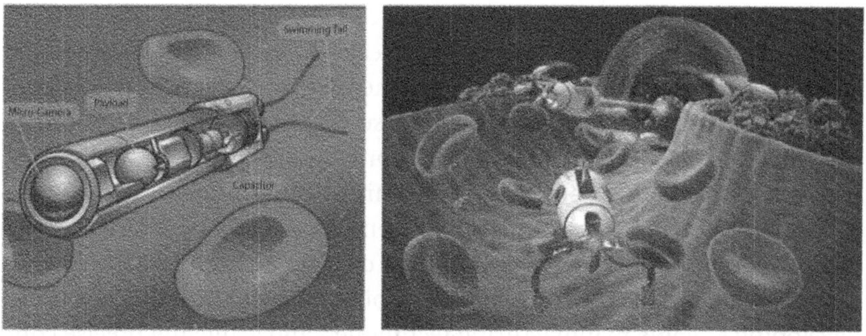

Figure 1.9 Nanorobots are theoretical microscopic devices measured on the scale of nanometers [25].

AI-powered algorithms facilitate targeted cancer therapy by analysing vast datasets. AI identifies the most effective nanoparticle characteristics—such as size, shape, and surface modifications—that enable precise delivery to tumour sites as shown in Figure 1.8. Additionally, AI aids in selecting optimal drug combinations and dosages for enhanced therapeutic efficacy. AI-based nano blood-swimming robots, often referred to as nanorobots or nanobots, represent a groundbreaking innovation in healthcare as shown in Figure 1.9. These microscopic devices are designed to navigate the bloodstream, perform medical tasks, and deliver targeted therapies with unparalleled precision.

Nanobots are typically constructed at the nanoscale (10–100 nanometers) using biocompatible materials such as metals (gold, silver), polymers, or even DNA-based structures. They are powered by a variety of propulsion systems, including chemical reactions, magnetic fields, or ultrasound waves, enabling them to move through the bloodstream efficiently. Some designs mimic biological structures like flagella or cilia for enhanced mobility.

AI algorithms embedded within the nanobots or externally controlling them allow real-time navigation and decision-making. These algorithms process data from onboard sensors to guide the nanobots toward target sites such as tumours or clots. Machine learning (ML) capabilities enable nanobots to adapt to dynamic biological environments, like adjusting their pathways to avoid obstacles or responding to changes in pH, temperature, or biomarker concentrations. Nanobots are designed to recognize specific biomarkers (proteins, antigens, or pH levels) associated with diseased cells, ensuring precise targeting. Some nanobots are designed to physically break down blood clots or deliver clot-dissolving agents. AI aids in identifying clot locations and assessing treatment success. Nanobots can also perform microscale repairs, such as sealing vessel leaks or repairing damaged tissue at the cellular level. Onboard sensors gather data on various physiological parameters, such as oxygen levels, glucose concentrations, or inflammatory markers. AI processes this data to provide insights and guide treatment adjustments. Nanobots can transmit imaging or biochemical data to external systems, allowing doctors to monitor their performance and adjust their missions as needed. While the concept is promising, challenges remain in areas like ensuring biocompatibility to avoid immune reactions, developing scalable manufacturing techniques, and achieving precise control and safe biodegradation. Future advancements in AI, materials science, and biomedical engineering are expected to overcome these hurdles, making AI-based blood-swimming robots a transformative tool in diagnostics, therapy, and preventive medicine.

- AI-based nanorobots designed to break down kidney stones are a cutting-edge innovation in urology. These microscopic devices navigate the urinary system to locate and fragment kidney stones into smaller, more easily excretable pieces, offering a minimally invasive alternative to traditional methods such as extracorporeal shock wave lithotripsy (ESWL) or surgery.
- Nanorobots are typically made from biocompatible materials ensuring they can operate safely within the human body. They are equipped with tools or mechanisms for mechanical or chemical breakdown of kidney stones, such as lasers, chemical payloads, or ultrasound-generating components. Nanorobots equipped with micro-tools or oscillating elements apply mechanical forces to break the kidney stones into smaller fragments. AI adjusts the force and frequency to optimize the breakdown process based on the stone's composition (calcium oxalate or uric acid).
- Some nanorobots use AI to guide tiny laser systems that emit focused energy beams, vaporizing or fragmenting the stone. The laser intensity and duration are controlled to prevent damage to surrounding tissues. AI-controlled nanorobots generate high-frequency ultrasound waves, creating microbubbles that collapse near the kidney stone. This process, known as cavitation, generates localized shock waves that break the stone into smaller pieces. Nanorobots can

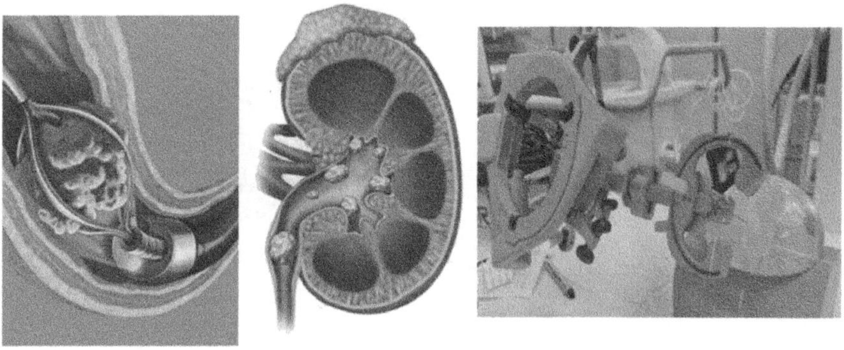

Figure 1.10 Removal of kidney stone [26].

release chemicals or enzymes that dissolve kidney stones. AI regulates the release rate to ensure the chemicals target only the stone, minimizing side effects on surrounding tissues as shown in Figure 1.10. After breaking the stone, nanorobots can guide fragments toward the urinary tract for natural excretion. AI systems optimize fragment size to ensure they can pass without causing blockages or discomfort. Some nanorobots are equipped to remove smaller particles of kidney stones themselves, either by suction or by attaching to them for easy transport out of the body [26].

AI-based organ printing as shown in Figures 1.11(a) and 1.11(b) combines advanced bioprinting techniques with artificial intelligence to create functional tissue or organ structures. This emerging field has the potential to revolutionize medicine by addressing the shortage of donor organs and enabling personalized therapies. Bioprinting is a specialized form of 3D printing that uses bio-inks (gel-like substances containing living cells) to create complex biological structures. Layers of bio-ink are deposited to mimic the architecture of natural tissues or organs.

AI algorithms analyze biological data (imaging scans from CT or MRI) to create precise 3D models of tissues/organs, optimize the internal structure, such as vasculature, to ensure functionality and simulate tissue mechanics and predict how printed structures will behave. Robots designed for tooth repair and dental cleansing combine robotics, AI, and precision engineering to improve oral health [27]. These devices offer significant potential for enhancing dental care accessibility, reducing discomfort, and improving procedural accuracy.

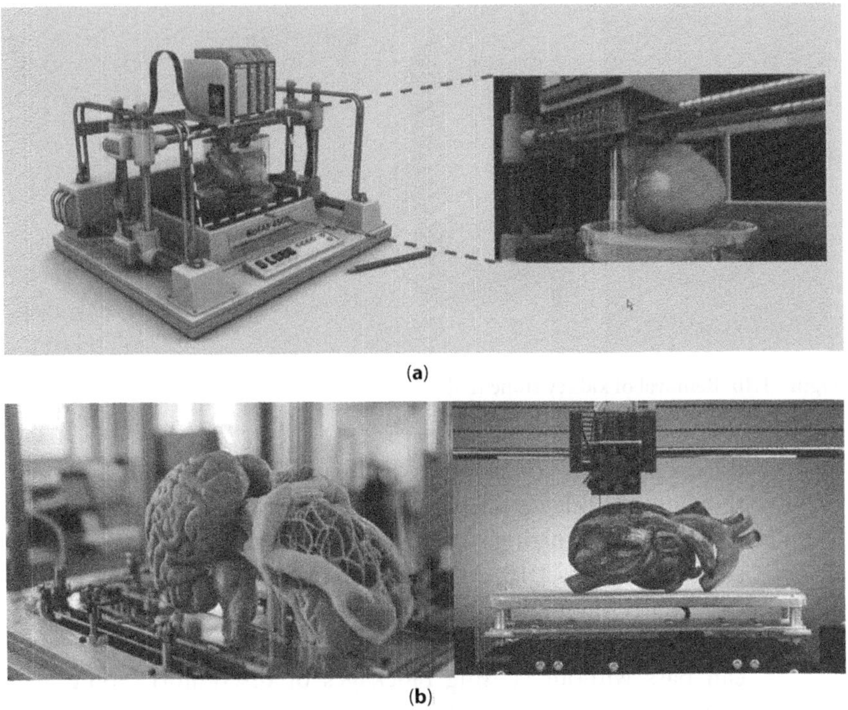

Figure 1.11 (a and b) 3D organ printing concept [27].

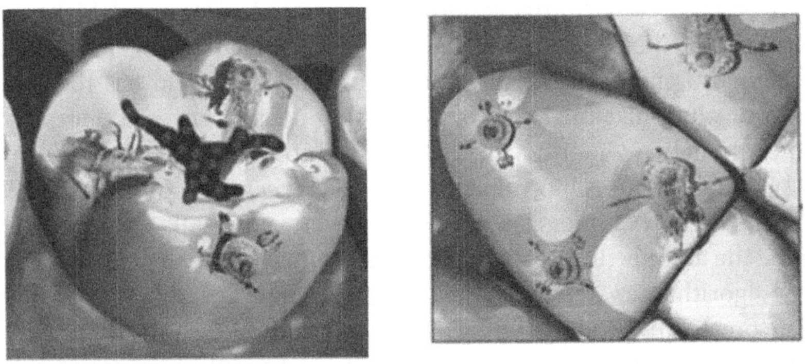

Figure 1.12 Tooth repair and dental cleansing by nano robots [28].

Tooth repair robots as shown in Figure 1.12 are systems that assist or autonomously perform dental procedures like cavity filling, enamel repair, or crown installation. It scans teeth using imaging (3D scans, X-rays) and identifies decay or damage. Precision drilling employs minimally invasive tools for removing decay while preserving healthy enamel. Such a robot deploys bio-compatible materials (e.g., resins or ceramics) for cavity fillings or small repairs. It reduces pain and anxiety through precise movements and shorter procedure times. Robots can use laser-assisted cavity preparation, which is less invasive and more accurate.

1.9 Dental Cleansing Robots

Dental cleansing robots are advanced devices designed for teeth cleaning, plaque removal, and periodontal health maintenance. These are small robots that clean teeth inside the mouth. For example, a micro-robotic swarm using magnetic fields to remove plaque and biofilms. Robots that assist dentists in deep cleaning procedures, such as scaling or root planning, provide real-time analysis of plaque distribution and oral health markers.

1.10 Challenges in Nanotechnology for Healthcare

The intersection of machine learning and nanotechnology is driving transformative advances in healthcare. By enabling smarter design, efficient data processing, and predictive insights, ML not only complements nanotechnology but also empowers it to address complex medical challenges effectively [28]. In spite of nanotechnology being a promising technology there are several challenges which have to be addressed by researchers to ensure that the technology is safe and effective to be used in healthcare. To mention a few challenges, the long-term effects of nanoparticles on human health are not fully understood. Ensuring that nanomaterials do not cause adverse immune reactions, toxicity, or organ damage is critical for their widespread adoption. While nanomaterials can be synthesized in small quantities for research purposes, scaling up production for commercial applications in healthcare can be challenging and costly. The use of nanotechnology in healthcare raises ethical questions regarding privacy, consent, and the potential for misuse. For example, nanosensors capable of detecting diseases in real time could also be used for surveillance purposes. Quality datasets at the nanoscale are limited, which can hinder ML model performance. Also, many ML models, especially deep learning, act as "black boxes," making it difficult to explain their predictions.

Nanotechnology in healthcare is a rapidly advancing field that holds the promise of revolutionizing medical treatments and diagnostics. The ability to manipulate materials at the molecular level allows for highly targeted and effective therapies, improved diagnostics, and real-time monitoring of diseases. While challenges such as toxicity, regulation, and scalability remain, the potential benefits of nanotechnology in healthcare far outweigh these obstacles. As research and development continue, the integration of nanotechnology with other advanced technologies, such as machine learning and artificial intelligence, will unlock new opportunities for personalized medicine, precision therapy, and improved patient outcomes.

1.11 Case Studies

Case studies illustrate the real-world impact of combining machine learning (ML) and nanotechnology in healthcare [29–32]. These examples highlight how these technologies address complex challenges, leading to significant advancements in diagnostics, therapeutics, and patient outcomes.

1.11.1 Smart Nanocarriers for Cancer Therapy

Cancer treatment often faces challenges such as off-target effects and drug resistance. Smart nanocarriers address these issues by delivering drugs directly to cancer cells while minimizing damage to healthy tissues. Machine learning optimizes the design and functionality of nanocarriers for cancer therapy by analyzing datasets of nanoparticle size, surface chemistry, and drug release profiles; predicting the best formulations for specific cancer types, and simulating cellular uptake and drug release kinetics to improve therapeutic efficacy.

- Researchers developed gold nanoparticles (AuNPs) functionalized with polyethylene glycol (PEG) for improved biocompatibility and targeting ligands specific to cancer cell receptors, AOptimize gold nanoparticle-based drug carriers for chemotherapy delivery.
- A dataset of nanoparticle parameters (size, ligand density, and drug loading efficiency) and their therapeutic outcomes *in vitro* and *in vivo* was generated. A neural network was trained to identify the optimal combination of these parameters to maximize drug accumulation in tumors. A neural network was trained on experimental data to predict the

behavior of nanoparticles under different conditions. The model identified nanoparticle configurations with enhanced tumor-targeting capabilities.

- The optimized nanocarriers demonstrated higher drug accumulation in tumors and reduced systemic toxicity compared to conventional methods. Drug accumulation in tumor tissues was 4x higher than in surrounding healthy tissues. Patients experienced significantly fewer systemic side effects compared to traditional chemotherapy. Preclinical trials showed a 30% increase in survival compared to untreated controls.

1.11.2 Early Diagnosis of Neurodegenerative Diseases

Neurodegenerative diseases like Alzheimer's and Parkinson's are challenging to diagnose early due to the subtle and gradual onset of symptoms. The objective was to detect Alzheimer's disease biomarkers in cerebrospinal fluid (CSF) using nanosensors. Nanotechnology-based diagnostics can detect early biomarkers, but interpreting complex sensor data requires advanced computational methods. Machine learning algorithms enhance the diagnostic potential of nanosensors by identifying patterns in biomarker levels that precede symptoms, classifying sensor signals into diagnostic categories and providing predictive insights into disease progression.

- Researchers employed gold nanoparticle-based sensors functionalized with antibodies targeting beta-amyloid (Aβ) peptides, a hallmark of Alzheimer's. A support vector machine (SVM) analyzed nanosensor data to distinguish between healthy individuals and early-stage Alzheimer's patients.
- Signal data from the nanosensors were fed into a support vector machine (SVM).
- The model was trained to distinguish between healthy individuals and those with early-stage Alzheimer's based on Aβ levels.
- The nanosensor-ML system achieved over 90% accuracy in early diagnosis, enabling earlier interventions that could slow disease progression.
- Principal component analysis (PCA) was used to reduce noise and highlight key patterns in the sensor data.

- The SVM classified samples into "healthy" and "Alzheimer's" categories with an accuracy of over 90%.
- Patients diagnosed in the early stages of Alzheimer's were able to begin treatment to slow disease progression.
- The nano sensor-ML system was cost-effective and suitable for large-scale screening programs.

1.11.3 COVID-19 Vaccine Delivery Optimization

The development and distribution of mRNA vaccines during the COVID-19 pandemic relied heavily on lipid nanoparticles (LNPs) for efficient delivery. These nanoparticles protect the fragile mRNA and ensure it reaches target cells. The objective was to optimize lipid nanoparticles (LNPs) for the delivery of mRNA vaccines during the COVID-19 pandemic. Machine learning improved vaccine delivery systems by optimizing the composition and size of LNPs, predicting stability under various storage and environmental conditions, and enhancing cellular uptake for increased vaccine efficacy.

- Deep learning models analysed experimental data to predict the most stable and efficient LNP formulations.
- A deep learning model trained on existing datasets of LNP formulations and their stability under different conditions was used.
- The model predicted LNP compositions with high stability and efficiency for mRNA delivery.
- The ML-optimized LNPs contributed to the rapid development and global rollout of mRNA vaccines, saving millions of lives.
- ML-guided design resulted in LNP formulations that significantly enhanced mRNA stability and delivery efficiency, contributing to the success of mRNA vaccines like Pfizer-BioNTech and Moderna.
- The use of ML shortened development timelines, enabling rapid vaccine rollout.

1.11.4 Nanobiosensors for Diabetes Management

Managing diabetes requires continuous glucose monitoring, which can be cumbersome with traditional devices. Nanotechnology-based biosensors offer a more convenient and precise solution. The objective was to develop a wearable nanobiosensor for non-invasive glucose monitoring. Graphene-based sensors were engineered to detect glucose levels in sweat, offering a non-invasive alternative to blood-based testing. ML enhances the functionality of nanobiosensors by improving sensor accuracy through noise reduction, predicting glucose trends based on historical data and enabling real-time alerts for abnormal glucose levels.

- A recurrent neural network (RNN) processed biosensor data to predict glucose fluctuations and provide personalized insulin recommendations.
- Patients using the ML-enhanced biosensor reported improved glycemic control and greater convenience compared to conventional devices.
- An RNN analyzed biosensor data to identify glucose trends and predict future levels.
- The model was trained on real-world data from diabetic patients to improve personalization.
- Each patient's historical data was used to tailor the model for individual glycemic patterns.
- Patients using the nanobiosensor experienced fewer episodes of hyperglycemia and hypoglycemia.
- The device reduced the need for invasive blood sampling, offering a more convenient monitoring solution.

1.12 Advantages and Challenges

The integration of machine learning (ML) with nanotechnology in healthcare has opened unprecedented opportunities, offering significant advantages. However, these innovations are also accompanied by several challenges that must be addressed to maximize their potential. This section outlines the key benefits and hurdles of this interdisciplinary approach.

1.12.1 Advantages

- ML algorithms optimize nanotechnology for highly specific applications, reducing trial-and-error approaches.
- The ability to process large datasets rapidly accelerates the discovery of novel nanomaterials.
- Predictive models minimize risks associated with nanotoxicology.

- Combining nanotechnology and ML enables therapies tailored to individual patients, increasing treatment efficacy while minimizing side effects.
- ML algorithms analyze large datasets to identify trends and predict outcomes, speeding up drug discovery and nanomaterial design.

1.12.2 Challenges

- Nanotechnology is a relatively new field, leading to limited datasets for training ML models.
- Generating and curating high-quality datasets for ML training remains a bottleneck.
- Modeling nanoscale phenomena often requires high computational resources.
- Issues of data privacy and the potential misuse of advanced healthcare technologies.
- Nanoparticles improve the sensitivity of imaging techniques like MRI or CT scans, while ML algorithms analyze these images for better diagnostic accuracy.
- The high dimensionality and complexity of data from nanotechnology experiments make processing and analysis challenging.
- Wearable nanobiosensors coupled with ML provide real-time insights into patients' health conditions, enabling timely interventions.
- Integrating ML and nanotechnology often involves handling sensitive patient data, raising privacy concerns.
- Successful integration of ML and nanotechnology requires close collaboration between fields with distinct terminologies, methodologies, and priorities.

By addressing these challenges head-on, the full potential of ML and nanotechnology in healthcare can be unlocked, paving the way for innovative and equitable solutions.

1.13 Future Trends

The convergence of machine learning (ML) and nanotechnology is expected to redefine healthcare in the coming decades. Emerging trends highlight advancements in technology, methodologies, and applications, pointing to a future where this interdisciplinary approach achieves even greater precision, scalability, and accessibility.

- Quantum computing promises to revolutionize ML by solving complex optimization problems in nanomaterial design.
- Future ML models will integrate real-time patient data with nanotechnology-enabled diagnostics to provide adaptive treatment plans.
- The widespread adoption of ML and nanotechnology could address disparities in healthcare by offering cost-effective diagnostic and therapeutic solutions.
- Adaptive drug delivery systems that alter dosage based on real-time patient data analyzed by ML models.
- Multi-functional nanoparticles that switch between diagnostic and therapeutic roles depending on ML-predicted disease progression.
- Reduction in production costs through automated processes.
- Improved scalability for mass production of ML-optimized nanodevices.
- Discovery of new nanomaterials with tailored properties for specific medical applications.
    - Faster response times for critical healthcare interventions.
    - Scalable systems for population-level health monitoring.
    - Extended lifespan of implantable devices.
    - Reduced maintenance for nanotechnology-enabled healthcare solutions.

1.14 Conclusion

The integration of machine learning with nanotechnology in healthcare represents a paradigm shift in medicine. By leveraging the predictive power of ML, researchers can harness the unique properties of nanomaterials to develop more effective diagnostics, treatments, and preventive measures. While challenges remain, the continued interdisciplinary collaboration between AI experts, nanotechnologists, and healthcare professionals will undoubtedly unlock new possibilities for improving global health outcomes.

AI disclosure statement: The author confirms that no artificial intelligence (AI)-generated content or AI-assisted writing tools were used in the preparation of this chapter (including summaries and the abstract). All text, analysis, and conclusions are the result of the author's original work.

References

1. Cid Vidal, X., Dieste Maroñas, L., Dosil Suárez, A., Modern machine learning: Applications and methods, in: *Machine Learning and Artificial Intelligence with Industrial Applications: From Big Data to Small Data*, Springer International Publishing, 2022.
2. Mukhamediev, R.I., Popova, Y., Kuchin, Y., Zaitseva, E., Kalimoldayev, A., Symagulov, A., Yelis, M., Review of artificial intelligence and machine learning technologies: classification, restrictions, opportunities and challenges. *Mathematics*, 10, 15, 2552, 2022.
3. Raffa, V., Vittorio, O., Riggio, C., Cuschieri, A., Progress in nanotechnology for healthcare. *Minim. Invasive Ther. Allied Technol.*, 19, 3, 127–135, 2019.
4. Wong, I.Y., Bhatia, S.N., Toner, M., Nanotechnology: emerging tools for biology and medicine. *Genes Dev.*, 27, 22, 2397–2408, 2013.
5. Conte, R., Foggia, R., Valentino, A., Di Salle, A., Kandsi, F., Calarco, A., Nanotechnology advancements transforming molecular diagnostics: Applications in precision healthcare. *Int. J. Nano Dimens.*, 15, 3, 2024.
6. Liu, W., Wu, Y., Hong, Y., Zhang, Z., Yue, Y., Zhang, J., Applications of machine learning in computational nanotechnology. *Nanotechnology*, 33, 16, 162501, 2022.
7. Singh, A.V., Ansari, M.H.D., Rosenkranz, D., Maharjan, R.S., Kriegel, F.L., Gandhi, K., Luch, A., Artificial intelligence and machine learning in computational nanotoxicology: unlocking and empowering nanomedicine. *Adv. Healthc. Mater.*, 9, 17, 1901862, 2022.
8. Kala, D., Gupta, S., Kaushal, A., Nanotechnology in healthcare, in: *Synthesis and Applications of Nanoparticles*, pp. 405–416, 2022.
9. Pramanik, P.K.D., Solanki, A., Debnath, A., Nayyar, A., El-Sappagh, S., Kwak, K.S., Advancing modern healthcare with nanotechnology, nanobiosensors, and internet of nano things: Taxonomies, applications, architecture, and challenges. *IEEE Access*, 8, 65230–65266, 2020.
10. BigOhTech, Benefits of machine learning in healthcare. Retrieved December 29, 2024, https://bigohtech.com/benefits-of-machine-learning-in-healthcare/.
11. Leong, Y.X., Tan, E.X., Leong, S.X., Lin Koh, C.S., Thanh Nguyen, L.B., Ting Chen, J.R., Ling, X.Y., Where nanosensors meet machine learning: Prospects and challenges in detecting Disease X. *ACS Nano*, 16, 9, 13279–13293, 2020.

12. Guo, Y., Shao, S., Gu, R., Hu, X., Zhao, M., Peng, Y., Zhang, W., Zhang, B., Ding, J., Wang, N., Peng, H., Han, J., Application of nanomaterials in early diagnosis of cancer. *Nanotechnol. Rev.*, 12, 1, 20230116, 2023.
13. Singh, A.P., Biswas, A., Shukla, A., Maiti, P., Targeted therapy in chronic diseases using nanomaterial-based drug delivery vehicles. *Signal Transduction Targeted Ther.*, 4, 1, 33, 2019.
14. Chandrakala, V., Aruna, V., Angajala, G., Review on metal nanoparticles as nanocarriers: Current challenges and perspectives in drug delivery systems. *Emergent Mater.*, 5, 6, 1593–1615, 2022.
15. Alghamdi, M.A., Fallica, A.N., Virzì, N., Kesharwani, P., Pittalà, V., Greish, K., The promise of nanotechnology in personalized medicine. *J. Pers. Med.*, 12, 5, 673, 2022.
16. Saallah, S. and Lenggoro, I.W., Nanoparticles carrying biological molecules: Recent advances and applications. *Kona Powder Part. J.*, 35, 89–111, 2018.
17. Wang, X., *et al.*, Comparative Study of Three Carbon Additives: Carbon Nanotubes, Graphene, and Fullerene-C60, for Synthesizing Enhanced Polymer Nanocomposites. *Nanomaterials*, 10, 5, 838, 2020.
18. Singh, S.P., Yadav, S.K., Mondal, K., What We Need to Know about Quantum Dots Nanoparticles, in: *Science and Applications of Nanoparticles*, pp. 29–53, Jenny Stanford Publishing, 2022.
19. Nasteski, V., An overview of the supervised machine learning methods. *Horizons*, 4, 51-62, 56, 2017.
20. Brune, H., Ernst, H., Grunwald, A., Grünwald, W., Hofmann, H., Krug, H., Wyrwa, D., *Nanotechnology: Assessment and Perspectives*, vol. 27, Springer Science & Business Media, 2006.
21. Rimban, E., Intersecting Horizons of Artificial Intelligence and Nanotechnology, 2024, Available at SSRN 4757599.
22. Doyle, M., Predictive Modeling, in: *Virtual Testing and Predictive Modeling: For Fatigue and Fracture Mechanics Allowables*, pp. 247–289, 2009.
23. Adir, O., Poley, M., Chen, G., Froim, S., Krinsky, N., Shklover, J., Schroeder, A., Integrating artificial intelligence and nanotechnology for precision cancer medicine. *Adv. Mater.*, 32, 13, 1901989, 2020.
24. National Programme on Technology Enhanced Learning (NPTEL), Preview: NOC24_BT72, 2024, Retrieved from https://onlinecourses.nptel.ac.in/noc24_bt72/preview.
25. Parmar, D.R., Sonia, J.P., Patel, A., Sen, D.J., Nanorobotics in advances in pharmaceutical sciences. *Int. J. Drug Dev. Res.*, 2, 2, 247–256, 2010.
26. Verma, J., Nath, M., Tripathi, P., Saini, K.K., Analysis and identification of kidney stone using Kth nearest neighbour (KNN) and support vector machine (SVM) classification techniques. *Pattern Recognit. Image Anal.*, 27, 574–80, 2017.

27. Ozbolat, I.T. and Yu, Y., Bioprinting toward organ fabrication: Challenges and future trends. *IEEE Trans. Biomed. Eng.*, 2023, https://doi.org/10.1109/TBME.2013.2243912.
28. Sudakar, J. and Nirmanik, S.M., Nanorobotics in medical field. *Int. J. Res. Appl. Sci. Eng. Technol.*, 10, 12, 2022.
29. Kishore, D.M., Bindu, S., Manjunath, N.K., Estimation of yoga postures using machine learning techniques. *Int. J. Yoga*, 15, 2, 137–143, 2022.
30. Kishore, D.M., Bindu, S., Manjunath, N.K., Smart yoga instructor for guiding and correcting yoga postures in real time. *Int. J. Yoga*, 15, 3, 254–261, 2022.
31. Tripathi, S.L., Prakash, K., Balas, V.E., Mohapatra, S., Nayak, J. (Eds.), *Electronic Devices, Circuits, and Systems for Biomedical Applications*, Elsevier, Amsterdam, The Netherlands, 2021.
32. Thillaiarasu, N., Lata Tripathi, S., Dhinakaran, V. (Eds.), *Artificial Intelligence for Internet of Things: Design Principle, Modernization, and Techniques*, 1st ed., CRC Press, Boca Raton, 2022, https://doi.org/10.1201/9781003335801.

2

FPGA-Based Hardware Accelerators for ANN and DNN

Abhishek Kumar

School of Electronics and Electrical Engineering, Lovely Professional University, Punjab, India

Abstract
The integration of hardware accelerator with field programmable gate array (FPFA) into a real-time system has shown to be crucial for improving neural network performance and efficiency. Hardware accelerators are specialized components made to relieve general-purpose central processing units (CPUs) of a computationally challenging task; they enhance system performance while handling specialized activities like matrix multiplications. These components expressly for real-time applications show high reliability, low latency, and deterministic behavior requirements. The usage of hardware accelerators for artificial neural networks (ANN) and deep neural networks (DNN), which are designed to function like the human brain, is covered in this chapter and implemented with Verilog programming language to construct the accelerators for ANN and DNN on the Kintex series of FPGA.

Keywords: Hardware accelerator, ANN, DNN, matrix multiplication, machine learning, healthy lifestyle

2.1 Introduction

Analyzing data from diverse physiological systems, including the brain, heart, and muscles, is a crucial component of the study field known as biomedical signal processing. The diagnosis and management of medical disorders, examination of signals can yield useful information. However,

Email: abkvjt@gmail.com

Abhishek Kumar and Suman Lata Tripathi (eds.) Smart Chips for Smart Devices: VLSI Design for Next-Generation IoT Solutions, (31–48) © 2026 Scrivener Publishing LLC

analyzing biological data is a computationally demanding endeavor that takes a lot of time and resources. This issue can be resolved by using hardware accelerators, which enable quicker processing speeds and use less power than conventional software-based techniques [1–4]. Applications for biomedical signal processing that utilize hardware accelerators include voice recognition, magnetic resonance imaging (MRI) reconstruction, and EEG and ECG analysis. Complete biomedical systems can be built using the accelerators in conjunction with other hardware elements like sensors and actuators. Processing the electrical impulses produced by the brain and heart, respectively, is done during EEG and ECG analysis [5]. The analysis of these signals can yield important data for the detection and management of diseases including epilepsy and arrhythmias. In comparison to conventional software-based techniques, hardware accelerators can offer quicker processing times and greater accuracy. Reconstruction in magnetic resonance imaging (MRI) is the processing of MRI scan data to produce images of the body's internal organs. This data processing is computationally demanding and takes a lot of time and resources. Hardware accelerators can offer more efficient computation and less power use.

An advanced computing device known as a hardware accelerator is created to speed up a particular computation or collection of computations, usually by offloading them from the main processor, as shown in Figure 2.1. Hardware accelerators can be employed in real-time systems to enhance the speed and effectiveness of time-sensitive processes. Real-time systems can carry out tasks in a predetermined amount of time. These systems can be employed in a variety of applications, including robots, automotive systems, and control systems. For real-time systems to ensure prompt

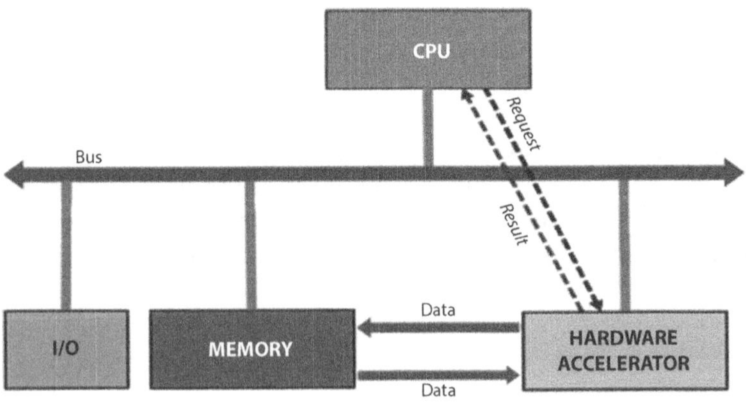

Figure 2.1 Communication between CPU and accelerator [4].

and accurate responses to events, high performance and low latency are essential. Real-time systems can operate more quickly by shifting computation-intensive activities from the primary processor to hardware accelerators [6]. Hardware accelerators are specialized computer parts created to speed up particular computations. Utilizing these devices can free up CPU resources for other operations by offloading computation-heavy tasks from the primary processor. Digital signal processors (DSPs), graphics processing units (GPUs), field-programmable gate arrays (FPGAs), and application-specific integrated circuits (ASICs) are a few types of hardware accelerators that can be utilized in real-time systems.

Machine learning algorithms are used more frequently than ever in a variety of fields, including data analytics, natural language processing, and computer vision. These algorithms' high computing requirements, however, might result in longer processing times and significant energy consumption on conventional processors. To address this challenge, hardware accelerators for machine learning have been developed to offload the computation of these algorithms from the general-purpose processors and accelerate the processing of the machine learning tasks [7].

The primary objective of hardware acceleration is to speed up processing by utilizing specially designed hardware to execute a certain process or algorithm; it shows the benefit of allowing the CPU to handle other data while the hardware co-processor handles the work necessary to speed up the process. A co-processor's setup phase involves the main processor computing the required parameters and waiting for the results. This stage has significance because it establishes a basis for the computations. The efficiency of this relationship is important; if the overhead of corresponding with the co-processor exceeds the time needed for computing, then the advantages of utilizing a co-processor are wasted [8]. During communication it is essential that the costs of communication, such as synchronization delays and data transfer times, be smaller than the computational savings from offloading jobs to the co-processor in order to obtain speed improvements using co-processors.

Latency – It is necessary to reduce the amount of time needed to transmit data to and receive output from the co-processor.
Bandwidth – To ensure proper communication between the processor and co-processor without any bottlenecks.
Task Offloading's Efficiency – The tasks that are designated for the co-processor should be those that are most advantageous to be executed in parallel processing.

A hardware accelerator is a specialized hardware component designed to offload and accelerate specific tasks from a general-purpose processor or a software application. Hardware accelerators can improve the performance and efficiency of a system by offloading computationally intensive tasks to dedicated hardware components that are optimized for the specific task [9]. Numerous applications, including machine learning, image and video processing, cryptography, and networking, use hardware accelerators. By shifting computationally demanding operations to specialized hardware components, hardware accelerators can greatly increase a system's performance and efficiency. The CPU's workload can be reduced and the system's overall performance can be enhanced. Machine learning hardware accelerators are pieces of specialized hardware created to speed up machine learning workloads. These accelerators are substantially quicker than conventional CPUs since they are designed to carry out the intricate matrix and vector operations that are frequently used in machine learning algorithms [10–12].

Examples of hardware accelerators [13]

Graphics Processing Units (GPU) are a popular option for accelerating machine learning. They possess a significant number of parallel processing units that can be used to execute matrix and vector operations concurrently. Deep learning models can be trained and inferred using GPUs.

Field-Programmable Gate Arrays (FPGA) are hardware components that can be programmed to carry out particular functions. Custom neural network designs can be implemented using FPGAs, which can also be configured to execute matrix and vector operations concurrently.

Application-Specific Integrated Circuits (ASIC) are specialized chips that are built and optimized for certain uses. ASICs like Google's Tensor Processing Unit (TPU) and Intel's Nervana Neural Network Processor (NNP) are made expressly for machine learning.

Digital Signal Processors (DSP) are specialized microprocessors made for applications involving signal processing. They can be used to speed up machine learning, especially in programs that analyze audio or video.

Neuromorphic Chips are specifically made for activities like speech and picture recognition; they are built to resemble the architecture of the human brain.

Applications for machine learning that leverage these hardware accelerators range from audio and picture recognition to natural language processing and recommendation systems.

2.2 Hardware Accelerator for FPGA

FPGAs are a common choice for accelerating machine learning (ML) operations since they are a type of hardware accelerator that can be designed to carry out particular tasks. FPGAs are an appealing alternative for creating unique ML hardware because of their great degree of flexibility and ability to be optimized for certain ML models. The Xilinx Virtex UltraScale+ FPGA is a well-liked FPGA architecture for machine learning [14]. This FPGA has memory, DSP, and programmable logic blocks that can be used to construct different machine learning (ML) algorithms. The SDSoC tool, for system-level design, the Vitis AI tool, for ML acceleration, and the Vivado HLS tool for high-level synthesis are tools that Xilinx offers to ML developers. This FPGA has memory blocks, DSP blocks, programmable logic, and a high-speed transceiver that can link other FPGAs together [15]. The OpenCL SDK for FPGA, the Intel FPGA Deep Learning Acceleration Suite, and the Intel DevCloud for FPGA, which enables developers to test their designs on distant FPGA hardware, are tools that Intel offers to ML developers.

The PYNQ platform, which is built on the Xilinx Zynq FPGA and comes with a Python-based programming environment for ML acceleration, is one of the open-source FPGA platforms for machine learning. The FINN framework additionally offers a collection of tools for creating FPGA accelerators for specific ML tasks. Overall, with a large selection of topologies and development tools, FPGAs offer a flexible and potent hardware accelerator for ML. For a variety of ML activities, especially those requiring high data throughput, like convolutional neural network (CNN) inference, FPGAs can significantly speed things up and increase energy efficiency. Designing and optimizing FPGA-based ML systems, however, can be difficult and calls for specialized training and experience. Figure 2.2 presents a block diagram of a hardware accelerator with input and weight matrix; the

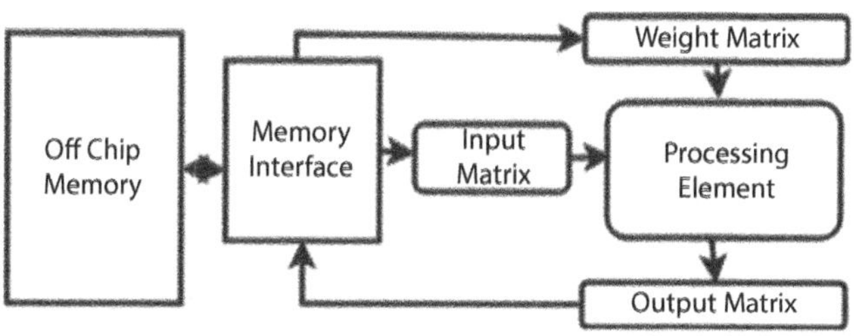

Figure 2.2 Neural network block diagram.

accelerator is created to carry out matrix operations for Artificial Neural Networks (ANNs) and Deep Neural Networks (DNNs) used in a variety of applications, including computer vision, natural language processing, and biomedical signal processing [16]. An FPGA device with specialized hardware programming makes up the hardware accelerator with input and weight matrices. A processing element block is included in the design that multiplies the input matrix and weight matrix. Systolic arrays, folding, and pipelining techniques are just a few of the ways the processing element block might be implemented. The performance of the hardware accelerator can be enhanced using these methods.

2.3 Neural Network Accelerator

A hardware accelerator for a neural network accelerator is created specifically to handle the computationally demanding task needed by ANN and DNN. It can speed up neural network operation, enabling quicker inference and training durations [17]. Dedicated AI chips like Google's Tensor Processing Unit (TPU), GPUs, FPGAs, and ASICs are a few examples of neural network accelerators. Verilog, VHDL, OpenCL, TensorFlow, and PyTorch are just a few of the several programming languages and tools that may be used to program these accelerators. For each input, these neural networks need to execute numerous computations, such as matrix multiplication, activation function calculations, and bias addition. These calculations can be quite time-consuming and together with energy costs can reduce the system's overall performance. The solution is to use hardware accelerators within machine learning applications, ANNs, and DNNs. DNNs can have numerous hidden layers and are more complicated than ANN, requiring more processing and storage, shown in Figure 2.3.

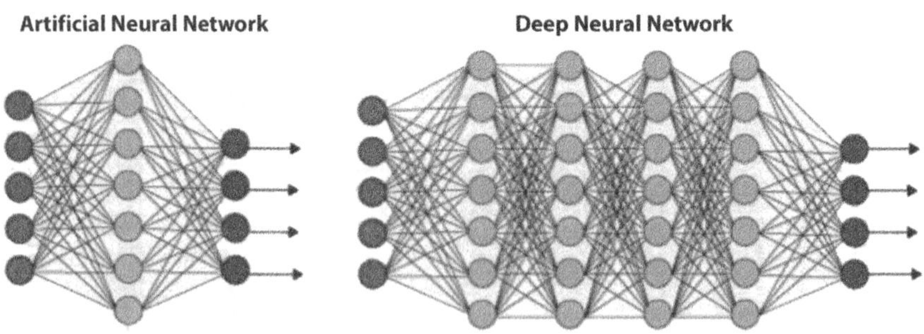

Figure 2.3 ANN and DNN architecture [18].

Hardware acceleration may therefore be more advantageous for DNNs than ANNs in terms of the following:

(a) ***Accuracy***: Particularly for applications requiring image and speech recognition, DNNs are typically more accurate than ANNs. This is so that more intricate patterns and correlations in the data may be captured by DNNs. A hardware accelerator can reduce the higher processing complexity that results from this enhanced accuracy.
(b) ***Hardware acceleration***: DNNs require higher levels of parallelism than ANNs. DNNs are better suited for hardware acceleration utilizing GPUs or specialized ASICs since they need more memory bandwidth and processing capacity.
(c) ***Training***: Even with hardware acceleration, training DNNs can take a long time. Because DNNs need big datasets and lots of iterations to update the entire network's various parameters. DNNs can quickly produce correct predictions after being training.
(d) ***Applications***: DNNs are better suited to complicated tasks like audio and picture recognition, natural language processing, and autonomous cars, while ANNs are frequently employed for straightforward classification and prediction tasks.

In summary, both ANNs and DNNs can be advantageous from hardware acceleration; DNNs can stand to gain more as a result of their increasing complexity and requirement for greater computational and memory bandwidth. However, the particular goal and the implementation resources at hand determine which neural network should be used.

2.4 Artificial Neural Networks (ANN) Accelerator

ANNs are a type of machine-learning algorithm that are inspired by the structure and functions of the human brain. They are utilized in many different applications, including natural language processing, predictive analytics, and picture and speech recognition. They are often implemented as specialized hardware that is better equipped than general-purpose processors to handle the intricate mathematical calculations and speed up operations needed by ANNs. Parallel processing techniques to make use of the inherent parallelism of neural networks and putting computations on parallel hardware is a popular method for constructing a hardware accelerator for ANNs. The memory hierarchy is a crucial factor; to store input features, weights, and intermediate outcomes, ANNs often need a lot of memory. The memory hierarchy must be carefully managed to reduce the number of memory accesses and maximize throughput to maximize performance. The hardware must be designed to efficiently execute the specific operations required by the neural network, while the software must be optimized to take advantage of the hardware architecture and minimize data movement and processing time.

Overall, hardware accelerators for ANNs have the potential to significantly improve the performance and efficiency of machine-learning applications. As the demand for machine learning continues to grow, hardware accelerators will play an increasingly important role in enabling these applications to run faster and more efficiently.

Verilog program of a hardware accelerator for ANN

```
module ANN_Accelerator( clk, reset, input_vector, output_vector);
parameter N = 10; // number of input neurons
parameter M = 5; // number of output neurons
   input clk;
   input reset;
   input [N-1:0] input_vector;
   output reg [M-1:0] output_vector;

// weights and biases of the perceptron
reg signed [N-1:0] weights [0:M-1];
reg signed [M-1:0] biases;

// internal variables
reg signed [M-1:0] weighted_sum;
reg signed [M-1:0] activation;
integer i,j;
// initialize weights and biases
initial begin
   weights [0] = {1, 2, 3, 4, 5, 6, 7, 8, 9, 10};
   weights [1] = {-1, -2, -3, -4, -5, -6, -7, -8, -9, -10};
   weights [2] = {0, 1, 0, -1, 0, 1, 0, -1, 0, 1};
   weights [3] = {1, 1, 1, 1, 1, 0, 0, 0, 0, 0};
   weights [4] = {-1, -1, -1, -1, -1, 0, 0, 0, 0, 0};
   biases = {0, 0, 0, 0, 0};
end
```

```
always @(posedge clk) begin
  if (reset) begin
    weighted_sum <= 0;
    activation <= 0;
  end else begin
    // compute weighted sum
    for (i = 0; i < M; i=i+1) begin
      weighted_sum[i] <= biases[i];
      for (j = 0; j < N; j=j+1) begin
        weighted_sum[i] <= weighted_sum[i] + weights[i][j] * input_
        vector[j];
      end
    end
    // apply activation function
    for (i = 0; i < M; i=i+1) begin
      if (weighted_sum[i] > 0) begin
        activation[i] <= weighted_sum[i];
      end else begin
        activation[i] <= 0;
      end
    end
    // assign output vector
    output_vector <= activation;
  end
end
endmodule
```

In the above program, N and M are parameters that define the number of input and output neurons, respectively. The input_vector is a vector of N input values that are multiplied by the weights of the perceptron and added to the biases. The result is passed through an activation function (ReLU) to produce the output_vector, which is a vector of M output values. Figure 2.4 presents the synthesized schematic obtained with Vivado Tool based on Kintex series FPGA family. The Figure 2.4 presents a schematic of ANN accelerator implemented with Kintex series of FPGA family. The primitive requirements for the implementation are given in Figure 2.5. The dynamic power requirement for the ANN accelerator is much higher than the static power consumption shown in Figure 2.6.

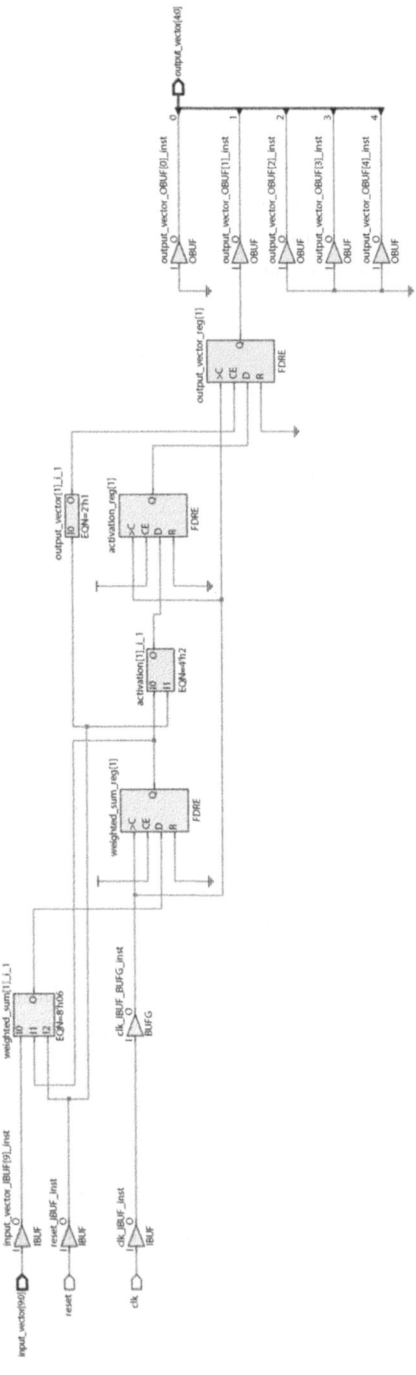

Figure 2.4 Synthesized schematic of ANN accelerator.

Primitives

Ref Name	Used	Functional Category
OBUF	5	IO
IBUF	3	IO
FDRE	3	Flop & Latch
LUT3	1	LUT
LUT2	1	LUT
LUT1	1	LUT
BUFG	1	Clock

Figure 2.5 Primitive of ANN.

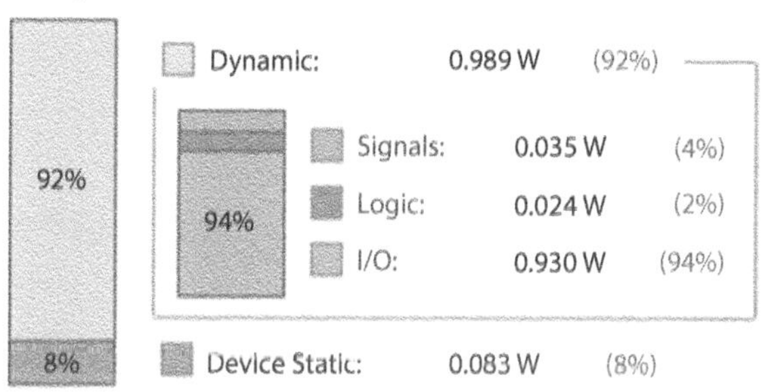

Figure 2.6 Power report of ANN.

2.5 Deep Neural Network (DNN) Accelerator

Hardware accelerators for Deep Neural Networks (DNNs) are becoming increasingly important as the demand for machine learning applications continues to grow. These accelerators can significantly speed up the processing time for DNNs by offloading computationally intensive operations from the CPU or GPU. One common approach for implementing a hardware accelerator for DNNs is to use parallel processing techniques to exploit the inherent parallelism of neural networks [19]. By mapping these computations onto parallel hardware, the overall processing time can be reduced. Another important consideration for hardware accelerators is the memory hierarchy. DNNs typically require large amounts of memory for storing input feature maps, weights, and intermediate results. To maximize performance, it is important to carefully manage the memory hierarchy to minimize the number of memory accesses and maximize data reuse.

Verilog program of a hardware accelerator for DNN

```
module DNNAccelerator(clk, reset, in_data, out_data);
   input [31:0] in_data; // input data
   input clk; // clock input
   input reset; // reset input
   output reg[31:0] out_data; // output data

   // define your neural network architecture here
   reg [15:0] hidden1_out, hidden2_out;
   reg [31:0] weight1[15:0], weight2[15:0], weight3[31:0], dot_prod;
   reg [15:0] bias1[15:0], bias2[15:0], bias3;
   integer i,j;

   always @(posedge clk, posedge reset)
   begin
   if (reset)
   dot_prod=0;
   else
   begin
      // first hidden layer
      for (i = 0; i < 16; i=i+1) begin
                        dot_prod = 0;
         for (j = 0; j < 32; j=j+1) begin
            dot_prod = dot_prod + in_data[j] * weight1[i][j];
         end
         hidden1_out[i] = dot_prod + (bias1[i] > 0) ? (dot_prod + bias1[i])
         : 0;
      end

      // second hidden layer
      for (i = 0; i < 16; i= i+1) begin
                               dot_prod = 0;
         for (j = 0; j < 16; j= j+1) begin
            dot_prod = dot_prod + hidden1_out[j] * weight2[i][j];
         end
         hidden2_out[i] = dot_prod + (bias2[i] > 0) ? (dot_prod + bias2[i])
         : 0;
      end
```

```
        // output layer
                                dot_prod = 0;
        for (i = 0; i < 16; i=i+1) begin
            dot_prod =dot_prod + hidden2_out[i] * weight3[i];
        end
        out_data = dot_prod + (bias3 > 0) ? (dot_prod + bias3) : 0;
    end
    end
    always @(posedge clk, posedge reset) begin
        if (reset) begin
            // reset logic goes here
            hidden1_out <= 0;
            hidden2_out <= 0;
        end else begin
            // update weights and biases here
        end
    end
endmodule
```

Figure 2.7 presents DNN accelerator architecture consisting of two hidden layers with 16 neurons each and an output layer with a single neuron. The ReLU activation function is used for each neuron. First-way block is used to compute the output of the neural network for the current input data. The weights and biases for each layer are stored in separate arrays. The dot product of the input data and the weights are computed for each neuron in the first hidden layer and then passed through the ReLU activation function. The same process is repeated for the second hidden layer, and then for the output layer. The final output is the dot product of the second hidden layer output and the output layer weights, passed through the ReLU activation function. The second always block is used to update the weights and biases of the neural network based on the error signal from the output layer. Figure 2.8 and Figure 2.9 present the primitive requirement and power report, respectively, for implemented architecture of the DNN accelerators.

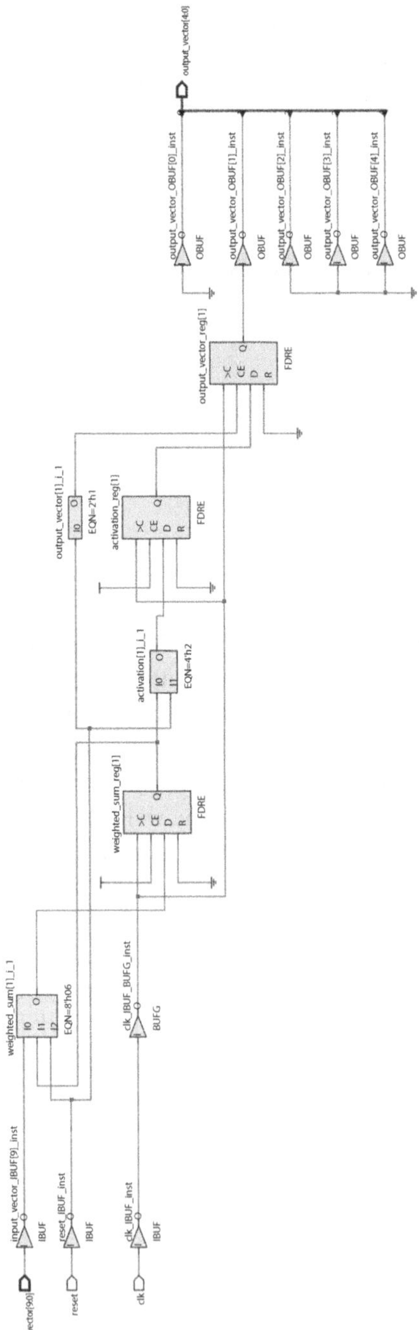

Figure 2.7 Synthesized schematic of DNN accelerator.

Primitives

Ref Name	Used	Functional Category
OBUF	32	IO

Figure 2.8 Primitive requirement.

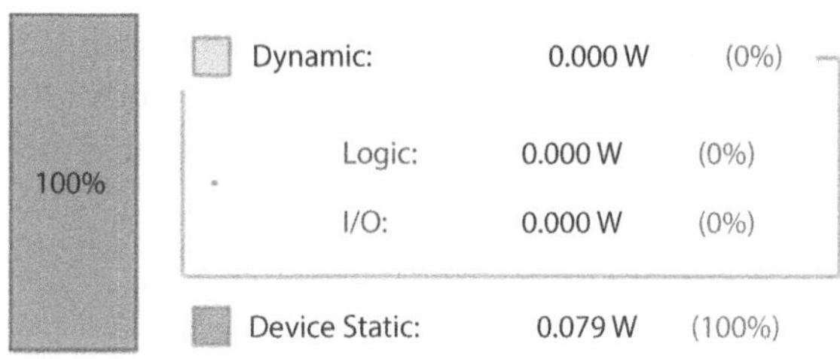

Figure 2.9 Power report.

2.6 Discussion

ANN and DNN are two widely used machine learning techniques that have different architectures and computational requirements. Hardware accelerators have been developed for both ANN and DNN implementations to

Table 2.1 ANN vs. DNN.

	ANN accelerator	DNN accelerator
Processing Time	Lower power and performance compared to the DNN accelerator	Higher power and performance compared to the ANN accelerator
Memory	Lower memory compared to DNN accelerator	Higher memory compared to the ANN accelerator
Flexibility	More flexible and easier to customize	Less flexible harder to flexible
Inference	Provides efficient inference for smaller models	Efficient inference for complex models
Cost	Lower	High
Application	Good for simple learning application	Good for complex learning applications

improve performance and energy efficiency. Here, we compare the hardware accelerators for ANN and DNN based on their architectures, performance, and energy efficiency. A study [19] compared the performance and energy efficiency of various accelerators, including GPUs, FPGAs, and TPUs, for both training and inference of machine learning models. According to the study, while GPUs were more effective for training, TPUs offered the best performance and energy economy for inference. The performance and energy efficiency of ASICs and FPGAs for the inference of deep neural networks were compared in a different study [20–24]. A comparative study of ANN and DNN accelerator is presented in Table 2.1.

2.7 Conclusion

Integration of FPGA-based ANN and DNN accelerator presents a transformative advancement in signal processing. Since these accelerators can execute complex operations in parallel, they are excellent at efficiently managing large datasets. Through the customization of hardware solutions based on machine learning algorithms, they can attain maximum efficiency. Additionally, efficiency is enhanced by eliminating the need for external memory and enabling high-throughput processing, on-chip memory streamlines operations. FPGA-based accelerators have great potential to neural network applications; selection of the best hardware solution requires machine learning application as well as limitations of available resources.

AI disclosure statement: The author confirms that no artificial intelligence (AI)-generated content or AI-assisted writing tools were used in the preparation of this chapter (including summaries and the abstract). All text, analysis, and conclusions are the result of the author's original work.

References

1. Abadi, M., Barham, P., Chen, J., Chen, Z., Davis, A., Dean, J., Kudlur, M., Tensorflow: A system for large-scale machine learning. *12th Symposium on Operating Systems Design and Implementation*, pp. 265–283, 2016.
2. Chen, T. and Du, Z., FPGA-Based Deep Learning Accelerators: A Comprehensive Overview. *ACM Trans. Reconfigurable Technol. Syst.*, 12, 4, 1–30, 2019.

3. Courbariaux, M., Bengio, Y., David, J.P., Binaryconnect: Training deep neural networks with binary weights during propagations. *Advances in Neural Information Processing Systems*, pp. 3123–3131, 2015.
4. Farabet, C., Couprie, C., Najman, L., LeCun, Y., Learning hierarchical features for scene labeling. *IEEE Trans. Pattern Anal. Mach. Intell.*, 35, 8, 1915–1929, 2013.
5. Wöhrle, H., Tabie, M., Kim, S.K., Kirchner, F., Kirchner, E.A., A hybrid FPGA-based system for EEG-and EMG-based online movement prediction. *Sensors*, 17, 7, 1552, 2017.
6. Ravi, M., Sewa, A., Shashidhar, T.G., Sanagapati, S.S.S., FPGA as a hardware accelerator for computation intensive maximum likelihood expectation maximization medical image reconstruction algorithm. *IEEE Access*, 7, 111727–111735, 2019.
7. Del Sozzo, E., *et al.*, A common backend for hardware acceleration on FPGA. *IEEE International Conference on Computer Design (ICCD)*, IEEE, pp. 427–430, 2017.
8. Reuther, A., Michaleas, P., Jones, M., Gadepally, V., Samsi, S., Kepner, J., AI accelerator survey and trends. *IEEE High Performance Extreme Computing Conference (HPEC)*, IEEE, pp. 1–9, 2021.
9. Krizhevsky, A., Sutskever, I., Hinton, G.E., Imagenet classification with deep convolutional neural networks. *Advances in Neural Information Processing Systems*, pp. 1097–1105, 2012.
10. LeCun, Y., Bengio, Y., Hinton, G., Deep learning. *Nature*, 521, 7553, 436–444, 2015.
11. Liu, Z., Li, J., Shen, Z., Huang, G., Yan, S., Zhang, C., Learning efficient convolutional networks through network slimming. *IEEE International Conference on Computer Vision*, pp. 2736–2744, 2017.
12. Russakovsky, O., Deng, J., Su, H., Krause, J., Satheesh, S., Ma, S., Berg, A.C., ImageNet large scale visual recognition challenge. *Int. J. Comput. Vision*, 115, 3, 211–252, 2015.
13. Szegedy, C., Ioffe, S., Vanhoucke, V., Alemi, A., Inception-v4, inception-resnet and the impact of residual connections on learning. *Proceedings of the Thirty-First AAAI Conference on Artificial Intelligence*, pp. 4278–4284, 2021.
14. Qi, L. and Yao, K., Artificial intelligence enterprise human resource management system based on FPGA high performance computer hardware. *Microprocess. Microsyst.*, 82, 103876, 2021.
15. Prabakaran, B.S., Mrazek, V., Vasicek, Z., Sekanina, L., Shafique, M., AproxFPGAs: Embracing ASIC-based approximate arithmetic components for FPGA-based systems. *57th ACM/IEEE Design Automation Conference*, IEEE, pp. 1–6, 2020.
16. Imani, M., Samragh, M., Kim, Y., Gupta, S., Koushanfar, F., Rosing, T., RAPIDNN: In-memory deep neural network acceleration framework, *arXiv preprint arXiv:1806.05794*, 2018.

17. Palumbo, A., Vizza, P., Calabrese, B., Ielpo, N., Biopotetial signal monitoring systems in rehabilitation: A review. *Sensors*, 21, 21, 7172, 2020.
18. Mostafa, B., El-Attar, N., Abd-Elhafeez, S., Awad, W., Machine and deep learning approaches in genome: Review article. *Alfarama J. Basic Appl. Sci.*, 1023, 2020, doi. org/10.21608/ajbas.
19. Jouppi, N.P., Young, C., Patil, N., Patterson, D., A domain-specific architecture for deep neural networks. *Commun. ACM*, 61, 9, 50–59, 2020.
20. Zhang, H., Nguyen, H., Bui, X.N., Nguyen-Thoi, T., Bui, T.T., Nguyen, N., Moayedi, H., Developing a novel artificial intelligence model to estimate the capital cost of mining projects using deep neural network-based ant colony optimization algorithm. *Resour. Policy*, 66, 101604, 2020.
21. Kumar, A., Sindhwani, M., Sachdeva, S., Facial Emotion Recognition (FER) with Deep Learning Algorithm for Sustainable Development, in: *Sustainable Engineering: Concepts and Practices*, pp. 415–434, Springer International, Cham, 2024.
22. Kumar, A. and Tripathi, S.L., SBOX under PVT variation. *Analog Integr. Circuits Signal Process.*, 105, 1, 73–82, 2020.
23. Kumar, A., Sindhwani, M., Sachdeva, S., Detection of Mastitis Disease in Cow with Machine Learning Classifiers. *Israa Univ. J. Appl. Sci.*, 7, 1, 112–127, 2023.
24. Tripathi, S.L. and Mahmud, M., *Explainable Machine Learning Models and Architecture*, Wiley-Scrivener Publishing, USA, 2023, https://doi. org/10.1002/9781394186570.

3

Analysis of Double-Gate Tunnel FET Using AI and ML

Siga Karthik* and Sanjeet Kumar Sinha

School of Electronics and Electrical Engineering, Lovely Professional University, Jalandhar, Punjab, India

Abstract

Shrinking conventional MOSFETs down to the sub-10 nm nodes is highly limited and requires advanced transistor technologies, which in this regard comprises the Tunnel Field Effect Transistors and Double-Gate T-FETs. These devices promise better low-power performance through characteristics such as a sub-threshold slope of less than 60 mV/dec and high ON/OFF ratios, which enables them to be used in applications that are sensitive to power. Optimization of the DG-TFET is much more challenging due to the complicated interaction between the design parameters: material properties, doping concentrations, and geometrical dimensions. Conventional optimization approaches face problems in tackling this non-linearity as well as the huge dimensionality of the problem factors; therefore, new suggestions are needed to design and analyse such devices.

AI and ML provide robust solutions to overcome the challenge of optimizing Double-Gate Tunnel FETs (DG-TFETs), which mainly include automatic analysis of data, advancement in prediction accuracy, and reduced simulation time. This paper discusses how AI and ML can be applied to DG-TFET analysis, where supervised learning models are ANN and SVM for performance prediction and unsupervised methods are k-means clustering and PCA for pattern recognition, and reinforcement learning for optimization. Case studies show that AI-based methods not only significantly speed up design but also reduce errors and enhance the performance of devices. This is an important breakthrough in low-power semiconductor technologies.

***Keywords*:** Double-gate T-FETs, tunnel field effect transistors, optimization of the DG-TFET

**Corresponding author*: sigakarthikroyal@gmail.com

Abhishek Kumar and Suman Lata Tripathi (eds.) Smart Chips for Smart Devices: VLSI Design for Next-Generation IoT Solutions, (49–64) © 2026 Scrivener Publishing LLC

3.1 Introduction

The relentless demand for the next high-performance, low-power electronic devices has led to the necessity of new transistor technology. Given that the scaling of conventional MOSFETs is reaching its limit at sub-10 nm nodes, it has drawn substantial attention towards alternative devices. Among these, Tunnel Field Effect Transistors (T-FETs) and their special case—Double-Gate Tunnel FETs (DG-TFETs)—are among the strongest contenders for next-generation low-power electronics. These transistors demonstrate excellent characteristics, with a sub-60 mV/dec subthreshold slope, to enable increased switching speed with power savings and improved leakage control, an essential need for power-sensitive applications [1–9].

There is a significant challenge in the optimization of the design and performance of DG-TFETs because of their complex structure and operation mechanism. Carrier transport in BTBT-based DG-TFETs mainly depends upon material properties, doping concentrations, and geometrical parameters such as gate length and dielectric thickness; hence, it will be very challenging to tune up the device for better performances using traditional design approaches. In this backdrop, AI and ML are emerging as powerful tools for handling such problems. Techniques of AI and ML allow efficient processing of the reams of data generated through simulations or experiments and make possible automation analysis and performance prediction, followed by optimization of DG-TFETs [10]. In particular, data-driven modelling in AI and ML allows not only the acceleration of design processes but also an insightful understanding of complicated interplays between design parameters and device performance. The next-generation transistor development will be increasingly dependent on AI-driven approaches towards the realization of low-power and high-performance electronics [11].

3.2 Double-Gate Tunnel FET (DG-TFET)

3.2.1 Basic Structure and Operation

A Double-Gate Tunnel FET consists of two gates that control a narrow channel. In Figure 3.1, the source area of the DG-TFET is highly doped with polarity opposite to that of the channel and charge carriers enter the channel through band-to-band tunnelling instead of the thermionic

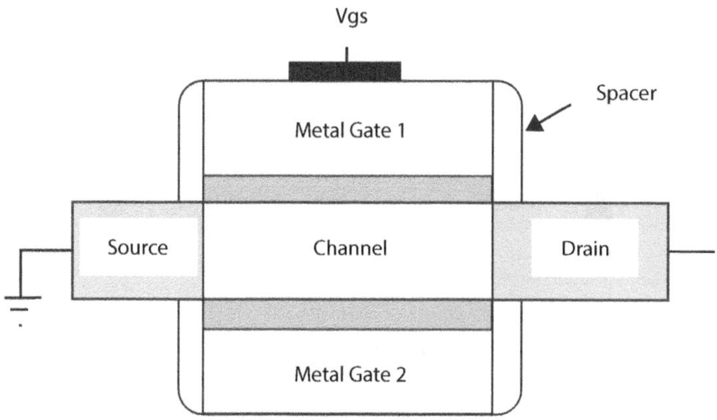

Figure 3.1 Basic structure of a double-gate T-FET [1].

emission. With two gates, electrostatic control can be enhanced which could therefore provide better channel regulation while suppressing short-channel effects [12].

3.2.2 Advantages and Challenges

Advantages:

- Low Sub-threshold Slope: T-FET's sub-threshold slopes are usually less than 60 mV/dec and hence can provide superior switching characteristics.
- High ON/OFF Ratio: In nature, DG-TFETs provide a high ON/OFF current ratio. This enables the usage of these devices in low-power applications [8].
- Reduced Leakage Current: The configuration consists of doubling the gate, thus reducing leakage current, which always remains a crucial challenge for traditional MOSFETs [10].

Challenges:

- Low ON-Current: The ON-current for the DG-TFET is relatively low, in comparison to the MOSFETs due to the tunnelling mechanism as shown in Figure 3.2 [10, 12].

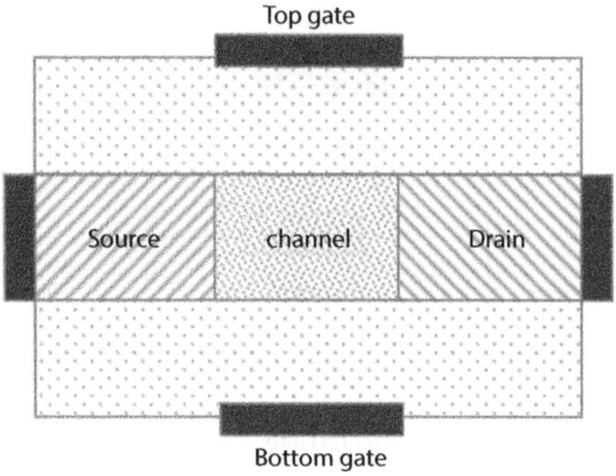

Figure 3.2 Tunnelling mechanism in DG-TFET [1].

- Complex Fabrication: The complex structure of DG-TFET makes it complicated to fabricate. Parameter Sensitivity: The performance parameters of DG-TFET are very sensitive to various factors such as material properties, doping concentration, and dimensions of the gate [18].
- Using the tunnelling mechanism, the carrier's tunnel across a potential barrier formed between the source and channel [3].

3.3 Role of AI and ML in DG-TFET Analysis

The interaction between design parameters and TFET device performance poses complexity, so the analysis of Double-Gate Tunnel FETs is quite challenging. Traditional approaches are inefficient in handling the high dimensional and strongly nonlinear nature of this data. The most important strengths are described below.

3.3.1 High-Dimensional Data Handling

AI models deal with complex relationships among numerous parameters by capturing the non-linear effect well, which is difficult to model by traditional methods. Pattern Recognition: The nature of ML algorithms makes it easy to identify unseen or hidden patterns and correlations in large datasets, thus optimum configurations of devices may be determined without exhaustive testing [6].

3.3.2 Better Predictions

It uses AI models, so it predicts the performance metrics accurately so not much physical experimentation is required. Optimization Process: It takes faster optimization of the design process and saves a lot of time and resources by finding optimal parameter configurations [16].

3.3.3 Reduced Simulation Time

AI-based surrogate models generate a quick approximation of high-fidelity simulations, which dramatically reduces the time required for computations and hence accelerates design [18]. Adaptive Learning: AI models learn from new data they encounter over time and constantly improve predictions and optimization strategies. These capabilities make AI and ML necessary for the improvement of DG-TFET technology by increasing the accuracy of performance predictions and hastening the process of design [15].

3.4 Applications of AI and ML in DG-TFETs

3.4.1 Current-Voltage Predictions

AI models can predict the current-voltage characteristics, threshold voltages and sub-threshold slopes based on the input parameters like gate length, oxide thickness, doping concentrations, etc., as shown in Figure 3.3.

3.4.2 Optimization of Device Parameters

ML algorithms can optimize those parameters affecting the performance of DG-TFET such as material choice, doping and geometrical dimensions, etc. Anomaly Detection: AI shall be able to detect anomalies and flaws in the device fabrication process or unusual patterns in the behaviour of devices [5].

3.4.3 Accelerating Simulations

Surrogate models developed using AI can closely follow the behaviour of DG-TFETs and therefore speed up device simulations, reducing computational time.

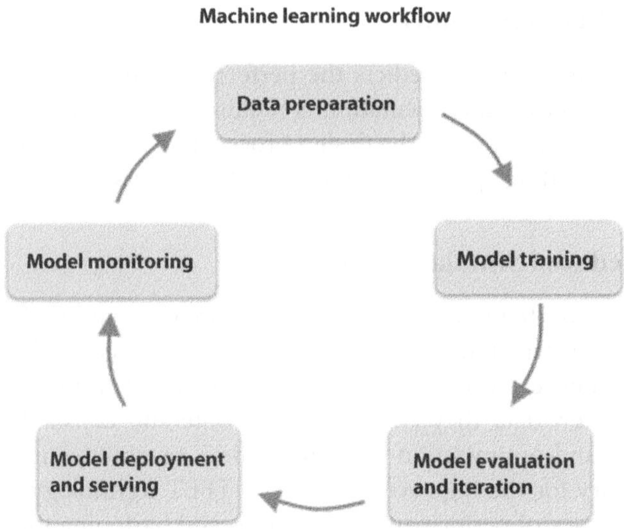

Figure 3.3 AI and ML workflow for DG-TFET analysis.

3.5 Machine Learning Models for DG-TFET Analysis

Supervised Learning for Performance Prediction

Generally, supervised learning techniques, where Artificial Neural Networks or Support Vector Machines or Decision Trees [5] and so on are learned from labelled data and used to predict the behaviour of DG-TFETs. Input features may include structural parameters such as gate length, doping concentration, etc. Output could be performance metrics such as ON-current, OFF-current, and threshold voltage, as shown Figures 3.4(a), 3.4(b) and 3.4(c). Recent works have also highlighted ML-facilitated design of TFETs using two-dimensional nanomaterials [17].

Artificial Neural Networks

This is highly capable of capturing non-linear relationships between input parameters and performance. ANNs can accurately predict the characteristics of devices with large datasets, as shown in Figure 3.5.

Support Vector Machines

SVMs can be used both for regression and classification. One can predict continuous variables, such as threshold voltage and determine whether a DG-TFET design meets the performance metrics.

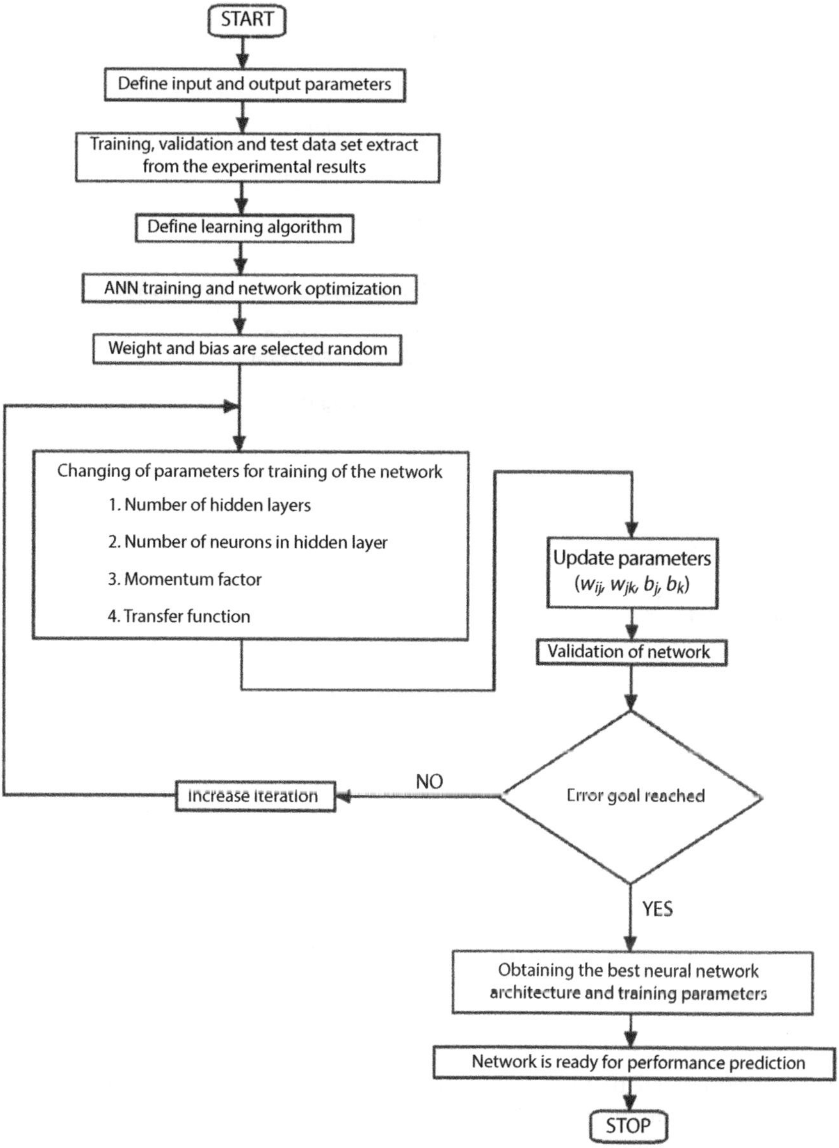

(a)

Figure 3.4 (a) ANN model for DG-TFET performance prediction.

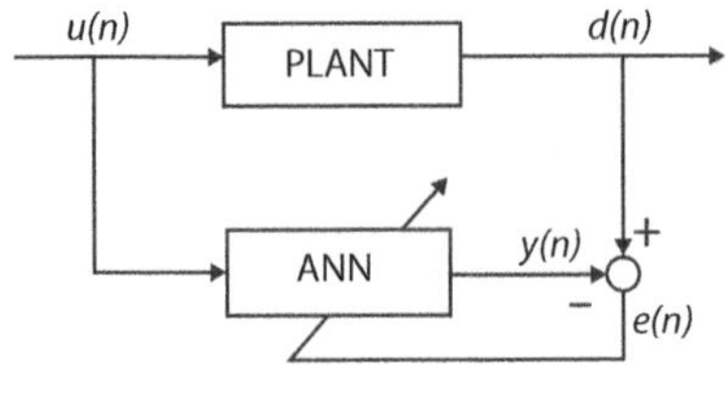

(b)

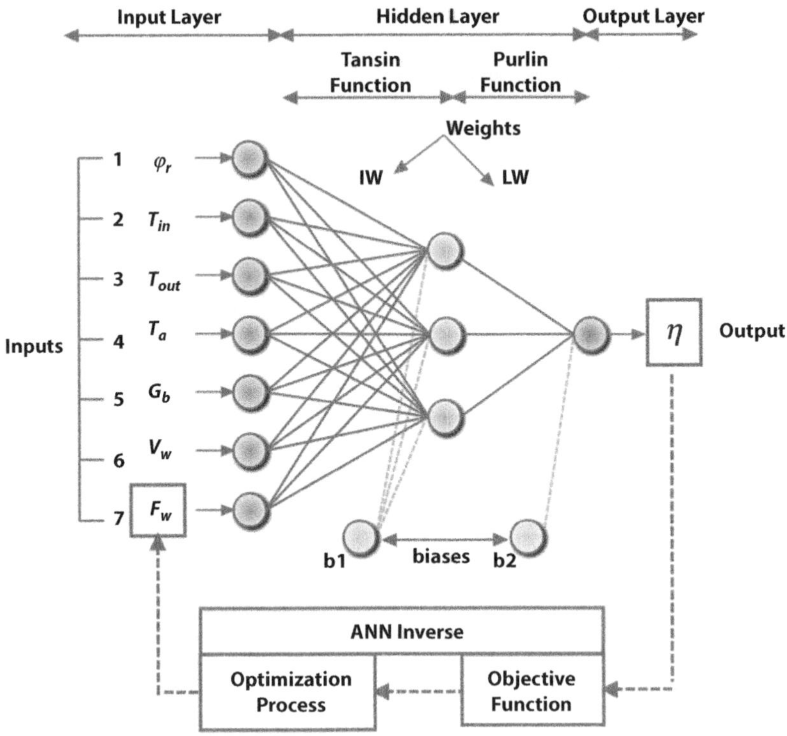

(c)

Figure 3.4 (Continued) (b) Block diagram of the developed modelling structure. (c) ANN architecture for the prediction of performance parameters of DG-TFET.

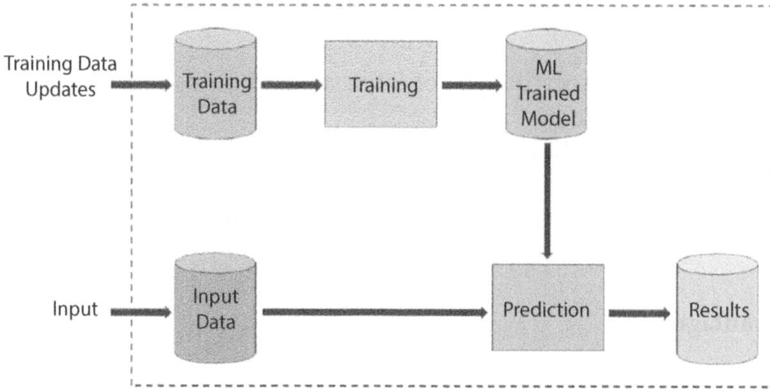

Figure 3.5a General workflow of AI & ML in DG-TFET analysis covering generation of the dataset, model training and optimization [1].

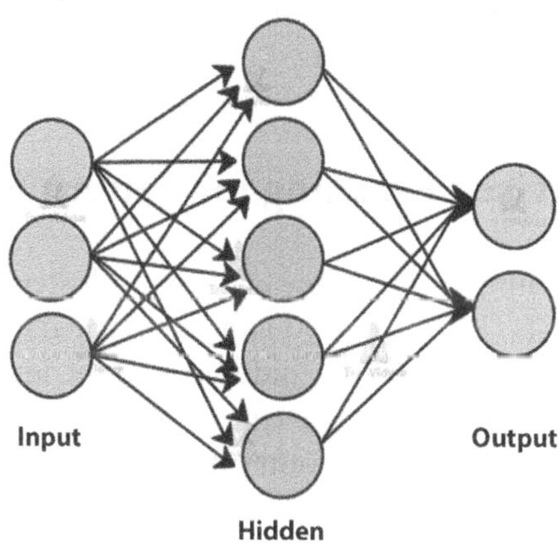

Figure 3.5b Architecture of artificial neural network.

Regression (SVR)

Goal: Predict metrics like threshold voltage and ON-current.
How: It maps input data into higher dimensions using kernels to handle non-linear relationships [14].
Advantages: Efficiently captures complex patterns, noise-robust.

Classification

Goal: Classify designs of interest based on certain performance criteria.
How: Find the optimum hyper-plane that separates different classes.
Application: It determines whether designs pass the thresholds concerning performance.

Advantages

Robustness: It is efficient even at high dimensions, and it would not over-fit.
Flexibility: Different data distributions are handled with different kernels.
Interpretability: Well-defined decision boundaries and insight into feature impacts.

3.6 Unsupervised Learning for Pattern Recognition

Unsupervised learning models like k-means clustering and Principal Component Analysis (PCA) are also applied to the data of DG-TFET for recognizing patterns that may be hidden without known labels. These models, on the one hand, recognize clusters of device designs that would behave similarly [13], while on the other hand, they reduce the dimensionality of the data set to make it easier to interpret.

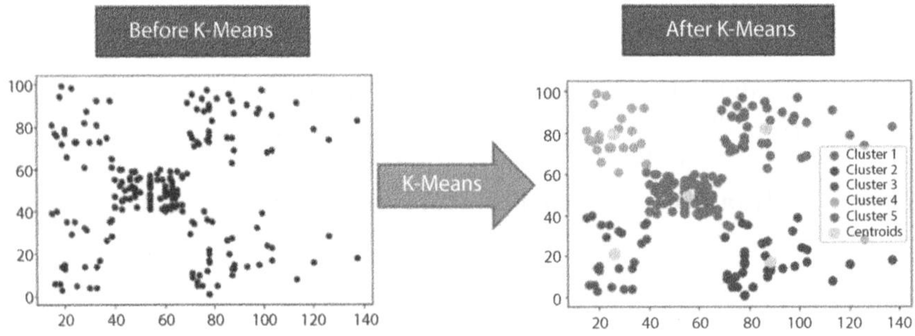

Figure 3.6 Clustering of DG-TFET designs using K-means [1].

Principal Component Analysis (PCA): PCA helps in the dimensionality reduction of the input features, and hence engineers may understand better the design parameters and their relationships with the performance metrics.

K-means Clustering: K-means clustering helps in grouping DG-TFET designs based on similarity, and thus this method may lead to easy identification of the best-performing designs [7] as shown in Figure 3.6.

3.7 Reinforcement Learning for Device Optimization

The models of Reinforcement Learning can also be considered to optimize the designs of DG-TFETs. In RL, the model iteratively tweaks the parameters of the devices based on the instantaneous reward received in response to every design implemented. In this regard, the model learns to maximize a reward function often defined as low power consumption, high ON-current, and minimal leakage. Figure 3.7 illustrates the reinforcement learning process used for DG-TFET optimization.

- Reward Function: The reward function can be designed to maximize desired performance metrics, such as the ON/OFF ratio or minimize power consumption [3].
- Exploration vs. Exploitation: RL algorithms need to balance exploring new design possibilities (exploration) with refining known successful configurations (exploitation).

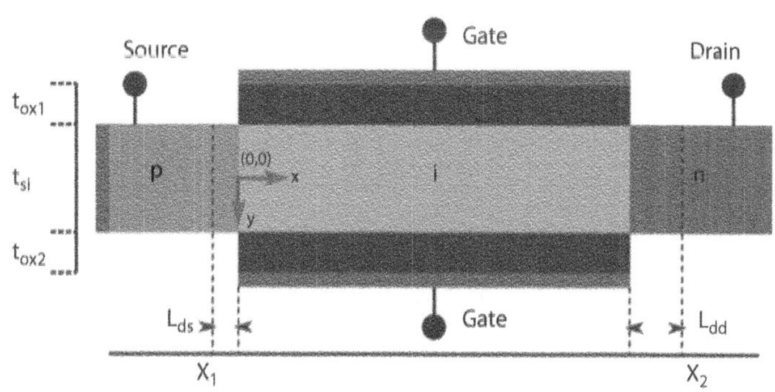

Figure 3.7 Reinforcement learning process for DG-TFET optimization [1].

3.8 Case Study: AI-Driven DG-TFET Optimization

3.8.1 Dataset Collection

The training dataset of AI and ML models is developed from the simulation process of TCAD. In addition, the different configurations of devices combined with relevant performance metrics such as:

- Input Features: Gate length, oxide thickness, doping levels, source/drain materials, etc.
- Output Metrics: ON-current I_ON, OFF-current I_OFF, sub-threshold slope and threshold voltage as shown in Figure 3.8 [7, 10].

3.8.2 Model Training and Validation

The trained models presented in Figure 3.9 are validated using unseen data for validating the accuracy during DG-TFET predictions. Further, Metrics such as Mean Absolute Error (MAE) and Root Mean Squared Error (RMSE) will be carried out to measure the performance of the model [3].

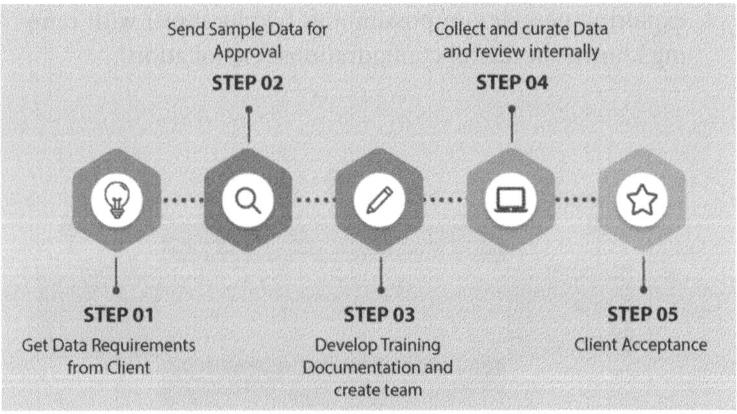

Figure 3.8 Data collection workflow.

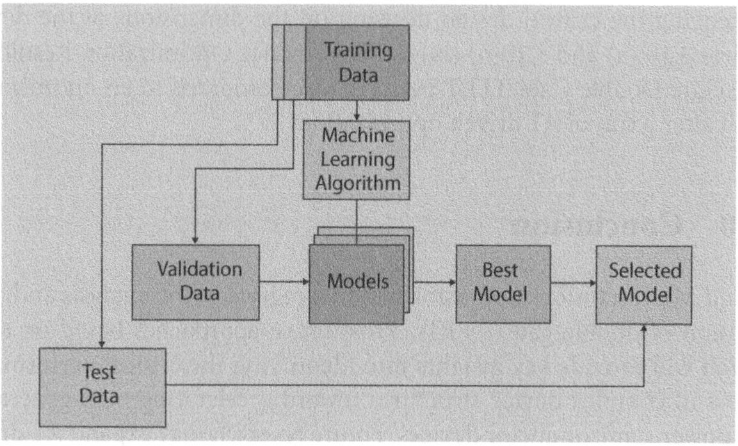

Figure 3.9 Model training and validation.

3.9 Results and Insights

A critical understanding of how the design parameters impact the performance of DG-TFET can be achieved by bringing the AI and ML models into play. For example, a trained model may indicate that the subthreshold slope is more sensitive to gate oxide thickness compared to gate length, thus

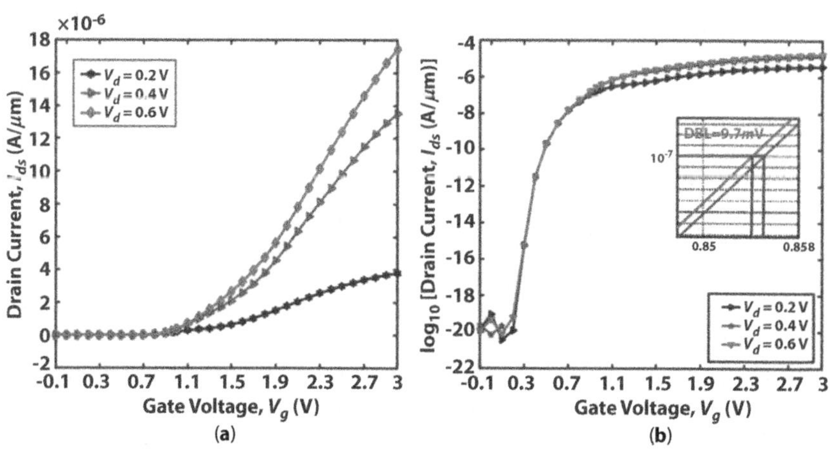

Figure 3.10 (a and b) Performance optimization results for a baseline double-gate TFET performance compared to an optimization result after a run of AI-driven optimization [1].

influencing the critical design decision on the dimensions of the device. Figures 3.10 (a) and 3.10(b) shows Performance Optimization Results for a baseline Double Gate-TFET performance compared to an optimization result after a run of AI-driven optimization.

3.10 Conclusion

AI and ML can unlock a transformational regime in the analysis and optimization of double-gate T-FETs. Prospective approaches based on automation will provide key insights into identifying the critical performance factors in complex device structures toward accelerating next-generation low-power semiconductor devices. Future research may expand AI-driven analysis to more advanced multi-gates or 3D T-FET architectures and advance the boundaries of semiconductor technology.

AI disclosure statement: The author confirms that no artificial intelligence (AI)-generated content or AI-assisted writing tools were used in the preparation of this chapter (including summaries and the abstract). All text, analysis, and conclusions are the result of the author's original work.

References

1. Bala., S. and Mamta, K., Design and simulation of nanoscale double-gate TFET/tunnel CNTFET. *J. Semicond.*, 39, 4, 044001_1 - 044001_5T, 2018.
2. Verreck, D., Groeseneken, G., Verhulst, A., The tunnel field-effect transistor, in: *Wiley Encyclopedia of Electrical and Electronics Engineering*, pp. 1–24, 2016.
3. Sen, S., Khosla, M., Raman, A., Design and Challenges in TFET, in: *Advanced Field-Effect Transistors*, pp. 23–45, CRC Press, 2023.
4. Potharaju, R. and Choudhuri, B., Evolution of Tunnel Field-Effect Transistor and Scope in Low Power Applications: A Detailed Review. *Micro and Nanoelectronics Devices, Circuits and Systems: Select Proceedings of MNDCS 2022*, pp. 141–158.
5. Bagga, N. and Sarkar, S.K., An analytical model for tunnel barrier modulation in triple metal double gate TFET. *IEEE Trans. Electron Devices*, 62, 7, 2136–2142, 2015.
6. Gupta, S.K. and Kumar, S., Analytical modelling of a triple material double gate TFET with hetero-dielectric gate stack. *Silicon*, 11, 3, 1355–1369, 2019.
7. Gholizadeh, M. and Hosseini, S.E., A 2-D analytical model for double-gate tunnel FETs. *IEEE Trans. Electron Devices*, 61, 5, 1494–1500, 2014.

8. Guha, S., Pachal, P., Ghosh, S., Sarkar, S.K., Analytical model of a novel double gate metal-infused stacked gate-oxide tunnel field-effect transistor (TFET) for low power and high-speed performance. *Superlattices Microstruct.*, 146, 106657, 2020.
9. Mukherjee, A., Debnath, P., Chanda, M., Machine learning-based output prediction of negative capacitance tunnel-FET. *Int. J. Electron.*, 111, 8, 1331–1345, 2024.
10. Suguna, M., Charumathi, V., Balamurugan, N.B., Hemalatha, M., Sriram Kumar, D., Machine learning-based multi-objective optimisation of tunnel field effect transistors. *Silicon*, 14, 17, 11109–11119, 2022.
11. Akbar, C., Li, Y., Thoti, N., Device-simulation-based machine learning technique for the characteristic of line tunnel field-effect transistors. *IEEE Access*, 10, 53098–53107, 2022.
12. Chong, C., Liu, H., Wang, S., Wang, Z., Algorithm prediction of single particle irradiation effect based on novel TFETs. *Nanotechnology*, 35, 45, 455201, 2024.
13. Turkane, S.M. and Kureshi, A.K., Review of the tunnel field effect transistor (TFET). *Int. J. Appl. Eng. Res.*, 11, 7, 4922–4929, 2016.
14. Nirschl, T., Fischer, J., Fulde, M., Bargagli-Stoffi, A., Sterkel, M., Sedlmeir, J., Schmitt-Landsiedel, D., Scaling properties of the tunnelling field effect transistor (TFET): Device and circuit. *Solid State Electron.*, 50, 1, 44–51, 2006.
15. Villalon, A., Le Carval, G., Martinie, S., Le Royer, C., Jaud, M.A., Cristoloveanu, S., Further insights into TFET operation. *IEEE Trans. Electron Devices*, 61, 8, 2893–2898, 2014.
16. Avci, U.E., Rios, R., Kuhn, K., Young, I.A., Comparison of performance, switching energy and process variations for the TFET and MOSFET in logic. *Symposium on VLSI Technology-Digest of Technical Papers*, IEEE, pp. 124–125, 2011.
17. Tsang, C.I., Pu, H., Chen, J., Multiscale Simulation and Machine Learning Facilitated Design of Two-Dimensional Nanomaterials-Based Tunnel Field-Effect Transistors: A Review, arXiv preprint arXiv:2409.18965, 2024.
18. Singh, S., Singh, N.K., Singh, S., Naughariya, A., Niharika, N., Design Optimization of Doping-less InGaAs TFET and GaAs/Si-Heterojunction Doping-less TFET for Potential Breast Cancer Sensing Applications, in: *Proceedings of International Conference on Communication and Artificial Intelligence: ICCAI 2020*, pp. 123–133, Springer Singapore.

4

NC-TFET Device and Its Applications in Future VLSI

Katuri Yeshwant* **and Sanjeet Kumar Sinha**

School of Electronics and Electrical Engineering, Lovely Professional University, Jalandhar, Punjab, India

Abstract

A Negative Capacitance Tunnel Field Effect Transistor (NC-TFET) is a novel semiconductor device that has attracted considerable attention from the scientific fraternity since it offers the possibility of relaxing the fundamental limitations of conventional complementary metal-oxide-semiconductor (CMOS) technology. The concept combines two concepts: ferroelectric materials for negative capacitance and tunnel field effect transistor (TFET) architecture. This device targets two major issues of Very Large Scale Integration (VLSI): reducing power consumption with proper performance at low voltage levels. Thus, the negative capacitance effect is introduced into the gate stack of the NC-TFET by adding a ferroelectric layer that amplifies the gate voltage and leads to a subthreshold swing below the thermal limit of 60 mV/decade. This would enable NC-TFETs to operate at lower supply voltages and significantly impact power dissipation. However, TFET architecture based on band-to-band tunneling allows one to achieve ultra-low leakage currents that contribute significantly to energy efficiency. We have simulated all this in T-CAD. NC-TFETs, being properly poised shortly for VLSI applications, will be a very critical enabler for ultra-low-power computing systems; they will become part and parcel of high-end wearable electronics, Internet of Things (IoT) devices, and advanced data centers. They will suit next-generation processors, memory technologies, as well as 3D integration, promising impressive energy efficiency while maintaining scalability. Nevertheless, NC-TFETs would offer an opportunity for more power-efficient and high-performance computing architectures in an emerging application space like neuromorphic computing and artificial intelligence hardware. At present with further device scaling into the sub-10nm

**Corresponding author*: katuri.yeswanth@gmail.com

Abhishek Kumar and Suman Lata Tripathi (eds.) Smart Chips for Smart Devices: VLSI Design for Next-Generation IoT Solutions, (65–76) © 2026 Scrivener Publishing LLC

regime, NC-TFETs are a promising route forward to extend Moore's Law as well as cater to the imminent challenges of future VLSI technology.

Keywords: Negative capacitance tunnel field effect transistor (NC-TFET), CMOS technology, ferroelectric, T-CAD, Internet of Things (IoT)

4.1 Introduction

Since the semiconductor industry is vigorously pushing toward greater performance and lower power consumption, new bottlenecks have emerged from the traditional MOSFETs. While in recent years low-power, high-efficiency devices have not only grabbed much importance in consumer electronics but also crucial applications like wearables, IoT, and energy-efficient processors, novel transistor designs that break away from this paradigm are emerging. The Tunnel Field Effect Transistor (TFET), using band-to-band tunnelling as a conduction mechanism, is one of the most promising alternatives and provides more advantages in ultra-low-power operation.

However, the subthreshold swing (SS) that TFETs experience causes them to have limitations in reducing this swing, which becomes a hindrance to their switching speed and efficiency at low voltages. A very recent breakthrough in this problem has been the development of a Negative Capacitance TFET (NC-TFET) that exhibits elements of TFET principles in combination with negative capacitance from ferroelectric materials. Integration will further improve TFETs, thus enabling quicker switching with less power consumption. Superior electrostatic control, very low voltage, and energy-efficient factors are guaranteed to make NC-TFETs highly prominent in the future of semiconductor devices.

This chapter provides a detailed perception of NC-TFETs and encompasses some parts of their working mechanisms, structure, material selection, difficulties in their manufacturing, and applications in the shifting scenario of the semiconductor landscape.

4.2 Basics of Tunnel Field Effect Transistor (TFET)

4.2.1 Working Principle of TFET

In a TFET, quantum mechanical band-to-band tunneling mediates current flow through the device, in contrast to thermionic emission over a barrier, which is what dominates conventional MOSFETs. This mechanism affords subthreshold swings less than 60 mV/decade, a limit present in MOSFETs due to the Boltzmann distribution [1].

4.2.2 Band-to-Band Tunneling (BTBT)

The BTBT mechanism at the heart of TFET operations involves electrons in the valence band of the source region, influenced by a strong electric field and hence tunnel into the conduction band of the channel region. Since the gate voltage controls the width of the tunnel barrier, it makes it narrow, thereby controlling the tunneling current through the drain [2] as shown in Figure 4.1.

4.2.3 TFET Advantages

- **Low Subthreshold Swing (SS):** The subthreshold swing of below 60 mV/decade achieved by TFETs makes them a reality for ultra-low power switching.
- **Low Off-State Leakage:** TFETs are observed to have significantly lower off-state leakage current since it is based on tunneling rather than thermal emission.

4.2.4 TFET Challenges

- **Low ON-Current (ION _{ON} ON):** Generally, MOSFETs suffer from an inherently low ON-current as compared to their TFET counterparts since their tunneling probability is

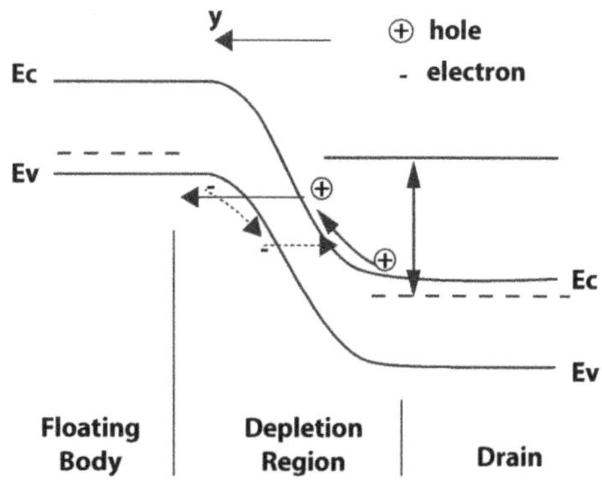

Figure 4.1 Band-to-band tunneling in TFET [10].

bounded by the width of the tunneling barrier and it cannot be easily reduced [3].
- **Material Constraints:** In contrast to MOSFET, TFETs rely on materials with very small band gaps and high tunneling efficiency, thereby limiting the scope of material choice.

4.3 Negative Capacitance Effect

Negative capacitance is the behavior of some ferroelectric materials that respond oppositely to the conventional expectation, that is, an increase in charge on the capacitor results in a reduction in voltage. This phenomenon is due to the intrinsic characteristics of ferroelectric materials, which yield non-linear polarization curves. Using Hafnium Zirconium Oxide (HZO), Lead Zirconate Titanate (PZT), or Barium Strontium Titanate (BST) as ferroelectric material, NC-TFETs take benefit from negative capacitance for the magnification of the gate voltage. In traditional transistors, the gate capacitance forms a voltage divider between the gate and the channel. By integrating a negative capacitance element into the gate stack, this divider is altered to provide voltage amplification at the channel. Essentially, the ferroelectric layer reduces the overall capacitance of the gate, which enhances the electrostatic control of the gate on the channel, thus enabling this device to switch at lower voltages. This is an important aspect in bringing down the subthreshold swing below the 60 mV/dec limit, thereby helping it to switch fast and utilize less power [4] as shown in Figure 4.2.

Such incorporation of negative capacitance into the gate stack of a TFET will lead to what is called NC-TFET—the enhancement in both power efficiency and switching speed.

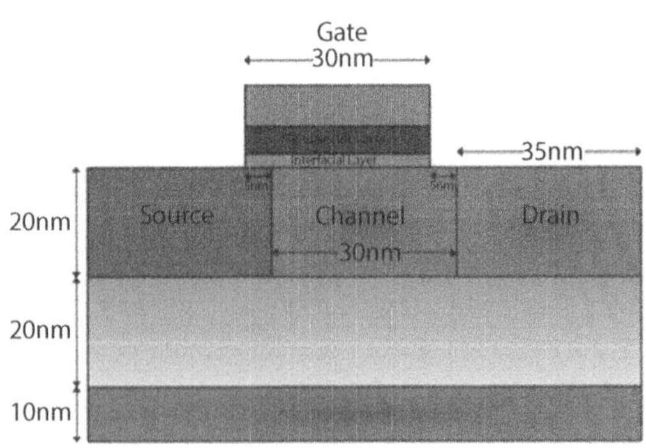

Figure 4.2 Negative capacitance effect in NC-TFET [11].

4.3.1 NC-TFET Device Architecture

The NC-TFET device structure is an extension of the conventional TFET with the significant difference being the inclusion of a ferroelectric material in the gate stack, which provides a layer that enables the generation of negative capacitance effects and thereby amplifies the control exerted by the gate on the channel [5] as shown in Figure 4.3.

4.3.2 Key Components of NC-TFET

1. **Source**: In the NC-TFET, the source region is heavily doped with p-type or n-type material, depending on the nature of the transistor being considered. Such heavy doping guarantees that the carrier will be available at the right energy level available for tunneling.
2. **Channel:** A thin channel of NC-TFET is formed using a narrow bandgap material to increase the probability of tunneling. Materials that can be used in the channel include:
   - Silicon-Germanium (SiGe) has a smaller bandgap compared to silicon, which makes it easier for easier tunneling.
   - III-V compounds such as Indium Arsenide (InAs) or Gallium Antimonide (GaSb) have low band gaps but high mobility and efficient tunneling.
3. **Drain**: The source to drain is highly doped with the opposite type of carrier compared with the source; n-type for p-TFETs, and p-type for n-TFETs. A large voltage difference

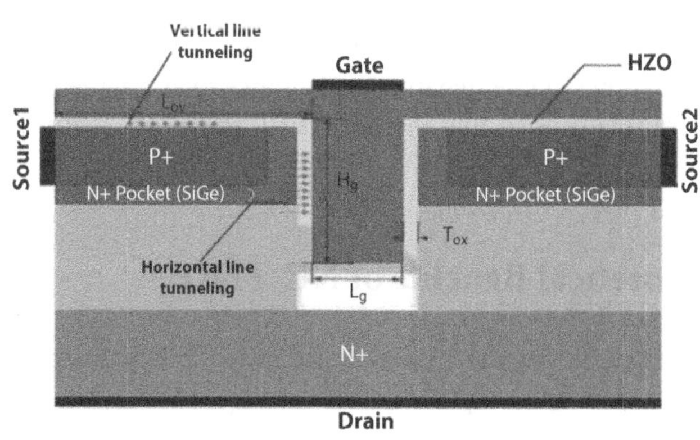

Figure 4.3 NC-TFET device structure [12].

is created between the source and the drain to enhance tunneling through the bandgap process.

4. **Gate**: An NC-TFET gate typically consists of a high-k dielectric layer, combined with a ferroelectric material. The negative capacitance due to the ferroelectric material enhances the gate voltage and thus the electrostatic control over the channel. The channel's conductance is modulated by the gate and consequently controls the tunneling of charge carriers from the source to the drain.

4.3.3 Materials for Negative Capacitance

The choice of ferroelectric material is critical in achieving the negative capacitance effect as desired in NC-TFETs. Among the commonly used ferroelectric materials are the following:

- **Hafnium Zirconium Oxide (HZO):** This is due to the reason that HZO retains all the benefits of being compatible with existing CMOS processes besides having the advantage of scaling down to smaller technology nodes.
- **Lead Zirconate Titanate (PZT):** PZT is one ferroelectric material that has been a focus for much research, even though it suffers from poor integration issues with more sophisticated semiconductor processes; it possesses good negative capacitance.
- **Barium Strontium Titanate (BST):** BST possesses good ferroelectric properties and therefore is used extensively in RF and memory applications.

Therefore, these materials should be engineered with great care in a manner that shows stable negative capacitance over a wide range of voltages and temperatures so that the NC-TFET works efficiently under various use cases.

4.4 Theoretical Benefits of NC-TFET

The integration of negative capacitance into TFETs introduces many advantageous concepts, especially concerning device performance and power efficiency.

4.4.1 Reduced Subthreshold Swing (SS)

It has been known that conventional MOSFETs are limited by the subthreshold swing of 60 mV/decade; this indicates that each drop in the gate voltage by 60 mV lowers the drain current by an order of magnitude. This restriction becomes a real challenge in designing devices that have to switch at lower voltages for low-power electronics. NC-TFETs can exploit negative capacitance to achieve subthreshold swings below 60 mV/decade. This finally leads to higher switching speeds and lower voltage operation, hence making NC-TFET suitable for ultra-low power applications; see Figure 4.4.

4.4.2 Lower Power Consumption

One of the most attractive features of NC-TFETs is their ability to be operated at ultra-low supply voltages, VDD_{DD}DD. The superior electrostatic control provided by the ferroelectric material enables a device to turn on or off at much lower gate voltages, which translates directly to less power consumption. As such, NC-TFETs are extremely attractive for application regimes in which energy efficiency is paramount, such as portable and battery-powered devices [6].

4.4.3 Enhanced Scalability

SCEs become a dominant issue as the semiconductor devices are scaled down further. SCEs degrade the transistor performance as the ability of the gate in the control of the channel reduces due to the diminution of the

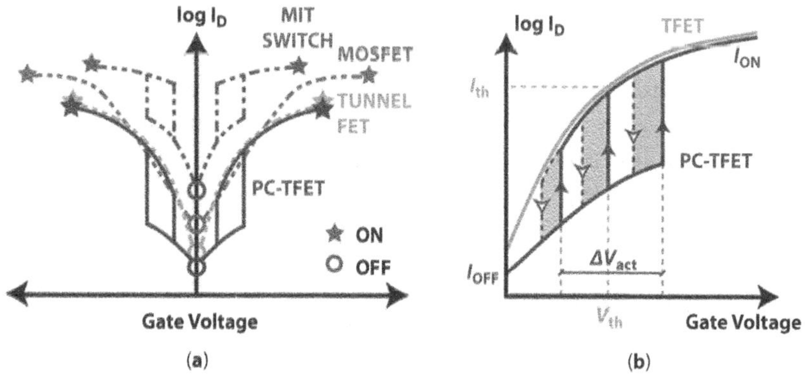

Figure 4.4 Subthreshold swing comparison between MOSFET, TFET, and NC-TFET [13].

length. NC-TFETs provide improved SCE immunity due to amplified gate control and hence have a brighter prospect for scaling into future technology nodes.

4.5 NC-TFET Fabrication and Process Integration

4.5.1 Challenges in Fabrication

Although theoretically attractive, the NC-TFET has various fabrication challenges that make its practical implementation difficult [7].

- **Ferroelectric Deposition:** High-precision deposition of a ferroelectric material is necessary when the ferroelectric material used in NC-TFET should be deposited uniformly and with reliable performance. Ferroelectric material thin films such as HZO need to be deposited using atomic layer deposition or other advanced techniques compatible with CMOS fabrication.
- **Material Stability:** The ferroelectric material must be stabilized in the negative capacitance values regardless of the long-term operation conditions such as temperature and voltage. Hence, this demands careful engineering so that the latter is long-term stable and reliable.

4.5.2 Integration with CMOS Processes

Another important aspect is the compatibility of NC-TFETs with traditional CMOS manufacturing processes. As the existing HZO material has been under strong development for advanced CMOS nodes, NC-TFETs can be added to the existing production lines with very minimal process modifications. They are a promising technology for future semiconductor products with less time and cost compared to adopting an entirely new device architecture [8].

4.6 Applications of NC-TFET

Since NC-TFETs can function at very low voltages and consume much less power, they are suitable for a wide range of applications, as shown in Figure 4.5.

- **Wearable Devices:** NC-TFETs could be implemented to increase battery life in wearable electronics, so these devices can operate longer with similar performance levels.
- **Internet of Things (IoT):** IoT devices are often placed in an energy-constrained environment, which will make the low-power operation of NC-TFETs highly beneficial in prolonging the lifetimes of devices.
- **Ultra-Low-Power Microprocessors:** NC-TFETs can be incorporated into next-generation processors. Those are designed for low-power consumption. Among them are those in mobile computing, sensors, and medical implants.
- **Medical Devices:** High power and significant heat dissipation are critical for biosensors and medical implants. The ability to operate very efficiently makes NC-TFETs optimal candidates for such applications wherein long-term performance and negligible interference with biological tissues are of great concern.

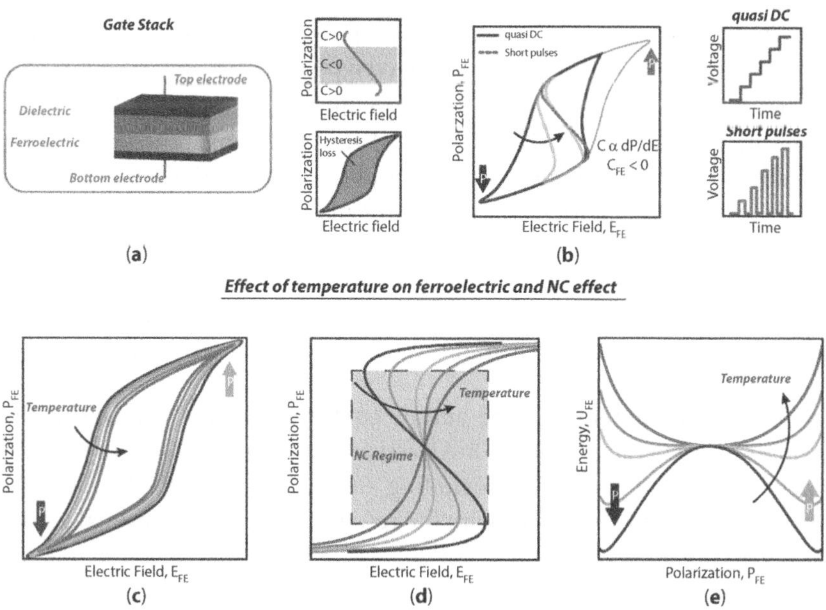

Figure 4.5 Power efficiency in NC-TFET applications [14].

4.7 Future Prospects and Research Directions

Although promising advancements in the NC-TFETs are indicated, several areas require more research and development [9].

- **Optimizing Ferroelectric Materials:** Research is still going on with ferroelectric material that has more robust negative capacitance properties, in particular with the scaling of devices down to even smaller technology nodes.
- **Device Modeling:** It requires more complex models to predict the behavior of NC-TFETs under different situations, assuring reliable and predictable functionality for various applications.
- **Thermal Stability:** For stable negative capacitance in NC-TFETs, the ferroelectric materials have to be proven to withstand a range of temperatures, which is one of the biggest hurdles for practical use in stressed applications.

4.8 Conclusion

NC-TFET devices are a breakthrough in the development of transistor science, holding the path for scalings of MOSFETs with fundamental limitations. The combination of quantum mechanical tunneling with the negative capacitance effect in the NC-TFET realizes superior subthreshold swing together with reduced power consumption in a qualitatively different way from limited scalabilities. These features make the NC-TFET device the frontrunner in ultra-low-power electronics for future small-portable to large-scale integrated circuits applications. As the study regarding ferroelectric materials and the method of their fabrication advances, NC-TFETs will be seen more frequently in future semiconductor technology.

AI disclosure statement: The author confirms that no artificial intelligence (AI)-generated content or AI-assisted writing tools were used in the preparation of this chapter (including summaries and the abstract). All text, analysis, and conclusions are the result of the author's original work.

References

1. Saurabh, S. and Kumar, M.J., *Fundamentals of Tunnel Field-Effect Transistors*, CRC Press, 2016.

2. Gundapaneni, S., Bajaj, M., Pandey, R.K., Murali, K.V., Ganguly, S., Kottantharayil, A., Effect of band-to-band tunneling on junctionless transistors. *IEEE Trans. Electron Devices*, *59*, 4, 1023–1029, 2012.
3. Lin, Z., Chen, P., Ye, L., Yan, X., Dong, L., Zhang, S., Chen, J., Challenges and solutions of the TFET circuit design. *IEEE Trans. Circuits Syst. I Regul. Pap.*, *67*, 12, 4918–4931, 2020.
4. Lin, C.I., Khan, A.I., Salahuddin, S., Hu, C., Effects of the variation of ferroelectric properties on negative capacitance FET characteristics. *IEEE Trans. Electron Devices*, *63*, 5, 2197–2199, 2016.
5. Anas, M., Amin, S.I., Beg, M.T., Anam, A., Chunn, A., Anand, S., Design and analysis of GaSb/Si-based negative capacitance TFET at the device and circuit level. *Silicon*, *14*, 17, 11951–11961, 2022.
6. Upadhyay, A.K., Rahi, S.B., Tayal, S., Song, Y.S., Recent progress on negative capacitance tunnel FET for low-power applications: Device perspective. *Microelectron. J.*, *129*, 105583, 2022.
7. Zhao, Y., Liang, Z., Huang, Q., Chen, C., Yang, M., Sun, Z., Huang, R., A novel negative capacitance tunnel FET with improved subthreshold swing and nearly non-hysteresis through hybrid modulation. *IEEE Electron Device Lett.*, *40*, 6, 989–992, 2019.
8. Nirschl, T., Wang, P.F., Weber, C., Sedlmeir, J., Heinrich, R., Kakoschke, R., Schmitt-Landsiedel, D., The tunneling field effect transistor (TFET) is an add-on for ultra-low-voltage analog and digital processes, in: *IEDM Technical Digest. IEEE International Electron Devices Meeting, 2004*, IEEE, pp. 195–198, 2004, December.
9. Reddy, N.N. and Panda, D.K., A comprehensive review on tunnel field-effect transistor (TFET) based biosensors: recent advances and prospects on device structure and sensitivity. *Silicon*, *13*, 9, 3085–3100, 2021.
10. Mamouni, F.E., Dixit, S.K., Schrimpf, R.D., Adell, P.C., Esqueda, I.S., McLain, M.L., Xiong, W., Gate-length and drain-bias dependence of band-to-band tunneling-induced drain leakage in irradiated fully depleted SOI devices. *IEEE Trans. Nucl. Sci.*, *55*, 6, 3259–3264, 2008.
11. Singh, A., Sinha, S.K., Chander, S., Effect of negative capacitance on heterojunction tunnel field effect transistor, in: *2021 5th International Conference on Electronics, Communication and Aerospace Technology (ICECA)*, IEEE, pp. 297–303, 2021, December.
12. Li, W., Jia, Q., Pan, Y., Chen, X.A., Yin, Y., Wu, Y., Wang, S., A T-shaped gate tunneling field effect transistor with negative capacitance, super-steep subthreshold swing. *Nanotechnology*, *32*, 39, 395202, 2021.
13. Vitale, W.A., Casu, E.A., Biswas, A., Rosca, T., Alper, C., Krammer, A., Ionescu, A.M., A steep-slope transistor combining phase-change and band-to-band-tunneling to achieve a sub-unity body factor. *Sci. Rep.*, *7*, 1, 355, 2017.
14. Kamaei, S., Saeidi, A., Gastaldi, C., *et al.*, Gate energy efficiency and negative capacitance in ferroelectric 2D/2D TFET from cryogenic to high temperatures. *npj 2D Mater. Appl.*, 5, 76, 2021.

5

Low-Power VLSI Design Techniques for High-Performance Applications

Ashutosh Kumar Yadav[1]*, Sweta Chander[1], Suman Lata Tripathi[2] and Sanjeet Kumar Sinha[1]

[1]School of Electronics and Electrical Engineering, Lovely Professional University, Phagwara, Punjab, India
[2]Department of Electronics and Telecommunication, Symbiosis Institute of Technology, Pune, Symbiosis International (Deemed University), Pune, India

Abstract
High-performance computers and portable devices continuously raise the demands of low-power integrated circuits. A thorough analysis of low-power design techniques that preserve or improve performance in sophisticated integrated circuits is presented in this research. The effects of important methods for lowering dynamic and leakage power are investigated, including multi-threshold, clock gating, power gating, DVFS, and transistor sizing. The design trade between power and silicon area are carefully examined to find the optimal application domains. In this chapter methodology to minimize the power consumption at the circuit and architecture levels, making them applicable to CPUs, signal processing units, and embedded systems without sacrificing the throughput has been analysed. Simulations and experimental data demonstrate that combining several power-saving techniques can result in significant increases in energy efficiency along with maintaining good performance in modern VLSI systems.

Keywords: Low-power VLSI, CMOS, clock gating, power gating, voltage scaling, multi-threshold design, energy efficiency, high-performance circuits

**Corresponding author*: aasrhiant.official@gmail.com

Abhishek Kumar and Suman Lata Tripathi (eds.) Smart Chips for Smart Devices: VLSI Design for Next-Generation IoT Solutions, (77–102) © 2026 Scrivener Publishing LLC

5.1 Introduction

In today's technology-driven world, the demand for portable and high-performance electronic devices necessitates efficient power management in Very Large Scale Integration (VLSI) designs. Power consumption in VLSI circuits directly impacts device longevity, performance, and thermal management. This chapter provides an in-depth exploration of low-power VLSI design techniques applicable to high-performance applications. It discusses the fundamental concepts of power consumption in VLSI circuits and elaborates on various low-power techniques at different abstraction levels, including technology, circuit, architecture, and algorithm levels. Additionally, advanced methods like approximate computing and machine learning-based optimizations are explored. This chapter synthesizes insights from numerous research studies to provide a comprehensive guide to state-of-the-art low-power VLSI design strategies.

The remarkable and exponential increase in the density of transistors, as initially posited by Moore's Law, has catalyzed profound advancements in the realm of computational power, facilitating the development of increasingly sophisticated technologies. Nevertheless, this significant surge in transistor density has concurrently engendered a plethora of challenges associated with elevated power consumption and the consequent thermal dissipation that these very challenges pose in the context of Very Large Scale Integration (VLSI) circuits. As the demand for high-performance applications escalates—especially within the domains of mobile and portable devices—there exists an intensified imperative for the implementation of low-power design methodologies that are capable of ensuring energy efficiency, all while maintaining a level of performance that does not suffer any degradation.

5.1.1 The Imperative for Low-Power Design

In light of the escalating demand for systems that not only deliver superior performance but also exhibit greater power efficiency, the domain of low-power Very Large Scale Integration (VLSI) design has emerged as an area of paramount significance within the realms of contemporary research and development. The power consumption associated with VLSI circuits has a direct and profound effect on various crucial aspects, including the overall performance of the system, the operational costs that are incurred, and the longevity of battery life in portable electronic devices. Furthermore, it is important to recognize that the heat produced by systems operating at high power levels can lead to a deterioration in performance, a reduction in reliability, and an increase in the costs associated with cooling solutions. The principal motivators driving the necessity for low-power design encompass several critical domains.

5.1.1.1 Portable Devices

The functionality of mobile phones, tablets, and laptops is heavily contingent upon their reliance on battery power, and thus, the implementation of low-power design principles is essential in ensuring that these devices can achieve prolonged battery life while simultaneously providing the level of performance required for the diverse range of modern applications that users expect.

5.1.1.2 Data Centers

The power consumption levels observed in data centers represent a significant portion of operational expenditures, and therefore, the reduction of power draw from servers and storage systems can yield considerable economic advantages as well as positive environmental impacts that cannot be overlooked.

5.1.1.3 IoT Devices

Devices that constitute the Internet of Things (IoT) are frequently powered by batteries or utilize energy harvesting methods, and as a result, they necessitate the development and deployment of ultra-low-power solutions that enable their operation over extended periods with minimal energy input, thereby enhancing their efficiency and applicability.

5.1.1.4 High-Performance Computing (HPC)

As the scale and complexity of computational tasks continue to expand, high-performance computing systems and supercomputers face the critical challenge of implementing efficient power management strategies that are aimed at minimizing both operational costs and the environmental ramifications associated with large-scale computing infrastructures.

5.2 Applications of Low-Power VLSI Design in High-Performance Systems

The techniques associated with low-power VLSI design are being employed across a diverse array of high-performance systems, which include but are not limited to the following sectors:

5.2.1 Mobile Computing

The contemporary landscape of smartphones and tablets necessitates a careful equilibrium between achieving high performance and maintaining extended battery life. The development of power-efficient processors, memory subsystems, and communication modules is of utmost importance in facilitating rich user experiences while simultaneously ensuring that power consumption remains at a minimal level.

5.2.2 AI and Machine Learning Accelerators

The computational demands associated with artificial intelligence workloads, especially those pertaining to deep learning, require an extensive array of computational resources. Dedicated accelerators, such as Google's Tensor Processing Unit (TPU) and NVIDIA's Graphics Processing Units (GPUs), have been specifically engineered to efficiently manage these intensive tasks; however, the issue of power consumption remains a critical concern within these domains. Consequently, low-power techniques in these areas are focused on optimizing computations at both architectural and circuit design levels to enhance efficiency.

5.2.3 Data Centers and Cloud Computing

Achieving power efficiency within data centers directly correlates to decreased operational expenditures and facilitates the capability to scale up computational resources without experiencing a proportional increase in energy consumption. The integration of low-power design techniques within processors, memory, storage solutions, and networking components is essential for optimizing performance in these high-stakes environments.

5.2.4 Internet of Things (IoT)

The devices that fall under the umbrella of the Internet of Things frequently operate in environments constrained by limited power availability, relying heavily on battery power or energy harvested from their surroundings. The application of low-power VLSI techniques is instrumental in enabling these devices to effectively perform tasks related to sensing, computation, and communication, all the while adhering to stringent limitations on power consumption, thereby enhancing their overall functionality and sustainability.

5.3 Importance of Low-Power VLSI Design

The burgeoning demand for portable electronic gadgets, which prominently includes smartphones, tablets, and an array of wearable technologies, has rendered power efficiency an absolutely critical parameter in the design process. The implications of high power consumption extend beyond merely shortening the lifespan of batteries; they also exacerbate challenges related to thermal management, consequently leading to potential device malfunctions and a reduction in overall reliability. In addition, within high-performance computing environments, such as expansive data centers that house numerous servers and critical infrastructure, the pursuit of power efficiency has emerged as a vital consideration for mitigating operational costs and minimizing the environmental impact associated with excessive energy consumption [1–3].

5.4 Scope of the Chapter

In this chapter, we will engage in a comprehensive exploration of the diverse array of techniques that are available for the purpose of designing low-power VLSI circuits that are adeptly tailored for high-performance applications. Our examination will encompass a wide spectrum of techniques that span various design levels, beginning at the technological and circuit design levels and extending through to architectural considerations and algorithmic strategies. Additionally, we will delve into advanced methodologies, including but not limited to approximate computing and the application of machine learning techniques, which present novel avenues for achieving significant power reduction within contemporary VLSI systems.

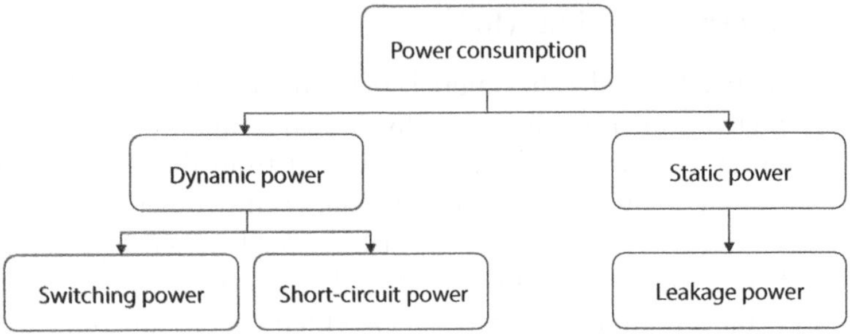

Figure 5.1 Types of power consumption.

5.5 Fundamentals of Power Consumption in VLSI Circuits

A comprehensive understanding of the constituents of power consumption in VLSI circuits is indispensable for the formulation of efficacious low-power design strategies. Power dissipation in VLSI circuits can be broadly classified into dynamic power, static power, and short-circuit power. A thorough understanding of the underlying components that contribute to power consumption in VLSI circuits is of paramount importance for the development of effective strategies aimed at low-power design provided in Figure 5.1. The phenomenon of power dissipation in VLSI circuits can be broadly classified into three primary categories: dynamic power, static power, and short-circuit power, each of which contributes uniquely to the overall power consumption profile [2, 3].

5.5.1 Dynamic Power Consumption

Figure 5.2 evidence that dynamic power presents the principal source of power consumption within Complementary Metal-Oxide-Semiconductor (CMOS) circuits and is intrinsically linked to the processes of charging and discharging capacitive loads inherent in the circuit design [4, 5].

The mathematical expression for dynamic power (Pdynamic) can be articulated as follows:

$$P_{dynamic} = \alpha .\ C_L\ V^2_{dd} f \tag{5.1}$$

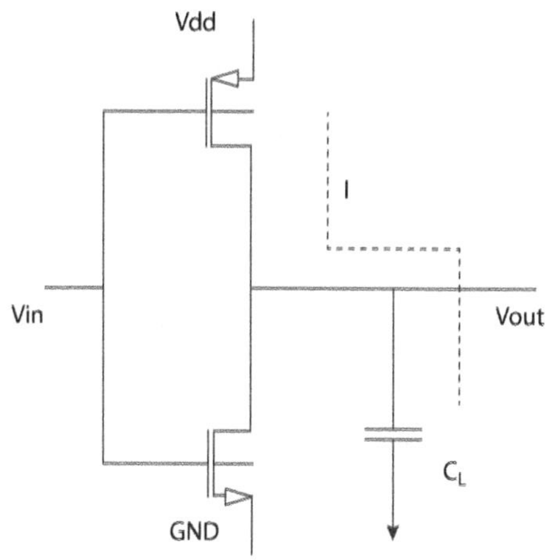

Figure 5.2 Dynamic power consumption circuit.

where: - α is the switching activity factor,
- C_L is the load capacitance,
- V_{dd} is the supply voltage, and
- f is the clock frequency.

In this equation, α signifies the switching activity factor, CL denotes the load capacitance, Vdd represents the supply voltage, and f corresponds to the clock frequency. In order to mitigate dynamic power consumption, it is essential to consider reducing any of these parameters; however, it is crucial to recognize that lowering Vdd and f may adversely affect circuit performance, thereby necessitating careful management of the requisite trade-offs involved.

5.5.2 Static Power Consumption

Static power, which is frequently referred to as leakage power, is generated due to leakage currents that persist even when the transistors are in a non-switching state. The significance of leakage power has become increasingly pronounced as the dimensions of transistors have continued to scale down. The total static power (Pstatic) can be represented mathematically as follows

$$P_{Static} = V_{dd} \times I_{leakage} \tag{5.2}$$

where $I_{leakage}$ denotes the leakage current. Various techniques have been developed to effectively reduce leakage currents, among which the utilization of high-threshold voltage transistors and the implementation of sleep transistors have proven to be particularly effective strategies [6–9].

5.5.3 Short-Circuit Power Consumption

Short-circuit power is the energy dissipated during the fleeting duration in which both PMOS and NMOS transistors are concurrently conducting as a result of a switching event. Although short-circuit power generally comprises a relatively minor component of the total power consumption, it can become a significant factor in circuits characterized by high load capacitances and rapid transition events that demand swift switching speeds.

5.6 Low-Power Design Techniques

A variety of techniques have been developed to reduce power consumption in VLSI circuits. These techniques span different levels of design abstraction.

5.6.1 Technology-Level Low-Power Design Approaches

Technology-level techniques are predicated upon modifications that occur at the material or device level, aiming to enhance the overall power efficiency of VLSI circuits in a manner that is both innovative and effective.

5.6.2 Multi-Threshold CMOS (MTCMOS)

The concept of Multi-Threshold CMOS, often abbreviated as MTCMOS, is a sophisticated technique that strategically utilizes transistors possessing varying threshold voltages within a singular integrated chip, which consequently facilitates a delicate and nuanced trade-off between operational speed and leakage power consumption. Specifically, transistors characterized by high threshold voltages are adept at significantly reducing leakage current in those paths of the circuit that are deemed non-critical in terms of timing, while conversely, transistors with low threshold voltages are strategically deployed in paths that are critically dependent on achieving high speed performance [10, 11].

5.6.3 Silicon-On-Insulator (SOI) Technology

The utilization of Silicon-On-Insulator technology, commonly referred to as SOI technology, serves to effectively diminish parasitic capacitance and consequently enhance switching speed by incorporating a thin layer of insulating material situated below the transistor structure. This innovative method of isolation plays a pivotal role in minimizing both dynamic power dissipation, which occurs during active switching, and static power dissipation, which arises when the circuit is not actively engaged, thus rendering SOI technology exceptionally suitable for applications that demand low power consumption alongside high-performance capabilities [9–11].

5.6.4 FinFETs and Gate-All-Around (GAA) Transistors

In light of the fact that traditional planar transistors are approaching their inherent scaling limitations, the advent of FinFETs in Figure 5.3 and Gate-All-Around transistors has emerged as a promising solution that offers superior control over short-channel effects while simultaneously mitigating leakage currents. The significance of these advanced transistor designs cannot be overstated, as they are integral to facilitating low-power operation, particularly in cutting-edge technology nodes that operate at dimensions below 10 nanometers, thereby pushing the boundaries of modern semiconductor technology.

5.7 Low Power Design Technique and Power Management

Circuit-level techniques optimize the design of basic logic gates and flip-flops to reduce power.

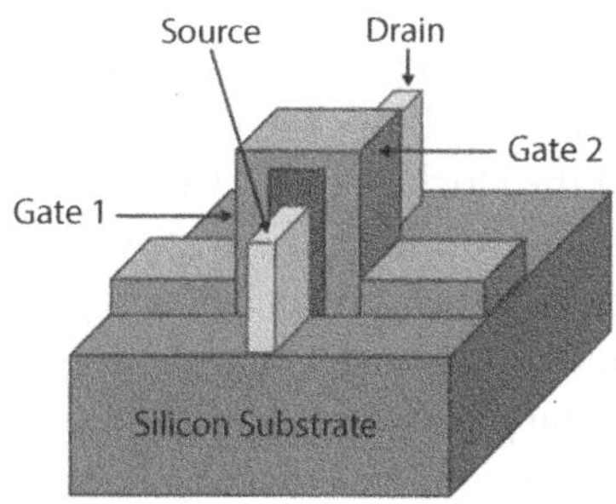

Figure 5.3 Structure of FinFET.

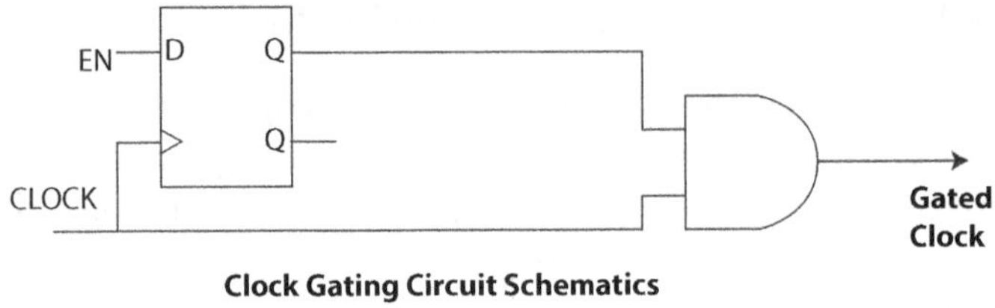

Figure 5.4 Schematics of clock gating circuit.

5.7.1 Clock Gating

The methodology presented in Figure 5.4 as clock gating is a highly effective technique aimed at reducing dynamic power consumption through the strategic disabling of the clock signal to specific segments of the circuit that are not actively engaged in switching operations. This technique proves particularly advantageous in the context of sequential circuits, where the clock network itself can account for a substantial portion of the overall power budget, thus enabling a more efficient allocation of power resources [12].

5.7.2 Power Gating

The power gating technique involves the disconnection of the power supply from those blocks within the circuit that are inactive, thereby resulting in a significant reduction in leakage power consumption. While this approach is undeniably effective in minimizing static power losses, it necessitates meticulous management of wake-up times and inrush currents to avert any potential degradation in circuit performance, which underscores the complexity involved in implementing this strategy effectively [12, 13].

5.7.3 Dynamic Voltage and Frequency Scaling (DVFS)

Dynamic Voltage and Frequency Scaling, often abbreviated as DVFS, is a sophisticated technique that allows for the real-time adjustment of both the supply voltage and the clock frequency in accordance with the specific demands imposed by the workload at any given moment, thus providing a finely tuned balance between power consumption and overall performance. This methodology shown in Figure 5.5 has garnered widespread adoption in contemporary processors, as it optimizes energy efficiency under fluctuating load conditions, making it a cornerstone of modern power management strategies [13].

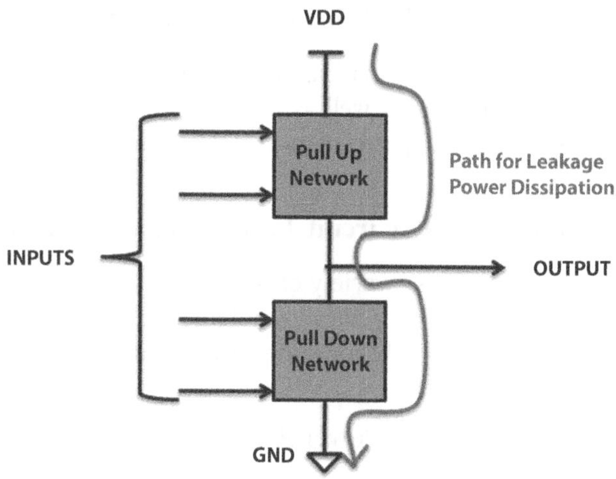

Figure 5.5 Power gating circuit.

5.7.4 Adaptive Body Biasing (ABB)

The implementation of Adaptive Body Biasing, or ABB, entails the dynamic adjustment of the body bias applied to transistors in order to effectively control their threshold voltage characteristics. By strategically applying forward body bias in paths that are critical to timing, thereby reducing delay, and conversely employing reverse body bias in non-critical paths to diminish leakage currents, ABB represents a highly flexible and adaptable approach to power optimization that can be tailored to the specific operational requirements of the circuit [14, 15].

5.8 Low-Power VLSI Design Methodologies

The intricate and multifaceted design of low-power Very Large Scale Integration (VLSI) systems necessitates a meticulous balancing act between enhancing performance metrics and minimizing power consumption levels, which are often at odds with one another. Within this complex domain, there exists a plethora of methodologies that traverse a wide spectrum, ranging from the fundamental transistor level all the way up to the higher echelons of algorithmic design. These sophisticated techniques can be systematically classified into four predominant categories, each with its unique focus and application:

5.8.1 Power-Efficient VLSI through Technology-Level Design

These are advanced strategies that capitalize on the latest advancements in semiconductor technology, as well as innovative transistor architectures that push the boundaries of performance while reducing power usage [20].

5.8.2 Low-Power VLSI: Circuit-Level Power Optimization

This category encompasses a variety of methods aimed at optimizing the actual circuit design itself to achieve a notable reduction in power consumption without substantially compromising performance [20].

5.8.3 Architecture-Level Power Reduction Strategies

These approaches are centered around system-level innovations that aim to enhance power efficiency through thoughtful architectural modifications and improvements that address energy utilization holistically [21].

5.8.4 Impact of Scaling on Leakage and Device Variability

This final category pertains to modifications made at the software and algorithmic level, which can substantially lead to computations that consume far less energy, thereby enhancing the overall efficiency of the system.

5.8.5 Technology-Driven Low-Power Design Strategies

The domain of technology-level techniques is primarily concerned with the enhancement of the semiconductor devices and materials that serve as the foundational building blocks for VLSI circuits. As the semiconductor industry has witnessed a rapid scaling of Complementary Metal-Oxide-Semiconductor (CMOS) technology, the issue of power consumption, particularly leakage power, has emerged as a formidable challenge that must be addressed with urgency and precision [20, 21].

5.8.6 Scaling and Moore's Law

For an extensive duration, the semiconductor industry has depended heavily on the predictive framework known as Moore's Law, which posits that the number of transistors that can be integrated onto a single chip will approximately double every two years. Historically, this exponential increase in transistor density has facilitated notable enhancements in both performance and energy efficiency, allowing for the development of increasingly sophisticated electronic devices. However, as the industry endeavors to scale technology down to smaller nodes, specifically those that are sub-20 nanometers, a myriad of challenges begins to surface that complicate the landscape [21, 22, 24]:

5.8.7 Increased Leakage Power

The trend toward smaller transistors often results in the utilization of thinner gate oxides, which inadvertently leads to elevated leakage currents that contribute significantly to an increase in static power consumption, thereby negating some of the performance gains achieved.

5.8.8 Increased Variability

As devices continue to shrink in size, process variations become increasingly pronounced, making it increasingly challenging to accurately predict power consumption and performance characteristics, which can result in unpredictable behavior in circuit operation. To effectively confront and surmount these challenges, the realm of semiconductor technology has seen a notable evolution beyond traditional planar Metal-Oxide-Semiconductor Field-Effect Transistors (MOSFETs) and Tunnel field effect transistors (TFETs), leading to the adoption of more advanced and sophisticated structures such as Fin Field-Effect Transistors (FinFETs) and Silicon-On-Insulator (SOI) devices, which offer promising solutions to the pressing issues of power consumption and variability.

5.9 Pipelining and Parallelism for Energy-Efficient VLSI Design

Architecture-level techniques focus on optimizing the overall design and operation of the circuit to reduce power consumption.

5.9.1 Pipelining and Parallelism

The architectural technique of pipelining entails the division of a computational task into multiple distinct stages, with each stage operating at a reduced frequency, thereby resulting in a substantial reduction in dynamic power consumption. In tandem with this method, the principle of parallelism involves the duplication of hardware resources to enable the concurrent execution of multiple tasks, which effectively reduces the switching activity associated with each individual operation and consequently lowers the aggregate power consumption across the entire system [16].

5.9.2 Voltage Islands and Power Domains

The concept of voltage islands refers to the practice of operating different segments of a chip at varying voltage levels, which allows for a more granular and precise control over power consumption. This strategy is particularly advantageous in large-scale System-on-Chip (SoC) designs, where disparate modules exhibit diverse power-performance requirements, thereby facilitating a more efficient utilization of energy resources across the entire integrated circuit [16, 17].

5.9.3 Low-Power Memory Design

The design of memory components, particularly static random-access memory (SRAM) and dynamic random-access memory (DRAM), represents a critical area of focus due to their significant contributions to overall power consumption within electronic systems. By employing techniques aimed at reducing bit-line capacitance, leveraging multi-level cell (MLC) architectures, and optimizing cache designs, it becomes possible to achieve substantial reductions in memory-related power consumption, thereby enhancing the overall energy efficiency of the system [17].

5.10 Efficient Processor Design

In the realm of computer architecture, one of the most significant advancements involves undertaking architectural optimizations at the processor level, which serves the dual purpose of crafting energy-efficient systems that not only deliver exceptional performance metrics but also do so while maintaining minimal power consumption levels, thus contributing to both economic and environmental sustainability [6, 7].

5.10.1 Out-of-Order Execution

In contemporary computing environments, modern processors have adopted the sophisticated technique of out-of-order execution, a method that enhances overall performance by enabling the execution of instructions as soon as the necessary operands are available for processing, rather than adhering strictly to the sequential order of instruction completion. This strategic approach to instruction execution not only increases the overall throughput of the processor but also circumvents the need for escalating clock speeds, which in turn leads to a more judicious and efficient utilization of power resources in the system [7, 8].

5.10.2 Superscalar Pipelines

The implementation of superscalar architecture within processors permits the simultaneous execution of multiple instructions within a single clock cycle, achieved through the deployment of multiple pipelines. This innovative design paradigm markedly augments performance levels without necessitating a corresponding rise in clock frequency, which subsequently translates to significant power savings, as the processor can perform more operations within the same energy constraints [9, 11].

5.10.3 Specialized Cores

An alternative and increasingly popular strategy in the domain of energy-efficient processor design is the incorporation of specialized cores that are tailored for executing specific tasks with optimal efficiency. For instance, a growing number of modern processors feature low-power cores specifically designated for managing background tasks, while simultaneously housing high-performance cores that are reserved for executing demanding applications. This heterogeneous architectural configuration empowers the system to transition dynamically between high-performance and low-power operational modes, thereby optimizing power consumption in accordance with varying workload demands.

5.10.4 Architectures and Specialized Processing Cores

At the algorithmic level, the pursuit of power-efficient computing is facilitated through the meticulous design of algorithms that strategically minimize the number of computational operations or leverage approximate computations, thereby reducing overall power consumption significantly and enhancing efficiency.

5.11 Energy-Efficient Algorithms and Approximate Computing

5.11.1 Energy-Efficient Algorithms

The design of energy-efficient algorithms is fundamentally focused on minimizing the overall number of computations, memory accesses, and data transfers, as all these factors are directly correlated with power consumption. For instance, sorting algorithms can undergo optimization processes aimed at reducing the frequency of comparisons made, while machine learning algorithms can be engineered to achieve convergence more rapidly, thereby decreasing the total number of iterations required to reach a solution, ultimately conserving energy.

5.11.2 Approximate Computing for Low-Power VLSI Design

The concept of approximate computing emerges as a powerful technique wherein the precision of computations is intentionally compromised to achieve substantial power savings. This methodology is predicated on the understanding that numerous applications, including image processing, machine learning tasks, and sensor data analysis, do not invariably demand exact results. By permitting minor inaccuracies in computations, approximate computing can lead to a remarkable reduction in power consumption, thereby enhancing the overall efficiency of the system.

5.11.3 Applications

The utility of approximate computing is particularly pronounced in scenarios where absolute accuracy is not a critical requirement, such as in multimedia processing applications, deep learning inference tasks, and the operation of Internet of Things (IoT) sensor nodes, where the trade-off between precision and power efficiency can yield significant benefits.

5.12 Advanced Techniques for Low-Power Design

In addition to the conventional methodologies previously discussed, a plethora of advanced techniques has emerged, each aimed at further optimizing power consumption within modern electronic systems, thereby contributing to the overarching goal of achieving greater energy efficiency.

5.12.1 Approximation-Aware Circuit Design for Power Optimization

The concept of approximate computing fundamentally capitalizes on the intrinsic capacity of numerous applications to withstand a certain degree of inaccuracies, particularly evident in domains such as multimedia processing and the burgeoning field of machine learning, thereby facilitating a substantial reduction in power consumption levels. By intentionally permitting systematic and controlled inaccuracies within computational processes, electronic circuits can function with diminished switching activities and a streamlined logic architecture, which consequently leads to remarkable and quantifiable power savings that can significantly enhance the overall energy efficiency of electronic systems.

5.12.2 Sub-Threshold Design

The methodology known as sub-threshold design entails the strategic operation of transistors within the sub-threshold region, specifically characterized by the condition where the gate-to-source voltage (Vgs) is less than the threshold voltage (Vth), in order to achieve an ultra-low power operational framework. Although it is acknowledged that this particular approach may result in a reduction in operational speed, it has proven to be exceedingly effective for applications constrained by energy limitations, such as sensor nodes deployed within Internet of Things (IoT) devices, where the imperative of minimizing power consumption takes precedence over the optimization of performance metrics [14, 15].

5.12.3 Machine Learning for Power Optimization

The application of machine learning methodologies presents a promising avenue for predicting power consumption patterns and dynamically optimizing various operational parameters, such as settings for dynamic voltage and frequency scaling (DVFS). For instance, reinforcement learning can be adeptly employed to adaptively modify power management strategies in real time, based on the continuously observed workload, thereby enabling an effective equilibrium between performance capabilities and energy efficiency, which is increasingly critical in the context of modern computational demands [11, 13, 15].

5.12.4 Hardware Acceleration for Low-Power Computing

Hardware accelerators represent a category of specialized processing units that have been meticulously engineered to execute specific computational tasks with a level of efficiency that surpasses that which is achievable by general-purpose processors. The strategic utilization of these accelerators allows for the offloading of compute-intensive operations, resulting in the attainment of substantial power savings that are increasingly crucial in the context of modern computing requirements.

For instance, in contemporary computing systems, various tasks, such as video encoding, data encryption, and deep learning inference, are frequently delegated to dedicated hardware accelerators, which may include Graphics Processing Units (GPUs), Tensor Processing Units (TPUs), and Field-Programmable Gate Arrays (FPGAs). These specialized accelerators have been meticulously optimized for their designated tasks, thereby enabling them to execute the necessary computations with a level of efficiency that notably exceeds that of a conventional general-purpose Central Processing Unit (CPU).

The introduction of specialized hardware into system architecture allows for the achievement of performance metrics that are comparable to those of traditional computing setups while simultaneously operating at significantly lower levels of power consumption. A prime example of this phenomenon can be observed in Google's Tensor Processing Unit (TPU), which has been specifically designed to cater to artificial intelligence workloads, thereby delivering not only enhanced performance but also superior energy efficiency when contrasted with the power consumption characteristics of standard general-purpose CPUs or GPUs in the context of deep learning inference tasks [17–20].

5.12.5 Optimizing Memory Hierarchies for Power and Performance

In numerous high-performance Very-Large-Scale Integration (VLSI) systems, the memory subsystems—including caches, main memory (Dynamic Random-Access Memory or DRAM), and non-volatile memory (NVM)—hold a pivotal role in determining the overall energy consumption profile of the system [21]. Given the escalating demands for data access in contemporary technological environments, the design of energy-efficient memory systems has emerged as a critical factor for minimizing power consumption in modern computing architectures. The architecture of main memory, caches, and various forms of storage are conceptualized with several key considerations aimed at achieving a harmonious balance between performance and power efficiency:

1. An emphasis is placed on minimizing both dynamic and leakage power expenditures within the memory cells and the associated memory controllers.
2. Memory architectures are optimized to ensure lower access times, which subsequently contributes to a reduction in energy consumption per access.
3. The strategic reduction of power consumption associated with the movement of data and the operations of the memory hierarchy is prioritized.

5.12.6 Sources of Power Dissipation in Memory Systems

Memory systems, particularly within the framework of large-scale designs, exhibit power consumption characteristics that arise from multiple sources. These sources encompass the following:

5.12.7 Dynamic Power

This term refers to the power consumed during memory access operations, specifically during read/write activities when data bits are altered or moved. The amount of dynamic power consumed is directly proportional to the switching activity, capacitance, and voltage levels of the memory cells involved in these operations [14, 15].

5.12.8 Static (Leakage) Power

This power is consumed when the memory subsystem remains idle, and its significance escalates dramatically in the realm of advanced memory technologies due to the presence of leakage currents [16].

5.12.9 Refresh Power

In the context of volatile memory types such as DRAM, there exists a necessity for additional power to be expended periodically in order to refresh the stored data, thereby maintaining data integrity [17].

5.12.10 Data Movement Power

The process of transferring data between various levels of memory, such as between caches, DRAM, and persistent storage, incurs power costs that are attributable to the energy consumed by interconnects, bus drivers, and memory controllers. In order to effectively mitigate power consumption within memory subsystems, it is imperative to address these sources of power dissipation at multiple hierarchical levels, extending from the individual memory cells to the overarching memory architectures [18].

5.12.11 Techniques for Low-Power Memory Design

An array of innovative techniques has been developed and implemented with the objective of reducing both dynamic and static power consumption within memory systems. These methodologies encompass a spectrum of strategies that range from the optimization of memory cells at the fundamental transistor level to the creation of more efficient memory hierarchies at the architectural level [21].

5.12.12 Memory Cell-Level Techniques

At the level of individual transistors, various techniques have been devised that specifically target the reduction of power consumption associated with individual memory cells, regardless of whether these cells are categorized as Static Random-Access Memory (SRAM), Dynamic Random-Access Memory (DRAM), or Non-Volatile Memory (NVM) cells, thereby ensuring a comprehensive approach to power efficiency in modern computing environments.

5.12.13 Low-Power SRAM Design

Static Random-Access Memory, commonly referred to as SRAM, is widely utilized within cache memory systems due to its rapid access speeds, which significantly enhance overall computing performance; however, it is important to note that this type of memory architecture is characterized by a considerable consumption of static power resulting from its inherent bistable latch structure that facilitates data storage. To mitigate the excessive power consumption associated with SRAM, several innovative techniques can be employed, including the strategic lowering of the supply voltage to decrease power usage, the implementation of high-threshold voltage transistors within non-critical pathways to minimize leakage, and the application of power gating strategies to deactivate sections of the SRAM that are not currently in use, thereby effectively reducing both dynamic power consumption and leakage power [22].

5.12.14 Reducing Leakage in SRAM

In contemporary designs of SRAM, there is a pronounced trend towards the utilization of stacked transistors in conjunction with multi-threshold voltage cells, which collectively serve to significantly minimize leakage power, a critical factor in enhancing energy efficiency. The employment of stacked transistors is particularly effective in this regard, as it establishes a higher effective threshold voltage that naturally limits the flow of leakage current; simultaneously, multi-threshold voltage designs facilitate the selective integration of both low-threshold and high-threshold transistors, allowing designers to meticulously optimize the delicate balance between power consumption and performance levels, thereby achieving superior operational efficiencies [22, 23].

5.12.15 Low-Power DRAM Design

Dynamic Random-Access Memory, otherwise known as DRAM, plays an essential role as the primary memory component in the majority of contemporary computing systems, providing the necessary storage for executing applications and tasks. However, it is crucial to recognize that the power consumption associated with DRAM is largely dictated by the necessity to periodically refresh memory cells to preserve stored data integrity, a process which can be quite power-intensive. Consequently, a primary objective for low-power DRAM design initiatives is to effectively reduce the power consumed during these refresh cycles, thereby achieving a more energy-efficient memory architecture [21–23].

5.12.16 Reducing Refresh Rate

In the realm of modern DRAM technologies, adaptive refresh mechanisms have been developed and implemented, which are designed to intelligently adjust the refresh frequency applied to different memory regions based on their susceptibility to data loss. By continuously monitoring the retention characteristics of individual DRAM cells and their performance metrics, it becomes feasible to extend the refresh intervals for those memory regions that exhibit lower susceptibility to data deterioration, thus resulting in a significant reduction of refresh power consumption and enhancing the overall energy efficiency of the DRAM design [25–30].

5.12.17 Sub-Array Activation

To further optimize power efficiency during read and write operations, a novel strategy known as sub-array activation has been introduced, which deviates from the traditional approach of activating entire memory banks. Instead, this method focuses on selectively activating only a limited number of rows within a memory bank, thereby substantially reducing the dynamic power requirements associated with read and write activities, and enabling a more judicious use of energy resources within the memory architecture [24, 25].

5.12.18 Low-Power Non-Volatile Memory (NVM) Design

Non-volatile memories, including but not limited to Flash memory, Phase Change Memory (PCM), and Spin-Transfer Torque Magnetic RAM (STT-MRAM), provide a distinct advantage in that they possess the ability to retain stored data without the need for a continuous power supply, which is a significant benefit for various applications. Nonetheless, it is imperative to acknowledge that these emerging memory technologies face their own set of challenges, particularly concerning issues related to high write energy consumption and limitations regarding endurance, which must be addressed to realize their full potential [24, 26].

5.12.19 PCM and MRAM

The advent of Phase Change Memory (PCM) and Magnetic RAM (MRAM) represents a promising shift in memory technology, as these innovative solutions demonstrate the capability of achieving lower static power consumption in comparison to both DRAM and SRAM. Specifically, PCM functions by altering the phase of a material from an amorphous state to a crystalline state to store data, while MRAM employs magnetic states to represent binary information; furthermore, PCM has been shown to consume considerably less power during data read operations, and STT-MRAM is particularly noteworthy for its high efficiency in managing both dynamic and static power consumption, thereby positioning these technologies as viable alternatives in the quest for energy-efficient memory solutions [25–31].

5.13 Conclusion

In this chapter, we have meticulously provided an extensive and comprehensive overview of contemporary low-power VLSI design techniques, while placing a significant emphasis on their relevance to high-performance applications that are increasingly prevalent in today's technology landscape. The relentless and ever-expanding demand for portable, efficient, and high-performance electronic devices has rendered power consumption a paramount concern within the sphere of VLSI design, necessitating innovative solutions to address this challenge effectively. Our exploration has encompassed several pivotal aspects. We undertook a thorough dissection of the primary sources of power dissipation that can be found in VLSI circuits—namely, dynamic power, static power, and short-circuit power—while elucidating their profound impact on overall power usage and the performance capabilities of the devices involved. Innovations that emerge at the technology level, such as Multi-Threshold CMOS technology, Silicon-On-Insulator (SOI) technology, and the advanced FinFET architecture, provide foundational improvements that are crucial for mitigating leakage currents and enhancing the intrinsic characteristics of transistors.

5.14 Future Directions

The trajectory of low-power Very Large Scale Integration (VLSI) design is poised to be profoundly influenced by the relentless progression of technological innovations and the advancements within the realm of materials science, in conjunction with the emergence of novel architectural frameworks and algorithmic methodologies. As electronic devices progressively diminish in size while simultaneously incorporating an increasing array of functionalities, this phenomenon will inevitably give rise to both new challenges and significant opportunities aimed at minimizing power consumption, all the while striving to maintain or even enhance overall system performance metrics. The advent of cutting-edge technologies, including but not limited to quantum computing, neuromorphic computing, and the implementation of three-dimensional (3D) integrated circuits, will serve to further transform and redefine the existing paradigms associated with low-power VLSI design, thereby establishing this field as one that is in a perpetual state of evolution and necessitating the implementation of innovative solutions alongside robust interdisciplinary collaboration among various scientific and engineering disciplines. In conclusion, the diverse array of techniques and methodologies thoroughly examined in this chapter lays down a robust and comprehensive foundation for the effective design of energy-efficient, high-performance VLSI systems that are increasingly relevant in today's technology-driven world. As we continue to progress into the future, the critical importance of power optimization in the design and development processes will only become more pronounced, thereby making it absolutely essential for both researchers and engineers to persist in their efforts to innovate, develop, and refine low-power design strategies that are capable of meeting the ever-growing demands and expectations of the fast-evolving electronics industry.

AI disclosure statement: The author confirms that no artificial intelligence (AI)-generated content or AI-assisted writing tools were used in the preparation of this chapter (including summaries and the abstract). All text, analysis, and conclusions are the result of the author's original work.

References

1. Pedram, M. and Nazarian, S., Thermal modeling, analysis, and management in VLSI circuits: Principles and methods. *Proc. IEEE*, 94, 8, 1487–1501, 2006.
2. Kang, S.M. and Leblebici, Y., *CMOS Digital Integrated Circuits: Analysis and Design*, 3rd, McGraw-Hill Education, 2003.
3. Roy, K., Mukhopadhyay, S., Mahmoodi-Meimand, H., Leakage current mechanisms and leakage reduction techniques in deep-submicrometer CMOS circuits. *Proc. IEEE*, 91, 2, 305–327, 2003.
4. Sakurai, T. and Newton, A.R., Alpha-power law MOSFET model and its applications to CMOS inverter delay and other formulas. *IEEE J. Solid-State Circuits*, 25, 2, 584–594, 1990.
5. Ye, Y. and Roy, K., QSERL: Quasi-static energy recovery logic. *IEEE J. Solid-State Circuits*, 36, 2, 239–248, 2000.
6. Colinge, J.P., *Silicon-on-Insulator Technology: Materials to VLSI*, 3rd, Springer, 2000.
7. Naeemi, A. and Meindl, J.D., Compact physical models for layout-dependent capacitance and resistance in graphene nanoribbon interconnects. *IEEE Electron Device Lett.*, 28, 5, 428–431, 2007.
8. Benini, L. and Micheli, G.D., Networks on chips: A new SoC paradigm. *Computer*, 35, 1, 70–78, 2000.
9. Mutoh, S., Douseki, T., Yamada, Y., *et al.*, 1-V power supply high-speed digital circuit technology with multithreshold-voltage CMOS. *IEEE J. Solid-State Circuits*, 30, 8, 847–854, 1990.

10. Brooks, D. and Martonosi, M., Dynamic thermal management for high-performance microprocessors, in: *Proceedings of the International Symposium on High-Performance Computer Architecture (HPCA)*, pp. 171–182, 2001.
11. Kulkarni, J.P. and Sylvester, D., High performance level conversion for dual V/sub DD/ design. *IEEE Trans. Very Large Scale Integr. VLSI Syst.*, 12, 9, 926–936, 2004.
12. Johnson, S.L. and Greenstreet, M.R., Low-power flip-flops with dual-threshold-voltage transistors. *IEEE Trans. Very Large Scale Integr. VLSI Syst.*, 5, 1, 96–103, 1997.
13. Singh, D. and Burleson, W., Power-aware CAD tools and methodologies: A survey. *IEEE Trans. VLSI Syst.*, 12, 6, 706–713, 2004.
14. Zhang, C., Vahid, F., Givargis, T., A self-tuning cache architecture for embedded systems. *ACM Trans. Embedded Comput. Syst. (TECS)*, 3, 2, 407–430, 2003.
15. Venkataramani, S., Ranjan, A., Roy, K., Raghunathan, A., Approximate computing: A survey. *IEEE Des. Test*, 32, 1, 83–99, 2015.
16. Calhoun, B.H. and Chandrakasan, A.P., Ultra-dynamic voltage scaling (UDVS) using sub-threshold operation and local voltage dithering. *IEEE J. Solid-State Circuits*, 39, 4, 775–785, 2004.
17. Huang, W. and Ren, X., Reinforcement learning for power management in computing systems: A survey. *ACM Comput. Surv. (CSUR)*, 52, 6, 1–36, 2019.
18. Kaur, K. and Noor, A., Strategies & Methodologies for Low Power VLSI Designs: A Review. *Int. J. Adv. Eng. Technol.*, 2011.
19. VLSI First, Best VLSI Seminar Topics, 2023.
20. Cadence, Low Power Design Techniques for Power Integrity in VLSI, 2001.
21. Synopsys, What is Low Power Design? – Techniques, Methodology & Tools, 2021.
22. Semiconductor Industry Association, The Future of Low Power VLSI, 2022.
23. International Technology Roadmap for Semiconductors, Low Power VLSI Design IEEE, Low-Power Design Techniques for High-Performance Applications, 2022.
25. Semiconductor Research Corporation, Research on Low-Power VLSI Design, 2022.
26. Low-Power VLSI Design Methodologies. *J. Circuits Systems Comput.*, 2016.
27. IEEE Computer Society, Low-Power VLSI Design Techniques for Emerging Applications, 2023.
28. Chander, S., Sinha, S.K., Chaudhary, R., *et al.*, Ge-Source Based L-Shaped Tunnel Field Effect Transistor for Low Power Switching Application. *Silicon*, 14, 7435–7448, 2022.
29. Chander, S., Sinha, S.K., Chaudhary, R., Prospects and Challenges of Different Geometries of TFET Devices for IoT Applications. *Nanosci. Nanotechnology-Asia*, 13, 4, 47–56(10), 2023.

30. Verma, S. and Tripathi, S.L., Effect of Mole Fraction and Fin Material on Performance Parameter of 14 nm Heterojunction Si1-xGex FinFET and Application as an Inverter. *Silicon*, 14, 8793–8804, 2022.
31. Bhattacharya, S., Tripathi, S.L., Nayana, G.H., Design Transmission Gates Using Double-Gate Junctionless TFETs. *Silicon*, 2024.

6

The Role of Artificial Intelligence and Machine Learning in Modern Healthcare Systems

Anudeep Goraya

School of Computer Science Engineering, Lovely Professional University, Punjab, India

Abstract

Artificial Intelligence (AI) and Machine Learning (ML) are revolutionizing healthcare by making patient care more efficient, data-driven, and personalized. Hospital operations are being optimized, patient outcomes are being improved, and medical decision-making is being transformed by these technologies. Rapid and precise analysis of large volumes of medical data is one of AI and ML's major advantages. AI can evaluate data from wearable health trackers, test findings, diagnostic pictures, and electronic health records to find trends that people would overlook. This makes it possible for medical professionals to identify illnesses earlier and with more accuracy, which improves treatment strategies and increases patient survival rates. AI-driven early illness detection is particularly effective for disorders like cancer, diabetes, and heart disease, where early intervention may drastically improve results. By analyzing patient records and recognizing tiny warning signals, AI helps clinicians make better educated and timely judgments. AI-powered solutions improve the interpretation of CT, MRI, and X-ray images in medical imaging, helping radiologists spot anomalies more accurately. AI-powered chatbots and virtual assistants are also transforming patient interactions by providing 24/7 support, answering medical queries, and reminding patients to take medication.

Email: anudeep.30953@lpu.co.in

Abhishek Kumar and Suman Lata Tripathi (eds.) Smart Chips for Smart Devices: VLSI Design for Next-Generation IoT Solutions, (103–122) © 2026 Scrivener Publishing LLC

Keywords: Artificial intelligence, machine learning, healthcare technology, AI diagnostics

6.1 Introduction

AI and ML are in the healthcare industry more and more. They are changing how healthcare is delivered, offering large opportunities to change healthcare systems and help improve patients' lives. Some tools and processes of these technologies are emerging for use in decision-making. AI will be used all across healthcare, from making accurate diagnoses and planning treatments to finding new medicine and managing public health. AI analyzes lots of data and sees and predicts patterns, thus helping doctors make the right decisions. This means that patient treatment and patient outcome improve [1].

AI is important in healthcare because it can look at massive amounts of data. Healthcare creates large amounts of information, but the established ways of looking at all this information might be slow and prone to error. AI does it fast, finding important elements that humans could miss. AI can predict what might happen to patients, detect sickness early, and even describe what medicine people need. Each person's health profile gets well analysed by AI, including X-rays, MRIs, and CT scans. With the help of AI, doctors may be able to detect diseases like cancer early, which helps people get better treatment. Rare sicknesses that doctors might miss can sometimes be spotted by AI. Regarding a patient's details, AI figures out what treatments suit them best. Each person gets exactly what they need so they don't get hurt and they feel much better through treatment. New medicine takes a long time to develop and is very expensive. In contrast, using AI, researchers find new medicines quickly and inexpensively, thus getting new drugs to people [2].

AI helps people find drugs that can be used for other diseases, resolving health problems faster. Helping with population health is another area in which AI makes a significant contribution. From data like medical records and other information about people's lives, AI finds patterns about health. This allows doctors to stop sicknesses from spreading, use resources better, and help communities stay healthier. AI can also help governments know what healthcare will be needed later, like new medicines, or predict diseases that might emerge [1, 2]. All over the world, hospitals have many problems—costs going up, people getting older, chronic diseases on the increase. AI is here to help. It makes workflows fast and reveals special secrets about health. With the assistance of AI, patients will get better and doctors won't be so overworked.

6.2 Overview of AI and ML Technologies in Healthcare

The introduction of artificial intelligence (AI) and machine learning (ML) technology has transformed the healthcare business. These breakthroughs alter how healthcare practitioners approach overall diagnosis, treatment, and patient care. AI aims to create computers that can do activities that previously required human intellect, such as learning, decision-making, and problem-solving [3]. We use machine-learning approaches to drive innovation and enhance patient outcomes. Let's look at the most essential ones as shown in Figure 6.1.

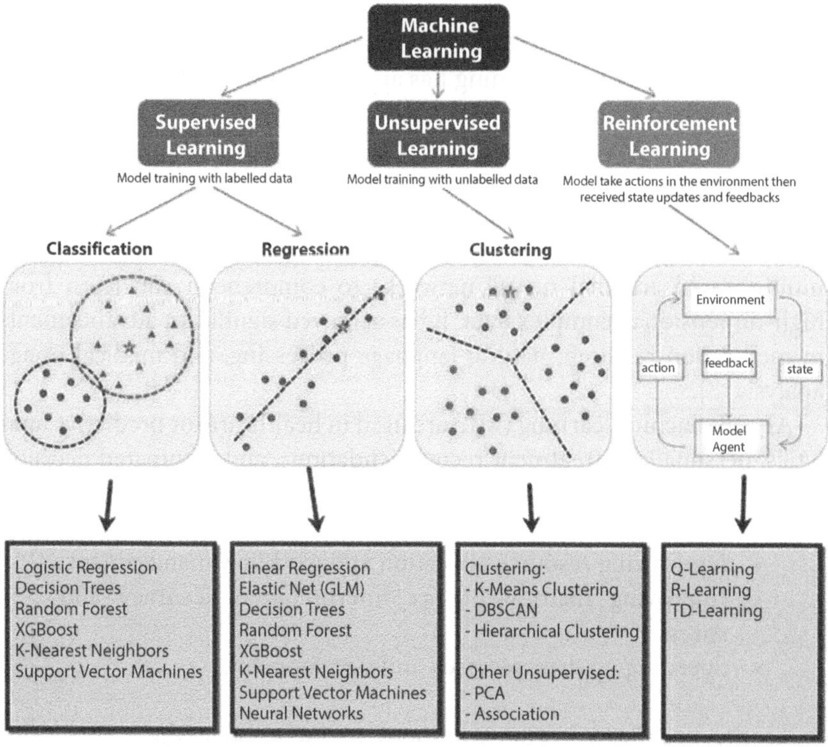

Figure 6.1 Classification of machine learning [3].

6.2.1 Supervised Learning

Algorithms are trained on labeled data to produce the desired output. In healthcare, supervised learning may be used to diagnose diseases, forecast risks, and prescribe treatments.

6.2.2 Unsupervised Learning

Unlike supervised learning, unsupervised learning algorithms search unlabeled data for hidden patterns, structures, and insights. This approach may be applied to patient stratification, anomaly identification, and population health management.

6.2.3 Reinforcement Learning

In this method, the agent interacts with a dynamic setting, receives input as a reward or penalty, and learns to do actions that maximize cumulative rewards. Reinforcement learning has applications in both robotic surgery and customized medicine.

6.2.4 Deep Learning

Deep learning, a powerful discipline of machine learning, employs multi-layered artificial neural networks to comprehend and learn from high-dimensional, complex data. It has achieved significant advancements in medication research, natural language processing, and medical image analysis [5].

AI and machine learning (ML) are used in healthcare for predictive analytics, personalized treatment recommendations, and automated decision support systems.

- Optimizing resource allocation and workflow management.
- Improving medical picture interpretation accuracy and speed.
- Speeding up drug research and development.

As the healthcare sector continues to incorporate these disruptive technologies, we can expect additional ground-breaking innovations to improve patient care and build a more efficient, data-driven healthcare system [4–11].

6.3 AI in Diagnostic Systems

Artificial intelligence (AI) is transforming the realm of diagnostic systems, as well as how clinicians manage patients. The use of artificial intelligence in healthcare is rapidly rising, with medical imaging being one of its most promising applications. AI-driven tools [12, 13] are enhancing the accuracy and efficiency of long-standing medical diagnostic methods such as CT scans, MRIs, and X-rays as shown in Figure 6.2.

AI-powered diagnostic devices utilize deep learning models and advanced algorithms to handle massive amounts of imaging data. These algorithms can detect minute patterns and anomalies that the human eye misses. Using this power, artificial intelligence can assist medical personnel in making earlier and more accurate diagnoses. For example, AI has demonstrated a remarkable ability to identify early illness indications that traditional approaches would miss in the detection of malignancies or retinal problems. Early identification is critical in cases of serious illnesses such as cancer because it allows for timely intervention, which improves patient outcomes and boosts survival rates [5].

Numerous case studies illustrate the efficacy of artificial intelligence in medical imaging. For example, AI systems have been found to outperform human radiologists in diagnosing early-stage lung cancer using CT images. These tools can swiftly evaluate photos and identify areas of concern for additional assessment by medical personnel. This not only

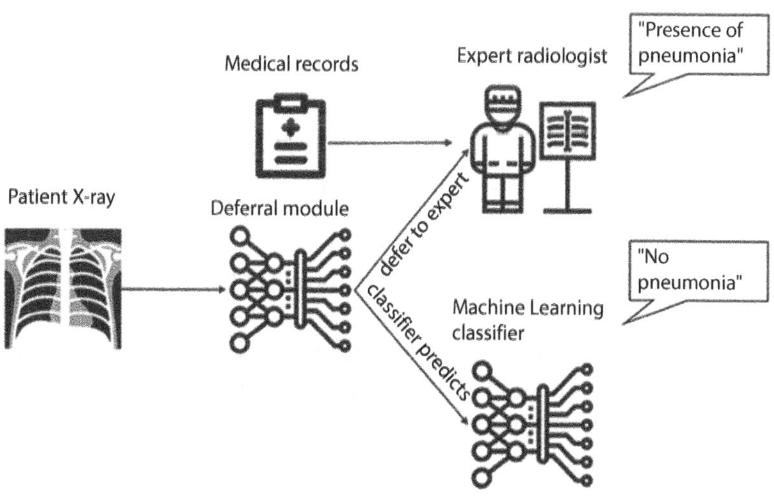

Figure 6.2 AI intervention in the diagnosis process.

improves diagnosis accuracy but also accelerates the procedure, decreasing healthcare practitioners' burden and allowing them to focus on more difficult duties. AI is not just altering medical imaging but also pathology and laboratory diagnoses. Traditionally, pathology involved manually evaluating tissue samples, which was time-consuming and labour-intensive. However, artificial intelligence (AI) is altering this by automating mundane operations and providing more precise and reliable outcomes. AI-powered systems can evaluate digitized slides of tissue samples for patterns that suggest the presence of illness. This automation lowers human error, guarantees consistency, and enables speedier delivery of findings [6]. AI's pattern recognition ability is already being used to identify biomarkers and more accurately predict when illnesses may develop. Biomarkers are proteins or genes that serve as crucial indicators of illnesses, guiding diagnosis and treatment. In comparison to traditional methodologies, AI systems can evaluate huge, complex information to identify biomarkers more rapidly and reliably. This is especially useful in fields such as cancer, where identifying genetic anomalies may help guide tailored treatment regimens.

The promise of AI in diagnostics goes beyond automation and efficiency. AI is driving a move toward more proactive and tailored treatment approaches. AI's capacity to assess patient-specific data and anticipate illness risk enables healthcare practitioners to propose individualized treatment alternatives. This personalized approach guarantees that patients receive care that is tailored to their particular requirements, which improves results while reducing unneeded treatments or interventions. In the future, diagnostic systems will be dependent on the seamless integration of artificial intelligence (AI) technologies. As AI advances, its role in healthcare will expand, providing increasingly more advanced tools for early illness identification, personalized treatment regimens, and improved patient outcomes. The ability of artificial intelligence to absorb large amounts of data will lead to continual improvements in diagnostic accuracy and efficiency, setting new standards for patient care.

6.4 AI in Treatment and Precision Medicine

Artificial intelligence (AI) is transforming the healthcare sector by changing how we approach precision medicine, pharmaceutical development, and therapy. The healthcare system is becoming more productive and efficient as AI is integrated into its operations, allowing for sophisticated prediction models and tailored care. One of the most visible benefits of artificial intelligence in healthcare is the development of tailored treatment strategies. Artificial intelligence (AI) systems may uncover individual trends and personalize treatment regimens to each patient's requirements by analyzing massive amounts of patient-specific data. This tailored strategy enhances the overall level of care by lowering the probability of unwanted side effects and enhancing patient outcomes. The diagram in Figure 6.3 depicts how Artificial Intelligence (AI), powered by big data, is transforming customized health promotion and chronic illness prevention.

The Left Side: The Foundation of Personalized Health.
The Person: The silhouette reflects the individual at the center of this individualized approach.
Factors influencing health: Several aspects influence a person's overall health. This includes:
Lifestyle factors, including sleep, stress, nutrition, and physical activity [6].
Big Data Analysis and Integration: The interconnected circles indicate the huge quantity of data obtained through deep phenotyping. The data is combined and analysed to provide patterns and insights.
AI-Driven Decision Making: The central processing unit with the brain emblem represents the function of artificial intelligence. AI systems use huge data to make informed judgments and predictions regarding a person's health [7].
Right Side: The Goal of Personalized Health.

The ultimate goal is to promote personalized health and prevent chronic diseases. Tailor-made plans are built based on AI analysis to boost

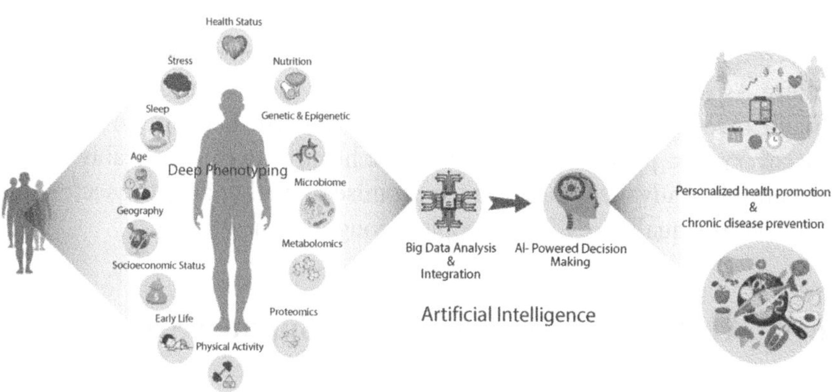

Figure 6.3 Lifestyle vs. disease prevention and health promotion.

individual health and prevent chronic illnesses. This might include tailored advice for nutrition, exercise, sleep, stress management, and even medicinal therapies.

Icons: The symbols on the right demonstrate individualized health:
Smartwatch: A wearable device for continuous health monitoring.
Person with Medical Staff: Represents customized healthcare interventions and assistance.
Healthy Food: Emphasizes the value of individualized nutrition.

The picture shows how artificial intelligence uses large data from numerous parts of a person's life to construct a highly tailored approach to healthcare, with the ultimate goal of improving health outcomes and preventing chronic illnesses. In the realm of drug research and development, AI is significantly accelerating the process. AI has the ability to minimize the time and costs associated with bringing new pharmaceuticals to market by finding promising drug candidates, forecasting their efficacy and safety, and expediting the clinical trials process. This efficiency is critical for addressing urgent health issues and making novel therapies available more rapidly. AI-powered prediction models are also changing the way healthcare practitioners handle patient care. These models can anticipate how patients will react to various therapies, allowing clinicians to make better-educated judgments. This technique is especially useful in the management of chronic illnesses, where early detection and tailored monitoring can have a major influence on patient health.

Another field where innovation is blossoming is robotics-assisted surgery, which has been strengthened by artificial intelligence. AI allows real-time data analysis and decision assistance, allowing robotic devices to help surgeons more precisely and accurately during difficult surgeries. This results in better surgical outcomes, more efficiency, and higher patient safety. As AI becomes more widely adopted in the healthcare industry, we may expect even greater advances in customized medicine, medication discovery, and patient care. The future of healthcare is inextricably linked to continued advances in AI, which promise to further alter the sector and improve the quality of life for people throughout the world [7].

6.5 AI Patient Management and Support Systems

Figure 6.4 portrays a smart healthcare system that uses networked gadgets and cloud technologies to provide remote healthcare services.

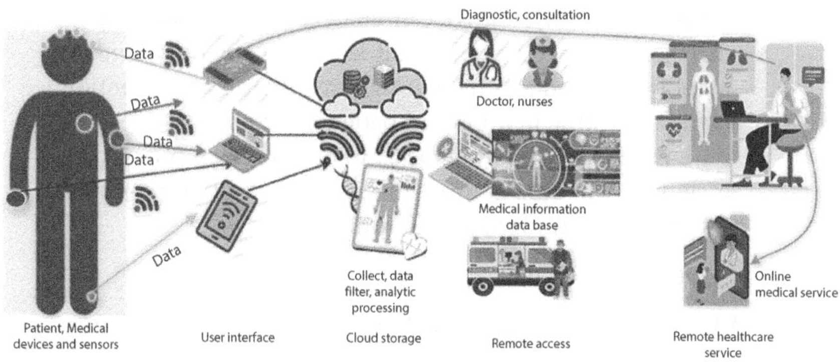

Figure 6.4 Smart healthcare system.

Data from a variety of sources, including patient-worn medical equipment and sensors, smartphone interfaces, and laptop entries, are sent to a cloud storage system. This consolidated repository, called the "medical information database," provides a complete source of patient data. Data is processed, including collecting, filtering, and analysis, before being used for diagnostic and consultation purposes [8]. Healthcare professionals, such as physicians and nurses, may access this information remotely, allowing them to give online medical services and assistance.

The system allows for a continuous flow of information, supporting efficient and accessible healthcare delivery regardless of geographical boundaries. This concept emphasizes the ability of technology to improve patient care.

6.6 AI in Electronic Health Records (EHR) Management

Automated Data Entry and Management: Artificial intelligence algorithms assist in the automation of clinical notes, lab findings, and patient histories, lowering the amount of time health-care personnel spend manually entering data.

- **Natural Language Processing (NLP)** can provide significant insights from unstructured data in Electronic Health Records (EHRs), including physician notes. This extraction procedure speeds up the retrieval of patient information and improves the quality of clinical decision-making.

- **Predictive Analytics:** AI can analyze patient data to predict health concerns. This skill facilitates the development of individualized treatment programs and early intervention tactics, resulting in improved patient outcomes.
- **AI-enhanced EHR systems** automate regular procedures including scheduling, invoicing, and prescription renewals, resulting in optimized clinical workflow.

6.7 AI for Virtual Health Assistants and Chatbots

Robotics-assisted surgery is a rapidly evolving subject that uses Artificial Intelligence (AI). AI enables immediate data analysis and decision assistance, allowing robotic systems to give surgeons greater precision and accuracy during complex surgeries. This leads to better surgical outcomes, increased efficiency, and greater patient safety. As artificial intelligence (AI) becomes more incorporated into the healthcare business, we may expect to see additional advances in customized medicine, medication development, and patient care. The future of healthcare is inextricably tied to the ongoing development of artificial intelligence (AI), which has the potential to transform the sector and enhance the general well-being of people globally [7, 8].

6.8 Remote Monitoring and Telemedicine

- **Continuous Patient Monitoring**: AI-integrated wearable devices continuously track vital signs, such as heart rate, blood pressure, and glucose levels. In case of abnormalities, healthcare providers are alerted in real time for timely intervention.
- **Telehealth Services**: AI-based platforms enable remote consultations through video calls, allowing patients to access healthcare professionals without in-person visits, which improves care accessibility in remote and underserved areas.
- **Disease Management**: For chronic conditions like diabetes and hypertension, AI telemedicine platforms offer personalized care plans, monitor treatment adherence, and adjust strategies based on real-time data.

- **Early Detection of Complications**: AI analyses data from remote monitoring devices to identify early signs of complications, enabling preventive actions and reducing the chances of hospital readmissions.

6.9 Machine Learning in Healthcare Data Analytics

With its revolutionary potential to improve patient outcomes, boost operational efficiency, and advance medical research, machine learning (ML) is fast emerging as a key component of healthcare data analytics. Figure 6.5 depicts a thorough summary of the use of machine learning in healthcare data analytics.

This figure depicts the many uses of machine learning (ML) techniques in the healthcare area. With the topic "ML algorithms in healthcare: app fields," the image highlights seven different domains in which these algorithms are having a major impact. These fields cover a broad spectrum of medical tasks, such as Diagnosis and Disease Identification, which recommends applying machine learning for prompt and precise diagnosis; Personalization of Medical Care, emphasizing the possibility of adjusting therapies to meet the requirements of specific patients; Medical Image Analysis, demonstrating how machine learning is used to comprehend intricate medical images; Medical Documentation Flow, indicating that paperwork is being automated and streamlined; predictive medicine, which focuses on using machine learning to predict health risks

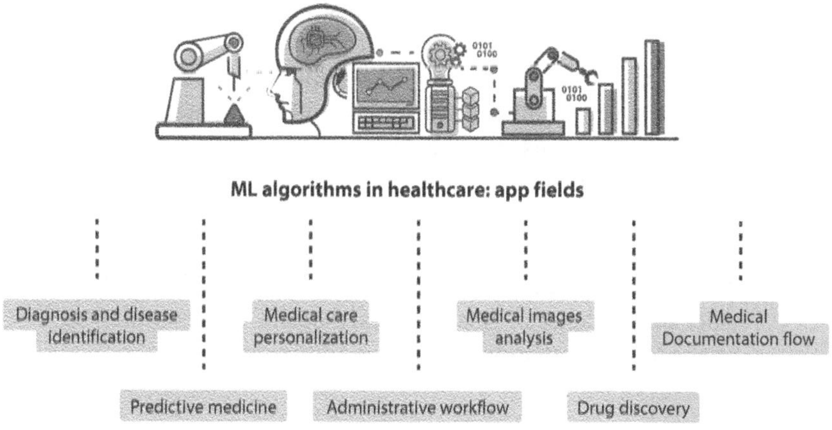

Figure 6.5 ML algorithms in healthcare.

and consequences; Administrative Workflow, which demonstrates how machine learning is used to optimize administrative processes; Predictive Medicine, emphasizing the use of ML to forecast health risks and outcomes; Administrative Workflow, showcasing the application of ML in optimizing administrative tasks; and Drug Discovery, highlighting the use of ML in accelerating the development of new treatments [8, 9]. The visual representation underscores the transformative potential of ML in revolutionizing various aspects of healthcare, from diagnostics and treatment to administration and research.

6.9.1 Importance of Data in AI and ML Models

Data is the cornerstone of any AI and ML model. Data from wearable technologies, clinical trials, electronic health records (EHRs), medical imaging, and patient-reported outcomes are used to train machine learning models in the healthcare sector. The accuracy and reliability of AI models are strongly influenced by the type and volume of data utilized for training. Having high-quality data that is reliable, diverse, and representative of many populations is essential for the model to generalize well and produce informative findings. Preprocessing and Data Cleaning: Errors, noise, and missing values are common in raw medical data. Proper data preparation, which includes data transformation, standardization, and cleaning, is necessary to build solid data.

6.9.2 Data Mining Techniques in Healthcare

Regression and classification are two supervised learning methods that are often applied in the medical field to forecast patient risk, treatment outcomes, and disease outcomes. Support vector machines (SVM) and logistic regression, for example, may categorize patients according to their risk of contracting a specific ailment. Clustering and Segmentation: Predictive models for patient readmission and emergency scenarios using unsupervised learning methods such as k-means clustering [10, 11].

- In order to predict readmission rates, AI systems examine patient data, such as previous hospital stays, comorbidities, and social factors. This enables medical professionals to make focused attempts to lower readmission rates.

- By anticipating demand, maximizing patient flow, enhancing resource allocation, and reducing wait times for urgent care, AI helps alleviate Emergency Department (ED) overcrowding.
- High-risk patients can be identified by AI systems for early monitoring and intervention, resulting in more effective treatment.

6.9.3 Predicting Deterioration

Algorithms driven by AI detect subtle shifts in a patient's health, promoting prompt intervention to prevent medical emergencies. AI is transforming patient support and management, improving patient outcomes, and healthcare efficiency, and providing more individualized care options. Hierarchical clustering assists in dividing patients into several groups according to similar traits. This may be used in personalized medicine, in which some patient subgroups receive customized therapies. Association Rule Mining: Methods like FP-Growth algorithms and Apriorism are used to find correlations and patterns among variables in big datasets. This can be utilized in the medical field to find correlations between various symptoms, comorbidities, and treatment results. Anomaly Detection: Algorithms for anomaly detection are used to find odd patterns in data, including finding uncommon diseases, detecting billing fraud, or discovering mistakes in medical imaging.

6.10 Predictive Analytics and Big Data

The diagram shown below outlines the predictive analytics process using big data. It begins with understanding the data, followed by employing big data technologies for data mining and text analytics. These steps lead to the core of predictive analytics: analysing the data through statistical methods, resulting in big data insights.

These insights inform the modelling phase, which is crucial for developing predictive models. Finally, the deployment stage utilizes these models for practical applications. Figure 6.6 visually represents the flow of information and the interconnectedness of each step in leveraging big data for predictive purposes, emphasizing the importance of a systematic approach [7–11].

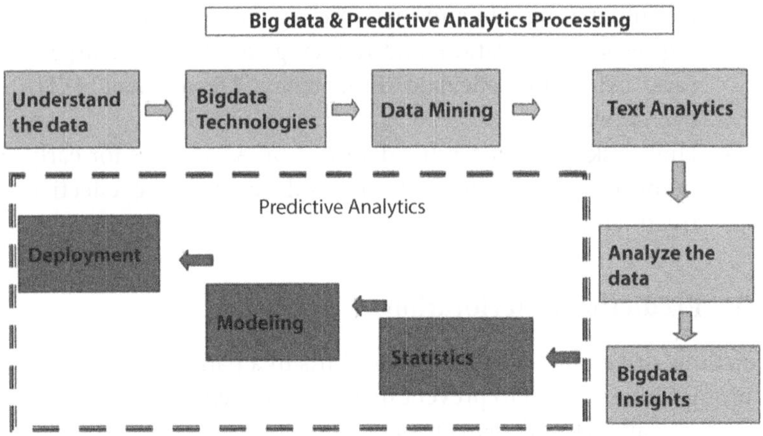

Figure 6.6 Big data predictive approach.

- **Combining Various Healthcare Data:** Electronic Health Records (EHRs), genetic sequencing, wearable technology, socioeconomic factors, and insurance claims are just a few of the many sources from which healthcare data is gathered. These disparate datasets may be integrated using big data analytics to provide a comprehensive picture of a patient's health.
- **Healthcare Predictive Modelling:** Disease progression, hospital readmission rates, and patient therapy responses are just a few of the health outcomes that may be predicted with the use of machine learning models such as random forests, gradient boosting machines, and deep learning approaches. Healthcare practitioners can lower the chance of unfavourable outcomes by using predictive analytics to take preventive action.
- **Population Health Monitoring:** By identifying trends and patterns in population health, big data analytics makes public health activities like health screening campaigns, vaccination drives, and monitoring of chronic diseases.
- **Improving Healthcare Operations:** By predicting patient admissions, planning staff schedules, and expediting the distribution of hospital resources, predictive analytics enhances operational effectiveness in the healthcare industry.

6.11 Data Privacy, Security, and Ethical Considerations

- **Regulatory Compliance**: Strict adherence to privacy and security rules is essential since healthcare information is sensitive. Patient data is protected by laws such as GDPR (in the EU), HIPAA (in the United States), and other local regulations.
- **Data Anonymization and De-identification**: Techniques including data masking, anonymization, and de-identification are used to protect patient identities while enabling research and analytics to access the data. These steps are crucial for preserving patient confidence and promoting data exchange.
- **Putting Security Measures in Place:** Protecting healthcare data from breaches and assaults requires robust cybersecurity measures, such as intrusion detection systems, access restrictions, and encryption. These safeguards guarantee that private data stays intact and protected.
- **Ethical Use of AI in Healthcare:** Fairness, transparency and the avoidance of biases are all ensured by applying AI and machine learning in healthcare in accordance with ethical norms. Getting informed permission and making sure AI-driven choices are transparent is crucial. Preventing disproportionate treatment of various patient populations requires addressing biases in the training data.

With precise diagnosis, individualized care, and improved service delivery, machine learning and big data analytics have the potential to dramatically improve healthcare. To build confidence and guarantee the appropriate use of AI in healthcare, it is crucial to address issues pertaining to privacy, security, and ethics.

6.12 Challenges and Limitations of AI and ML in Healthcare

Workflows, diagnostics, and patient care might all be completely transformed in the healthcare industry by combining artificial intelligence (AI) with machine learning (ML). To reach their full potential, these technologies must, however, overcome a number of obstacles and constraints. The main obstacles and restrictions of AI and ML in healthcare are outlined below.

- **Health information and medical background:** It is quite challenging to guarantee the security and privacy of this data while simultaneously avoiding unwanted access. The General Data Protection Regulation (GDPR) of the European Union and the Health Insurance Portability and Accountability Act (HIPAA) standards are challenging to achieve and can pose legal and regulatory issues when employing high-quality data for AI and machine learning applications [1–5].
- **Data ownership and informed permission:** Before using a patient's medical information for intellectual property purposes, the patient must provide their approval. This entails being explicit about who will have access to their data, how it will be used, and why. Another degree of complication is introduced by the continuous discussions around data ownership and patients' rights to manage or profit from their medical information. Either biased or unbiased cognitive models might be inferred from this data. AI systems may perform poorly for certain users due to a lack of diverse data, which raises the possibility of biased judgments and keeps healthcare services from being uniform and clean [6–11].

6.13 AI Models' Interpretability and Explainability

- **The Black-Box Problem:** A lot of advanced AI models, especially deep learning algorithms, operate as "black boxes," which makes it challenging to understand how they make decisions. The absence of transparency is a major obstacle to broad adoption in the healthcare industry, as choices have a direct impact on patient outcomes.
- **Transparency Regulation Requirements:** Demand for AI models that can clearly and understandably explain their decision-making processes is growing in highly regulated industries like healthcare. In order for healthcare practitioners to trust and confirm AI-driven recommendations—which are

crucial for patient safety and clinical acceptance—regulatory authorities frequently demand openness.

- **Increasing Trust Among Healthcare Providers:** Instead of making ambiguous forecasts, medical practitioners are increasingly inclined to use AI solutions that offer clear and useful insights. Enhancing the interpretability of AI models promotes cooperation between human specialists and AI systems and helps build physician confidence.
- **Methods to Improve Explainability:** Several methods have been created to improve the interpretability of AI models. Examples that shed light on how AI models make decisions include attention mechanisms and Local Interpretable Model-Agnostic Explanations (LIME), which were created to improve the interpretability of the model. These methods are still developing, though, and might not always offer the degree of clarity needed for practical use.

6.14 Conclusion

In contemporary healthcare systems, artificial intelligence (AI) and machine learning (ML) highlight the revolutionary effects these technologies have on patient care, diagnostic, and treatment procedures. Looking ahead, it is evident that AI and ML are radically changing the healthcare industry rather than merely improving current procedures. By facilitating more individualized and effective patient care, AI and ML are completely changing the way healthcare is delivered. Early illness identification, individualized treatment regimens, and better patient outcomes are made possible by the capacity to analyze enormous volumes of data. For example, research has demonstrated that AI may greatly increase medical imaging diagnostic accuracy, surpassing conventional techniques in the detection of diseases like lung cancer. Furthermore, AI's contribution to predictive analytics is essential for handling public health issues. AI can assist in forecasting disease outbreaks and enhance resource allocation in healthcare systems by examining trends in health data. To meet the growing healthcare demands brought on by aging populations and the incidence of chronic diseases, this proactive strategy is crucial. The use of AI in drug discovery is another example of how it may speed up the creation of novel treatments. Artificial intelligence (AI)-powered systems can evaluate biological and chemical data at previously unheard-of rates, cutting down on the time and expense involved in launching new medications. To respond to pressing medical emergencies and provide prompt access to efficient therapies, this breakthrough is essential.

As AI and ML technologies develop further, their incorporation into healthcare systems is anticipated to result in ground-breaking innovations that improve patient care and operational effectiveness. A future where healthcare is not only more accessible to a wider range of people but also more effective is promised by the continuous research and development in this area. To achieve these advantages and solve the intricate problems facing contemporary healthcare, sustained investment in AI technology will be essential.

AI disclosure statement: The author confirms that no artificial intelligence (AI)-generated content or AI-assisted writing tools were used in the preparation of this chapter (including summaries and the abstract). All text, analysis, and conclusions are the result of the author's original work.

References

1. Esteva, A., Robicquet, A., Ramsundar, B., Kuleshov, V., DePristo, M., Chou, K., Cui, C., Corrado, G., Thrun, S., Dean, J., A guide to deep learning in healthcare. *Nat. Med.*, 25, 1, 24–9, 2019 Jan.
2. Jiang, F., Jiang, Y., Zhi, H., Dong, Y., Li, H., Ma, S., Wang, Y., Dong, Q., Shen, H., Wang, Y., Artificial intelligence in healthcare: past, present and future. *Stroke Vasc. Neurol.*, 2, 4, 2017 Dec 1.
3. Topol, E.J., High-performance medicine: the convergence of human and artificial intelligence. *Nat. Med.*, 25, 1, 44–56, 2019 Jan.
4. Rieke, N., Hancox, J., Li, W., Milletari, F., Roth, H.R., Albarqouni, S., Bakas, S., Galtier, M.N., Landman, B.A., Maier-Hein, K., Ourselin, S., The future of digital health with federated learning. *npj Digital Med.*, 3, 1, 1–7, 2020 Sep 14.
5. Mudgal, S.K., Agarwal, R., Chaturvedi, J., Gaur, R., Ranjan, N., Real-world application, challenges and implication of artificial intelligence in healthcare: an essay. *Pan Afr. Med. J.*, 43, 1, 2022 Sep 2.
6. Holzinger, A., Langs, G., Denk, H., Zatloukal, K., Müller, H., Causability and explainability of artificial intelligence in medicine. *Wiley Interdiscip. Rev.: Data Min. Knowl. Discovery*, 9, 4, e1312, 2019 Jul.
7. Ahmad, M.A., Eckert, C., Teredesai, A., Interpretable machine learning in healthcare, in: *Proceedings of the 2017 ACM International Conference on Bioinformatics, Computational Biology, and Health Informatics*, pp. 559–560, 2017 Aug 15.

8. Esteva, A., Kuprel, B., Novoa, R.A., Ko, J., Swetter, S.M., Blau, H.M., Thrun, S., Dermatologist-level classification of skin cancer with deep neural networks. *Nature.*, 542, 7639, 115–8, 2017 Feb.
9. Rajpurkar, P., Irvin, J., Ball, R.L., Zhu, K., Yang, B., Mehta, H., Duan, T., Ding, D., Bagul, A., Langlotz, C.P., Patel, B.N., Deep learning for chest radiograph diagnosis: A retrospective comparison of the CheXNeXt algorithm to practicing radiologists. *PLoS Med.*, 15, 11, e1002686, 2017 Nov 20.
10. Choi, E., Bahadori, M.T., Schuetz, A., Stewart, W.F., Sun, J., Doctor ai: Predicting clinical events *via* recurrent neural networks, in: *Machine Learning for Healthcare Conference*, PMLR, pp. 301–317, 2016 Dec 10.
11. Bender, A. and Cortés-Ciriano, I., Artificial intelligence in drug discovery: what is realistic, what are illusions? Part 1: Ways to make an impact, and why we are not there yet. *Drug Discovery Today*, 26, 2, 511–24, 2021 Feb 1.
12. Tripathi, S.L. and Mahmud, M., *Explainable Machine Learning Models and Architecture*, Wiley-Scrivener Publishing, USA, 2023, https://doi.org/10.1002/9781394186570.
13. Thillaiarasu, N., Tripathi, S.L., Dhinakaran, V. (Eds.), *Artificial Intelligence for Internet of Things: Design Principle, Modernization, and Techniques*, 1st, CRC Press, Boca Raton, 2022, https://doi.org/10.1201/9781003335801.

7

CMOS 90nm-Based Two-Stage Differential Amplifier

Lingampalli Abhijith[1], Hakke Mayuri Shahuraj[1], Sobhit Saxena[1]* and Suman Lata Tripathi[2]

[1]*School of Electronics and Electrical Engineering, Lovely Professional University, Phagwara, Punjab, India*
[2]*Department of Electronics & Telecommunication, Symbiosis International (Deemed University), Pune, India*

Abstract
In CMOS technology, differential amplifiers are fundamental components that can serve important purposes, such as rejecting noise, providing maximum input impedance and often offering maximum common mode rejection. This amplifier is configured in various types suited to perform different applications. This chapter presents the two-stage differential amplifier which comes under the category of Op-Amp differential amplifier. This chapter shows the design and analysis of a two-stage differential amplifier with common mode and differential mode in 90nm CMOS technology.

Keywords: CMOS, two-stage, one-stage, differential amplifier, CMRR, GAIN, bandwidth

7.1 Introduction

Differential Amplifiers are essential components in analog circuit design and play a crucial role in various applications like in Op-Amps, Analog-to-Digital converters (ADCs), Communication Systems, Amplifiers [1]. They are designed to amplify the two signals difference between them while rejecting common mode noise. This principle makes the differential

**Corresponding author*: sobhitsaxena16@gmail.com

Abhishek Kumar and Suman Lata Tripathi (eds.) Smart Chips for Smart Devices: VLSI Design for Next-Generation IoT Solutions, (123–140) © 2026 Scrivener Publishing LLC

amplifier more effective in applications where signal integrity is crucial [2]. A high-performance differential amplifier is calculated by its characteristics like gain, bandwidth, common mode rejection ratio (CMRR) and gain band width product (GBP) [3].

Practically, differential amplifiers will enhance the minute differences between the two voltages, even in the presence of a much more common-mode signal. This makes it highly useful in sensor interfaces, audio processing, and communication circuits where the signal is covered in noise. Mostly performance of differential amplifier is measured by their gain, bandwidth, and most important is common-mode rejection ratio (CMRR), by which we can know how much ability is there to reject common-mode signals. There are three types of stage differential amplifier; likewise, a single-stage differential amplifier will perform basic functioning like differential amplification, which consists of a single differential pair, and it has limitations for gain, bandwidth, and CMRR. The two-stage differential amplifier offers high gain, better CMRR [4, 5] than single stage, making it more suitable for high-performance applications. This is the reason that designers often opt for a two-stage differential amplifier rather than single stage [6].

7.2 Theory of Differential Amplifiers [7, 8]

7.2.1 Basic Principles of Operation

A differential amplifier is designed in a way that amplifies the voltage difference between two input signals while rejecting the common voltage between the two inputs such as common-mode noise. This is totally different to a single-ended amplifier, which amplifies the voltage at a single input relative to a fixed reference point like ground. The unique feature of a differential amplifier is to reject common-mode signals, as we discussed earlier, which makes it crucial in applications where signal integrity is necessary. To understand the operation of a differential amplifier we must consider the two modes—differential mode and common mode.

7.2.2 Differential Mode

In this mode, the inputs to the differential amplifier are out of phase, meaning one input is positive while the other is negative relative to a common reference. The output of the amplifier is proportional to the difference between these two inputs. The key advantage of this mode is that it enhances the signal of interest (the difference between the inputs) while ignoring noise that may be present equally on both inputs [9].

7.2.3 Common Mode

In common mode, both inputs are in phase, meaning they rise and fall together. Optimally, a perfect differential amplifier may completely reject these common-mode signals, resulting in zero output. However, during practical implementation of differential amplifier circuits some of the common-mode signal can leak through the output due to imperfections in the amplifier's design [10].

7.2.4 Differential Gain and Common-Mode Rejection Ratio (CMRR)

The differential gain of a differential amplifier is the gain which is applied to the difference between the input signals.

$$A_{dm} = \frac{V_{out}}{V_1 - V_1}$$

Where A_{dm} is the differential gain, V_{out} is the output voltage, V_1 and V_2 are the input voltages.

Common mode gain which we will apply to average of the input signals is

$$A_{cm} = \frac{V_{out}}{V_{CM}}$$

The ability of the differential amplifier which is to reject the common-mode signals relative to its ability to amplify differential signals can be measured by an important parameter is CMRR.

In a case where an amplifier has CMRR value high that will indicate that it is effective at rejecting noise and maintains the integrity of the differential signal. In real-world applications, achieving a high CMRR is challenging due to mismatches in components, parasitic capacitances, and other non-idealizes in the circuit [11].

7.3 Differential Amplifier

7.3.1 Single-Stage Differential Amplifier

A single-stage differential amplifier is the simplest form of a differential amplifier, as shown in Figure 7.1, typically consisting of a pair of transistors (BJTs or MOSFETs) with their emitters or sources, in the case of MOSFET connected to a common current source. The bases (gates) of the transistors will be given the differential inputs, and the outputs are taken from their collectors (drains). The bias current will be provided by common current source to the transistors and ensures that the total current remains constant, regardless of the differential input voltage.

The key components of a single-stage differential amplifier include:

Transistor Pair: The two transistors form the core of the differential amplifier. Their matching is crucial for maintaining symmetry and achieving a high CMRR.
Current Source: The current source provides a stable bias current to the transistor pair. It is often implemented using a current mirror to ensure that the current remains constant even as the input voltage changes.
Load Resistors: The load resistors are connected to the collectors (drains) of the transistors. They convert the current changes caused by the differential input voltage into a voltage output.

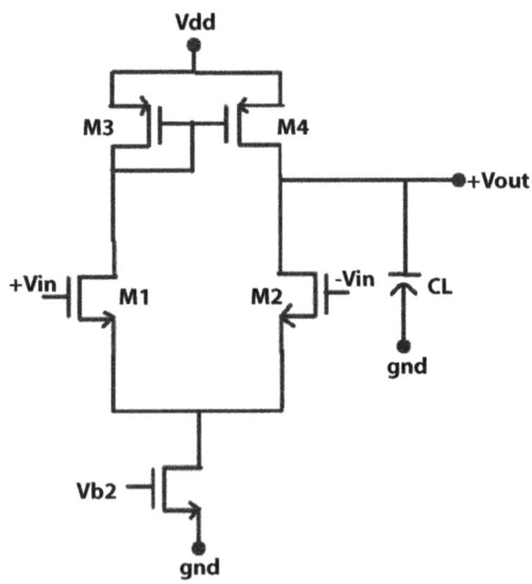

Figure 7.1 Schematic of single-stage amplifier.

In differential mode, the input signals are 180 degrees out of phase, meaning that when one input increases, the other decreases by the same amount. This differential input causes one transistor to conduct more while the other conducts less, resulting in a change in the voltage across the load resistors. The output voltage is the difference between the voltages at the collectors (drains) of the two transistors. The differential gain of the amplifier is given by:

$$Ad=gm.RL$$

Where g_m is the transconductance of the transistors, and R_L is the load resistance.

In common mode, both inputs move in the same direction, causing the transistors to conduct equally. Ideally, this should result in no change in the output voltage, as the voltage drops across the load resistors would be identical. However, in practice, due to mismatches in the transistors and the finite impedance of the current source, some common-mode signal may appear at the output. The common-mode gain is typically much smaller than the differential gain, but it is not zero. This non-zero common-mode gain reduces the CMRR and can degrade the performance of the amplifier in noisy environments.

7.3.2 Limitations of a Single-Stage Amplifier

While the single-stage differential amplifier is simple and effective for basic differential amplification, it has several limitations:

***Limited Gain*:** The gain of a single-stage amplifier is limited by the transconductance of the transistors and the load resistance. For applications requiring high gain, a single stage may not be sufficient.

***Limited Bandwidth*:** The bandwidth of the amplifier is also limited by the parasitic capacitances of the transistors and the load. At high frequencies, these capacitances can cause the gain to roll off, limiting the amplifier's performance.

***Finite CMRR*:** The CMRR of a single-stage differential amplifier is constrained by mismatches in the transistors and the current source. Achieving a high CMRR requires careful matching of components and attention to layout.

***Limited Output Swing*:** The output swing of the amplifier is limited by the voltage headroom available at the collector (drain) of the transistors. In low-power applications, this can be a significant constraint.

7.3.3 Two-Stage Differential Amplifier

To address the limitations of a single-stage differential amplifier, a two-stage configuration is often employed as shown in Figure 7.2. In a two-stage differential amplifier, the first stage provides the initial differential amplification, while the second stage acts as a gain or load stage to further amplify the signal and improve the overall performance. Multi-stage amplification offers several advantages, including the following:

***Higher Gain*:** By cascading multiple stages, the overall gain of the amplifier can be increased significantly.
***Improved CMRR*:** The second stage can be designed to enhance the CMRR by further rejecting common-mode signals.
***Greater Design Flexibility*:** The two stages can be optimized independently for gain, bandwidth, and noise performance, providing greater flexibility in design.

In a typical two-stage differential amplifier, the first stage is a differential pair, like the single-stage amplifier described earlier. The second stage is usually a common-source (for MOSFETs) or common-emitter (for BJTs) amplifier that amplifies the output of the first stage.

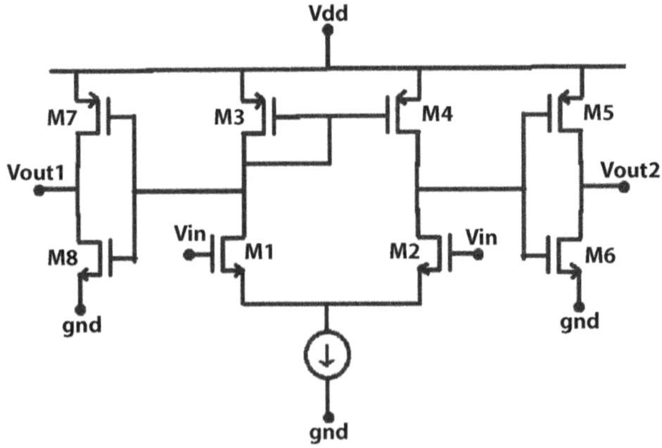

Figure 7.2 Two-stage differential amplifiers [12].

Stage 1: Differential Amplifier

The stage 1 is designed to provide the initial differential gain. It consists of a matched pair of transistors with a common current source and load resistors. The goal of this stage is to maximize the differential gain while minimizing the common mode gain, thereby setting the foundation for the overall CMRR.

Stage 2: Gain Stage/Load Stage

The second stage amplifies the differential signal produced by the first stage. This stage can be implemented as a common-source or common-emitter amplifier, depending on the type of transistors used. The second stage provides additional gain and can be designed to reject any residual common-mode signals that may have leaked through the first stage. The design of the second stage is critical for achieving the desired overall gain and bandwidth. It is important to consider factors such as transistor sizing, load impedance, and biasing to ensure optimal performance.

7.3.4 Advantages of Two-Stage Over Single-Stage Amplifiers

The two-stage differential amplifier offers several key advantages over a single-stage design:

***Higher Gain*:** The cascaded stages provide a much higher overall gain, making the two-stage amplifier suitable for applications requiring high signal amplification.
***Improved CMRR*:** The second stage enhances the CMRR by further rejecting common-mode signals, resulting in better noise rejection and signal integrity.
***Better Bandwidth*:** By carefully designing the second stage, the bandwidth of the amplifier can be extended, allowing it to operate effectively at higher frequencies.
Greater Flexibility: The two-stage design provides more design flexibility, allowing the designer to optimize each stage independently for different performance criteria.

7.3.5 Common-Mode Topology in Two-Stage Differential Amplifier [13]

In common-mode operation, the input signals to the differential amplifier are identical, causing both transistors in the first stage to conduct equally. Ideally, this should result in no change in the output voltage, as the differential signal (difference between the two inputs) is zero. However, in practice, due to mismatches in the transistors and the finite impedance of the current source, some common-mode signals may appear at the output. The first stage of the two-stage amplifier plays a crucial role in determining the overall CMRR. If the first stage has a high common-mode gain, the overall CMRR will be reduced, even if the second stage is designed to reject common-mode signals. To improve the common-mode rejection of the first stage, common-mode feedback (CMFB) circuits are often employed. ACMFB circuit senses the common-mode voltage at the output of the first stage and adjusts the biasing of the transistors to minimize the common-mode gain. CMFB circuits can be implemented in several ways, including the following:

Resistive Feedback: This type of feedback network can be used to recognize the common-mode voltage and adjust the biasing accordingly.
Active Feedback: In this type of feedback an additional amplifier can be used to recognize the common-mode voltage and provide feedback to the current source or transistor gates.
Capacitive Feedback: In this feedback capacitors can be used to recognize the high- frequency common-mode signals and provide feedback to improve CMRR at high frequencies.

CMFB circuits are essential in high-performance differential amplifiers, as they help maintain a high CMRR across a wide range of operating conditions. The CMRR of the first stage is critical to the overall performance of the two-stage differential amplifier. A high CMRR in the first stage ensures that common-mode signals are effectively rejected, allowing the differential signal to be amplified with minimal interference. This is particularly important in applications where the input signals are small and susceptible to noise.

Common-Mode Signal Rejection in the Second Stage

The second stage of the two-stage amplifier further amplifies the differential signal produced by the first stage. It also plays a role in rejecting any residual common-mode signals that may have leaked through the first stage. The design of the second stage, including the choice of transistor sizing and load impedance, is critical for maintaining a high overall CMRR.

In some designs, the second stage may include additional CMFB mechanisms to further enhance the common-mode rejection. This is particularly important in high-precision applications where even small common-mode signals can degrade the performance of the amplifier. CMFB circuit can fix the output common mode [14].

7.3.6 The Differential-Mode Topology in Two-Stage Differential Amplifier [15]

In differential mode, the input signals are out of phase, meaning that when one input increases, the other decreases by the same amount. This differential input causes one transistor in the first stage to conduct more while the other conducts less, resulting in a change in the voltage across the load resistors. The output to the first stage is the difference between the voltages at the collectors (drains) of the two transistors.

The differential gain of the first stage is proposed by the transconductance of the transistors and the load resistance. The goal in differential-mode operation is to maximize the differential gain while minimizing the common-mode gain, thereby enhancing the overall CMRR. The differential gain of the first stage is a key parameter that determines the overall performance of the two-stage amplifier. The gain is given by $Ad=gm.RL$, where gm refers to transconductance of the transistors, and RL refers to load resistance. The transconductance gm is proportional to the transistor's characteristics and bias current, meaning that the differential gain can be controlled by adjusting the bias current and the load resistance. The second stage amplifies this differential signal further, resulting in a much higher overall gain than a single-stage amplifier.

7.3.7 Importance of Matching and Symmetry in Differential-Mode Operation

In differential-mode operation, the matching of the differential pair transistors is crucial for achieving more (high) differential gain as well as less (low) common-mode gain. Mismatches in the transistor parameters, such as threshold voltage and transconductance, can cause the reduction of differential gain and increase the value of common-mode gain, which can reduce the CMRR [16]. Symmetry in the circuit layout is also important for maintaining the performance of the differential amplifier. Symmetrically out ensures that parasitic capacitances and resistances are matched, minimizing their impact on the differential and common-mode gains. In high-performance differential amplifiers, careful attention is paid to the matching of components and the symmetry of the layout to maximize the differential gain and CMRR.

7.3.8 Impact of the Second Stage on Differential Gain

The second stage of the two-stage amplifier amplifies the differential signal produced by the first stage. The design of the second stage, including the choice of transistor sizing, load impedance, and biasing, has a direct impact on the overall differential gain and bandwidth of the amplifier. In some designs, the second stage may include additional gain stages to further increase the overall gain. The trade-off between gain and bandwidth is an important consideration in the design of the second stage, as increasing the gain typically results in a reduction in bandwidth. The second stage also plays a crucial role in shaping the frequency response of the amplifier, with the overall bandwidth being determined by the combined response of the two stages. In high-frequency applications, the design of the second stage is critical for maintaining a wide bandwidth while achieving the desired gain.

7.4 Design Considerations

7.4.1 Biasing Strategies for Two-Stage Amplifiers

Proper biasing is essential for the stable operation of a two-stage differential amplifier. Biasing strategies ensure that the transistors work in the desired region of their characteristic curves, providing consistent gain, bandwidth, and CMRR across a range of operating conditions. Common biasing strategies include [17] the following:

- ***Current Mirror Biasing***: A current mirror is used to provide a steady bias current to the differential pair. This technique will help to maintain bias current remains constant; even as the input voltage changes it will still provide the consistent gain and CMRR.
- ***Self-Biasing***: In self-biasing, the bias current is set by the transistor itself, using feedback from the output. This technique provides good stability but may be less accurate than current mirror biasing.
- ***Cascade Biasing***: A cascade biasing is a type of biasing by which we can improve the gain and the amplifier bandwidth by reducing the impact of parasitic capacitances and improving the separation between the input and output stages.

The biasing strategy will depend on the specific requirements of the application, including the desired gain, bandwidth, and power consumption. In the Figure 7.3 design we used the current source biasing, which is the most common biasing strategy used in this type of circuit. The symbol at the bottom (between the sources of M1 and M2) is a current source,

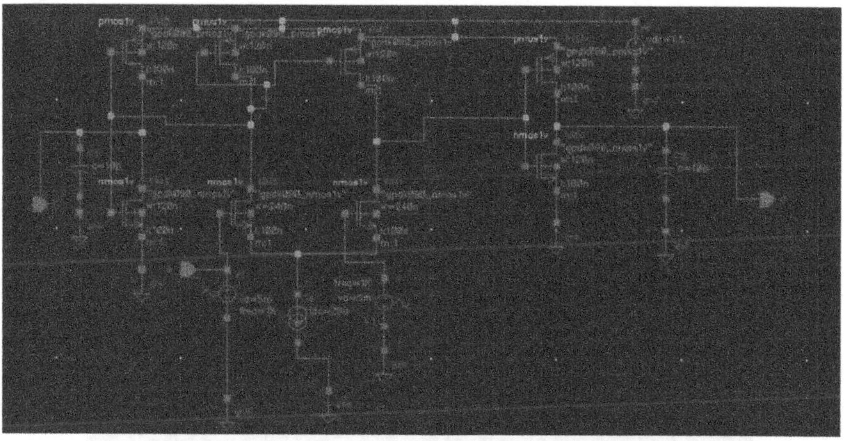

Figure 7.3 Schematic design of differential mode of two-stage differential amplifier.

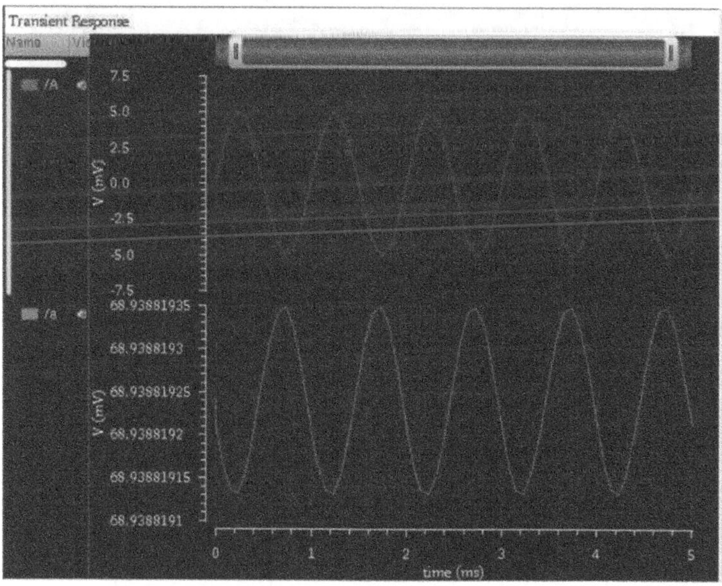

Figure 7.4 Transient response of differential mode differential amplifier.

which is used to bias the differential pair. This current source ensures a constant current flow through the differential pair (M1, M2), allowing the circuit to operate properly over different supply voltages. In this design we used Tail Current Source, which plays a crucial role in these differential amplifier configurations like enhancing the linearity of differential pair which ensures the accurate output signal. By maintaining a constant current, tail current source helps to reduce the distortion in the output signal

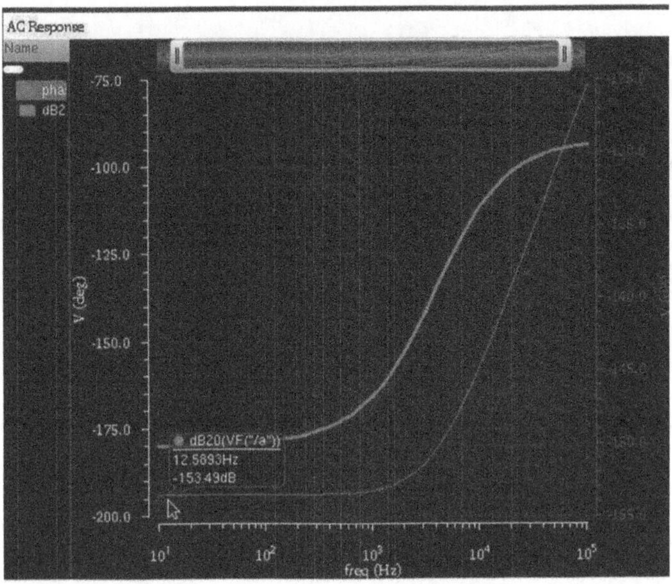

Figure 7.5 AC response of differential mode differential amplifier.

Table 7.1 Performance parameter of two-stage differential amplifier.

Parameter	**Figure**
Gain	80-120db
Phase Margin	45°-60°
Supply Voltage	1.8v–5v
Power Consumption	<5mv
CMRR	>80db
Technology	90nm

and improves CMRR by providing constant current. Figures 7.4 and 7.5 present the transient and AC response of two-stage amplifiers, and their performance is listed in Table 7.1.

7.4.2 Applications of Two-Stage Differential Amplifiers

Two-stage differential amplifiers [18] are fundamental blocks in many operational amplifier designs, in which the first stage provides the initial differential amplification, while the second stage provides additional gain and drives the output. This type of configuration offers high gain, better CMRR, and wideband width, making it ideal for precision analog applications.

Operational amplifiers are used in a wide range of applications:

- ***Signal Conditioning***: Op-Amps are used to amplify and filter signals in sensor interfaces, data acquisition systems, and other analog circuits. Op-Amps are also used in bio-signal processing and a detailed analysis is carried out by using gdpk90nm [3].
- ***Analog Computing***: Op-Amps are used in analog computers to perform mathematical operations, such as addition, subtraction, integration, and differentiation.
- ***Instrumentation Amplifiers***: These amplifiers are used to amplify the small differential signals in the presence of large common-mode noise; often two-stage differential amplifiers

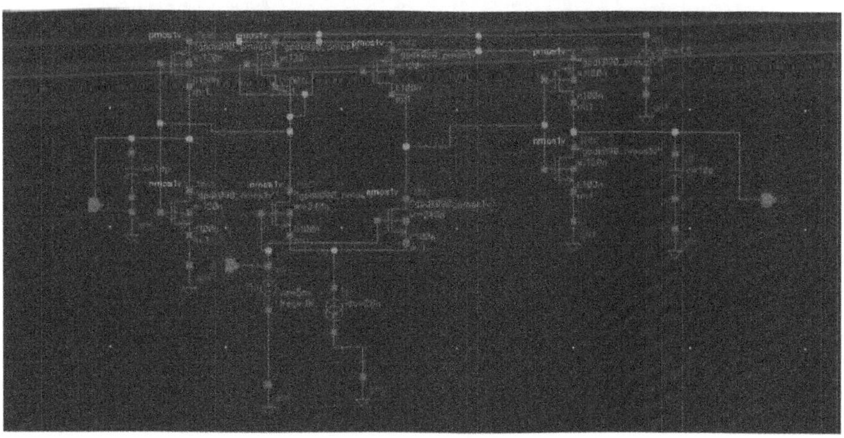

Figure 7.6 Schematic design of common mode of two-stage differential amplifier.

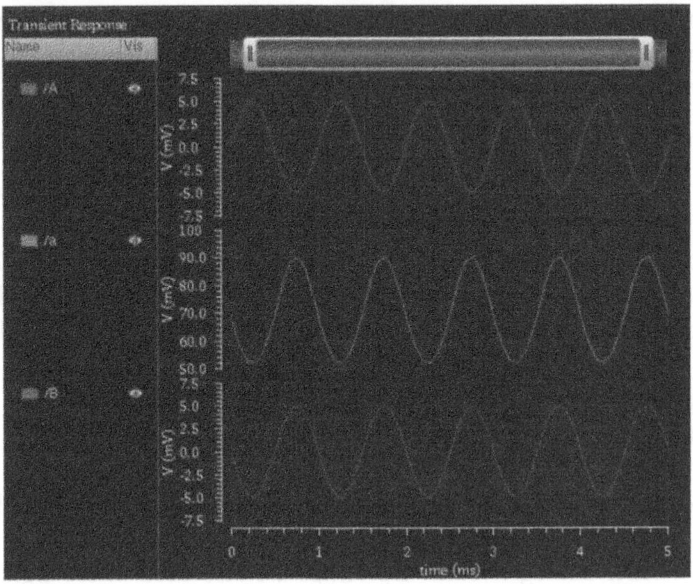

Figure 7.7 Transient response of common mode differential amplifier.

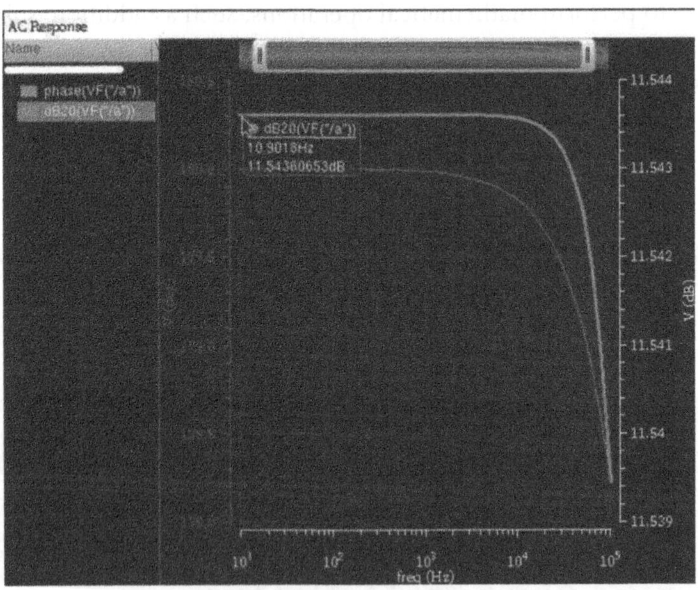

Figure 7.8 AC response of common mode differential amplifier.

will be used to achieve high gain and CMRR. The versatility and performance of two-stage differential amplifiers make them indispensable in the design of high-performance Op-Amps. The schematic of a two-stage amplifier is shown in Figure 7.6 and their simulation result is presented in Figures 7.7 and 7.8.

7.5 Conclusion

The two-stage differential amplifier offers significant advantages over single-stage designs, including higher gain, better CMRR, and greater design flexibility. Understanding the common-mode and differential-mode topologies is essential for optimizing the performance of these amplifiers. Practical implementation of a two-stage differential amplifier involves challenges such as transistor matching, parasitic management, stability, and power consumption. Simulation, prototyping, and iterative design are essential for overcoming these challenges and achieving optimal performance.

7.6 Future Trends and Considerations

As technology advances, the design of two-stage differential amplifiers continues to evolve. Emerging trends include the use of advanced semiconductor processes, such as silicon-on- insulator (SOI) and FinFET, to improve the performance of differential amplifiers. Additionally, the integration of two-stage differential amplifiers into system-on-chip (SoC) designs is becoming increasingly common, enabling more compact and power-efficient analog systems. In the future, the demand for high-precision, low-noise amplifiers is expected to grow, particularly in applications such as biomedical instrumentation [19], wireless communication, and quantum computing. The design of two-stage differential amplifiers will continue to play a crucial role in meeting these demands, with ongoing research focused on improving their performance, stability, and integration into complex analog systems.

AI disclosure statement: The author confirms that no artificial intelligence (AI)-generated content or AI-assisted writing tools were used in the preparation of this chapter (including summaries and the abstract). All text, analysis, and conclusions are the result of the author's original work.

References

1. Sharma, D. and Nath, V., CMOS Operational Amplifier Design for Industrial and Biopotential Applications: Comprehensive Review and Circuit Implementation. *Results Eng.*, 102357, 2024.
2. Safari, M.M. and Pourrostam, J., The role of analog signal processing in upcoming telecommunication systems: Concept, challenges, and outlook. *Signal Process.*, 109446, 2024.
3. Matthus, C.D., *et al.*, High gain and high bandwidth fully differential difference amplifier as current sense amplifier. *IEEE Trans. Instrum. Meas.*, 70, 1–11, 2020.
4. Tripathy, D. and Bhadra, P., A high speed two stage operational amplifier with high CMRR, in: *3rd IEEE International Conference on Recent Trends in Electronics, Information & Communication Technology (RTEICT)*, IEEE, 2018.
5. Lavanya, K. and Janamani Chandram, A., A high gain high CMRR two-stage fully differential amplifier using gm/I d procedure for bio-clinical applications. *Int. J. Sci. Eng. Technol. Res.*, 9, 14–18m, 2020.
6. Jain, P. and Mahesh Joshi, A., Low leakage and high CMRR CMOS differential amplifier for biomedical application. *Analog Integr. Circuits Signal Process.*, 93, 71–85, 2017.
7. Alzaher, H. and Ismail, M., A CMOS fully balanced differential difference amplifier and its applications. *IEEE Trans. Circuits Syst. II Analog Digital Signal Process.*, 48, 6, 614–620, 2001.
8. Ahn, H.-T. and J. Allstot, D., A 0.5-8.5 GHz fully differential CMOS distributed amplifier. *IEEE J. Solid-State Circuits*, 37, 8, 985–993, 2002.
9. Hornero, G., Casas, O., Pallàs Areny, R., Common mode response effects in differential measurements. *AEU Int. J. Electron. Commun.*, 128, 153510, 2021.
10. Ballo, A., Dario Grasso, A., Pennisi, S., CMOS Differential Stage with Improved DC Gain, CMRR and PSRR Performance, in: *26th IEEE International Conference on Electronics, Circuits and Systems (ICECS)*, IEEE, 2019.
11. Zhang, S., *et al.*, A 130-dB CMRR instrumentation amplifier with common-mode replication. *IEEE J. Solid-State Circuits*, 57, 1, 278–289, 2021.
12. Hamzah, M.H., Bahari Jambek, A., Hashim, U., Design and analysis of a two-stage CMOS op-amp using Silterra's 0.13 μm technology, in: *IEEE Symposium on Computer Applications and Industrial Electronics (ISCAIE)*, IEEE, 2014.
13. Massarotto, M., Carlosena, A., J. Lopez-Martin, A., Two-stage differential charge and transresistance amplifiers. *IEEE Trans. Instrum. Meas.*, 57, 2, 309–320, 2008.
14. Guerrero, F.N., *et al.*, Double-differential amplifier for sEMG measurement by means of a current-mode approach. *IEEE Access*, 10, 45870–45880, 2022.

15. Figueiredo, M., *et al.*, A two-stage fully differential inverter-based self-biased CMOS amplifier with high efficiency. *IEEE Trans. Circuits Syst. I Regul. Pap.*, 58, 7, 1591–1603, 2011.
16. Lee, C.-H., *et al.*, Balanced band-notched UWB filtering circular patch antenna with common-mode suppression. *IEEE Antennas Wirel. Propag. Lett.*, 16, 2812–2815, 2017.
17. Gómez-García, R., *et al.*, Multi-stub-loaded differential-mode planar multi-band bandpass filters. *IEEE Trans. Circuits Syst. II Express Briefs*, 65, 3, 271–275, 2017.
18. Tripathi, S.L., Raj, B., Verma, S., Narula, V., Saxena, S., High Gain Multistage CMOS Amplifier Design at 45nm technology node, in: *2023 IEEE Devices for Integrated Circuit (DevIC)*, Kalyani, India, pp. 237–242, 2023, doi: 10.1109/DevIC57758.2023.10134863.
19. Kumar, A., Tripathi, S.L., Verma, C., Raboaca, M.S., Enescu, F.M., Mihaltan, T.C., Design and Analysis of Low Power Bio-amplifier with Current Mirror Topology at CMOS 45nm Technology Node, in: *2022 14th International Conference on Electronics, Computers and Artificial Intelligence (ECAI)*, Ploiesti, Romania, pp. 1–5, 2022, doi: 10.1109/ECAI54874.2022.9847492.

8

Smart Self-Powered Wireless Sensor Network for Industrial Monitoring

G. Thirumalaiah[1] and Mudassir Khan[2]*

[1]Department of ECE, Annamacharya University, Rajampet, Andhra Pradesh, India
[2]Department of Computer Science, College of Computer Science, Applied College Tanumah, King Khalid University, Abha, Saudi Arabia

Abstract

The proposed project, Smart Self-Powered Wireless Sensor Network for Industrial Monitoring using Machine Learning, aims to create a self-sustaining industrial monitoring system powered by a 12V battery, which is charged *via* renewable energy sources like a solar panel and piezoelectric sensor. The system seamlessly switches between battery and AC mains power using relays to ensure uninterrupted operation. The core processing unit is a Raspberry Pi, which gathers information from a variety of sensors, such as a DHT11 for measuring humidity and ambient temperature, a DS18B20 for measuring the temperature inside the machine, and a MQ-135 gas sensor for identifying dangerous chemicals. The system also incorporates a motor driver to control industrial equipment speed, with two switches used to increase and decrease the motor speed. Additional components include an LCD for data display, a buzzer for audible alerts, and a CPU fan that activates when internal temperatures exceed safe limits. A GSM module is included for sending critical alert messages. The incorporation of machine learning, which analyzes sensor data to forecast and automate system reactions like setting off alerts or turning on actuators, is the basis of the system. This smart monitoring solution ensures efficient, real-time environmental and equipment condition tracking, making it highly suitable for industrial automation and safety.

Keywords: Raspberry Pi, IoT, motor driver, GSM

**Corresponding author*: mkmiyob@kku.edu.sa

Abhishek Kumar and Suman Lata Tripathi (eds.) Smart Chips for Smart Devices: VLSI Design for Next-Generation IoT Solutions, (141–178) © 2026 Scrivener Publishing LLC

8.1 Introduction

A Smart Self-Powered Wireless Sensor Network (WSN) for Industrial Monitoring represents a transformative evolution in the way industrial environments are monitored and managed. These networks utilize self-powered sensors that harvest energy from environmental sources, such as vibrations, light, or temperature variations, within the industrial setting. This self-sufficiency removes the need for external power sources and the necessity for regular battery replacement, two issues that standard wireless sensor systems frequently face. The network of interconnected sensors communicates wirelessly, enabling continuous and real-time monitoring of critical industrial parameters like temperature, humidity, pressure, equipment performance, and environmental conditions. This advancement ensures the seamless operation of industrial processes without disruptions due to power failures or battery depletion, contributing to enhanced productivity [1], safety, and operational efficiency. The integration of these self-powered sensors has also paved the way for a more sustainable approach to industrial monitoring, reducing reliance on non-renewable energy sources and minimizing the environmental footprint of traditional monitoring systems.

In the past, industrial monitoring systems were largely dependent on wired infrastructure and external power sources, both of which posed significant challenges in terms of installation, maintenance, and scalability [2]. The reliance on wires often resulted in complicated setups, higher costs, and limited flexibility, especially in large or hazardous environments [3]. The introduction of Wireless Sensor Networks (WSNs) marked a pivotal shift, offering greater scalability and ease of deployment in diverse industrial applications [4]. However, these early WSNs still required regular battery replacements or recharges, creating long-term maintenance issues and reducing the reliability of the systems [5]. Despite the promise of wireless monitoring, energy constraints continued to be a limiting factor, often requiring frequent service and intervention [6]. The evolution toward self-powered sensors represented a significant leap, eliminating the dependency on external power sources and allowing for truly autonomous, maintenance-free systems that could operate indefinitely by harvesting energy from their surroundings [7]. This breakthrough has not only extended the life cycle of these networks but has also made them more cost-effective and reliable for industries [8].

Today, the integration of smart data analytics with self-powered WSNs is driving significant advancements in industrial monitoring. These systems are no longer just passive data collectors; they are equipped with predictive analytics capabilities that enable real-time analysis, process optimization, and early detection of anomalies or potential failures [9]. By utilizing data from self-powered WSNs, industries can perform predictive maintenance, reducing unplanned downtime and increasing operational efficiency [10]. Furthermore, these networks provide valuable insights that enable better decision-making, improving the overall performance and safety of industrial operations. Looking to the future, advancements in energy harvesting technologies, wireless communication protocols, and data processing capabilities will continue to evolve, making self-powered WSNs even more efficient, intelligent, and adaptable. Smart Self-Powered WSNs will be essential to the creation of more intelligent, sustainable, and fully automated industrial systems, revolutionizing industries and resulting in a more interconnected, resource-efficient manufacturing ecosystem. They have the ability to integrate with cutting-edge technologies like artificial intelligence and the Internet of Things.

8.2 Related Works

A Smart Self-Powered Wireless Sensor Network (WSN) for industrial monitoring has evolved to provide an energy-efficient, sustainable, and autonomous solution. These networks utilize energy harvesting techniques, such as vibrations, light, or temperature variations, to power sensors without requiring external power sources or frequent battery replacements [11]. By wirelessly transmitting data, these systems ensure continuous operation in environments where traditional power sources are impractical or costly, contributing to the efficiency and sustainability of industrial processes [12]. With advancements in smart data analytics, these networks also support predictive maintenance and process optimization, making them key components in Industry 4.0. Self-powered WSNs have transformed industrial monitoring by offering reliable, cost-effective solutions for reducing downtime and improving overall operational productivity [13].

1. "Energy-Efficient Self-Powered Wireless Sensor Networks for Industrial Applications" explores the methods in [14].
2. "Harvesting Ambient Energy for Self-Powered Wireless Sensor Networks in Smart Manufacturing" [15] discusses integrating ambient energy harvesting to power WSNs in smart manufacturing, addressing energy storage and optimization challenges.

3. "Wireless Sensor Networks Powered by Ambient Energy for Remote Industrial Monitoring" [16] explores self-powered WSNs for remote industrial monitoring, emphasizing energy harvesting techniques for long-term sustainability.
4. "A Survey of Energy Harvesting Techniques for Self-Powered Wireless Sensor Networks" [17] provides an in-depth review of various energy harvesting methods, including piezoelectric, thermoelectric, and solar energy, for powering WSNs in industrial environments.
5. "Smart Factory Implementation Using Self-Powered Wireless Sensor Networks" [18] examines the role of self-powered WSNs in smart factories, focusing on real-time monitoring and predictive maintenance to optimize industrial operations.
6. "Self-Powered IoT Systems for Industrial Monitoring: Challenges and Opportunities" [19] discusses the integration of IoT and self-powered WSNs in industrial monitoring, addressing technical challenges and opportunities for IoT scalability.
7. "Energy Harvesting Wireless Sensor Network for Industrial Application in Harsh Environments" [20] explores the deployment of energy-harvesting WSNs in harsh industrial environments, providing solutions for reliable operation despite challenging conditions.
8. "Design and Analysis of Self-Powered Wireless Sensor Networks for Machine Condition Monitoring" [21] presents the design of self-powered WSNs for monitoring machine conditions, such as vibrations and temperature, to enable predictive maintenance.
9. "Vibration-Based Energy Harvesting for Industrial Wireless Sensor Networks" [22] investigates vibration-based energy harvesting for powering sensors in industrial environments, focusing on machinery health and environmental monitoring.
10. "Self-Powered Wireless Sensor Networks for Environmental and Structural Health Monitoring" [23] examines how self-powered WSNs can monitor environmental and structural health, offering sustainable solutions for industrial applications.

11. "Integrating Energy Harvesting and Wireless Sensor Networks for Smart Industrial Monitoring" [24] discusses the integration of energy harvesting with wireless sensors to enhance predictive maintenance and environmental monitoring in industrial settings.
12. "Battery-Free Wireless Sensors for Industrial Applications: A Review" [25] provides an overview of self-powered technologies used in wireless sensors for industrial monitoring, highlighting recent advancements in energy harvesting techniques.
13. "Adaptive Power Management for Self-Powered Wireless Sensor Networks in Industrial Applications" [26] explores adaptive power management strategies to optimize the performance of self-powered WSNs in industrial environments.
14. "Self-Powered Wireless Sensor Networks for Real-Time Industrial Data Collection and Analysis" [27] investigates the use of self-powered WSNs for continuous industrial data collection, enabling better decision-making and operational efficiency.

8.2.1 Cloud Computation with ThingSpeak

ThingSpeak enables cloud computation through MATLAB-based analytics, allowing users to process and analyze IoT data in real time. Key features include:

1. **Data Collection and Storage**
   - Supports HTTP, MQTT, and TCP for sensor data uploads.
   - Stores up to 8 fields per channel with timestamps.
2. **Real-time Visualization**
   - Built-in widgets for graphs, gauges, and maps.
   - Public and private dashboards for monitoring.
3. **MATLAB Analytics and Processing**
   - Custom scripts for filtering, predictive analytics, and AI models.
   - Fourier transforms, regression analysis, and anomaly detection.
4. **Triggering and Alerts**
   - Web hooks for integration with third-party services.
   - Email/SMS alerts based on sensor thresholds.

5. **IoT Integration**
   - Compatible with ESP8266, Raspberry Pi, Arduino, and other IoT devices.
   - Supports MQTT for low-power applications.

8.3 Existing System

Currently, industrial monitoring systems are predominantly based on traditional wired sensors and centralized control units. These systems often require considerable manual intervention and frequent maintenance due to their rigid, static design. The monitoring process in these existing systems usually involves basic techniques like simple alarms or local data logging, which do not capitalize on advanced technologies such as real-time data analysis or machine learning for predictive maintenance [20]. While some systems attempt to address power supply issues, they often fail to manage energy transitions efficiently. Many of these systems continue to rely heavily on mains electricity, creating a vulnerability to power outages and causing potential downtime, which impacts operational efficiency [19]. Additionally, integrating energy-harvesting technologies and optimizing power management between battery storage and mains power remains a significant challenge, leading to suboptimal energy use in these systems [3]. These limitations restrict the ability to achieve continuous, autonomous operation in environments where reliability is critical. The drawbacks of current systems are further compounded by the limited range of sensors used, such as the temp sensor DS18B20 (temperature sensor), and MQ-135 (air quality sensor), which fail to provide comprehensive data on various industrial parameters. This limits the scope of environmental monitoring and inhibits a holistic approach to industrial operations [5]. Furthermore, the existing models often lack the flexibility and scalability needed to adapt to changing operational conditions, making them unsuitable for modern industrial environments that require dynamic, real-time responses [6]. The absence of seamless integration between sensors and communication systems means that the data collected is often fragmented, preventing the formation of a cohesive monitoring solution. With limited automation and intelligence, these systems do not offer an intelligent response mechanism that could trigger predictive or corrective actions autonomously, leaving the system largely reactive instead of proactive [8]. This combination of inefficiencies, reliance on manual intervention, and inability to leverage real-time analytics significantly hampers the overall performance and effectiveness of traditional industrial monitoring systems, highlighting the need for innovative, self-powered, and intelligent solutions that can provide more reliable, efficient, and automated operations in modern industries [22].

8.4 Proposed System

The proposed model for industrial monitoring introduces a Smart Self-Powered Wireless Sensor Network (WSN) that revolutionizes traditional monitoring systems by integrating renewable energy sources, advanced sensors, and machine learning algorithms for enhanced functionality and operational efficiency. This model is powered by a 12V battery, which is charged using both solar panels and piezoelectric sensors that harvest energy from mechanical vibrations within the industrial environment. This dual energy harvesting system ensures uninterrupted operation by seamlessly switching between the stored battery power and the AC mains electricity *via* relays, eliminating the risk of downtime during power outages. The system is designed to collect comprehensive data from various sensors, including a DHT11 sensor for measuring temperature and humidity, a DS18B20 for monitoring internal machine temperature, and an MQ-135 sensor for detecting harmful gases. These sensors provide real-time insights into environmental and operational parameters, enabling timely adjustments and interventions in industrial processes.

The data collected by the sensors is processed by a central unit, a Raspberry Pi, which integrates machine learning algorithms to analyze the incoming data for trends, anomalies, and potential risks. Real-time monitoring and predictive maintenance are made possible by this cognitive processing, which reduces downtime and increases equipment longevity by anticipating possible system breakdowns before they happen. The system also features a motor driver, which allows precise control over motor speeds using dedicated switches for increasing and decreasing the speed, ensuring optimal performance under varying conditions. An LCD is included for visual data representation, while a buzzer provides audible alerts in case of critical events, such as temperature surges or gas leaks. Additionally, a GSM module is integrated to send out critical notifications and alerts to operators or maintenance personnel, further improving the system's responsiveness.

Automating decisions based on sensor data ensure more efficient operations and improved safety standards, particularly in complex and dynamic industrial environments. This holistic approach marks a significant step towards more autonomous, sustainable, and intelligent industrial monitoring systems.

The system integrates a variety of components to create a multifunctional and efficient platform. Key features include an MQ135 sensor for air quality monitoring, UV LEDs for targeted lighting applications, and a voltage sensor for tracking circuit parameters. The design incorporates a relay module for controlling high-power devices like 100W bulbs, and a CPU fan to manage heat dissipation. Solar panels provide the primary energy source, complemented by a 12V battery for energy storage and managed by an Arduino UNO microcontroller. The system is further enhanced with an LCD display for user interaction and an inverter to convert DC to AC power, enabling the operation of standard household appliances. Additionally, the project integrates a hydroponics system, where a soil moisture sensor triggers a DC pump motor for irrigation when moisture levels are low. It also sends SMS alerts *via* a GSM module and uploads data to ThingSpeak for remote monitoring and analysis.

In terms of operation, the system starts with energy harvesting from the solar panel and piezoelectric sensor, which store the generated power in a 12V battery. Sensors like the DHT 11 and Dallas Temperature sensor continuously monitor environmental conditions, while the MQ135 sensor keeps track of air quality. The Arduino UNO microcontroller acts as the central processing unit, collecting data from these sensors and managing various components such as the relay module, CPU fan, and motor driver. The LCD display provides real-time data to the user, and the GSM module sends alerts for critical conditions, ensuring constant communication. The inverter converts DC power to AC, making it suitable for household appliances. Additionally, the hydroponics system ensures proper irrigation by activating the DC pump motor when soil moisture levels are low, with data being uploaded to ThingSpeak for remote monitoring. This comprehensive approach ensures robust performance, energy efficiency, and advanced environmental and plant management, making it suitable for diverse applications, including industrial monitoring.

8.5 Hardware Components

8.5.1 12V Battery

Rechargeable batteries are energy storage devices that can be recharged after being discharged; they are a cost-effective option compared to single-use batteries. They offer multiple uses, reducing waste and providing a better long-term investment despite the higher initial cost and the need for a charger. They function similarly to conventional batteries but are recharged by applying DC current to their terminals. While rechargeable batteries reduce waste and offer long-term savings, they do have some drawbacks; they develop a memory effect if only partially discharged, which reduces their performance and overall lifespan [2]. Despite these challenges, rechargeable batteries are commonly used in cars, consumer electronics, and off-grid power storage systems.

8.5.2 Piezoelectric Sensor

A piezoelectric sensor operates based on the principle of piezoelectricity, where mechanical stress applied to certain materials generates electricity. These sensors are commonly used to measure physical quantities like acceleration and pressure [25]. Piezoelectric sensors can be made from materials like quartz and PZT ceramic, and often include an acceleration compensation element to reduce false readings caused by unwanted vibrations.

Piezoelectric sensors have several key characteristics, including the range of measurement, sensitivity, and reliability. The range of measurement is subject to specific limits depending on the sensor's design and application. Sensitivity (S) is defined as the ratio of the change in output signal (Δy) to the change in the signal causing the change (Δx), expressed as $S = \Delta y/\Delta x$. Reliability refers to the sensor's ability to maintain its characteristics within set operational conditions. Other specifications include a threshold of reaction, error rates, and indication time.

These sensors typically have an impedance value $\leq 500\Omega$ and operate in a temperature range of -20°C to +60°C, though they should be stored between -30°C and +70°C to prevent degradation. They also feature low soldering temperatures and a strain sensitivity of 5V/με. Quartz is often the preferred material for piezoelectric sensors due to its high flexibility.

8.5.3 Solar Panel

Photovoltaic solar panels, commonly used in both commercial and residential applications, convert sunlight into electricity through the photovoltaic effect. These panels consist of a series of interconnected photovoltaic cells, typically arranged in a 6x10 configuration, and can either be rigid or semi-flexible. The structural integrity of these panels is maintained by the top or back layer, protecting the cells from mechanical damage and moisture. Electrical connections within the modules are made in series to achieve desired voltage or in parallel for current capability [14]. Most solar modules are made from crystalline silicon cells, either monocrystalline or multicrystalline, though thin-film technologies using cadmium telluride, CIGS, and amorphous silicon are also prevalent. Innovations such as multi-junction cells and concentrator photovoltaics (CPV) are emerging, particularly for applications like spacecraft due to their high efficiency and power-to-weight ratio [17]. Solar modules also integrate features like bypass diodes to maximize output under partial shading and use weatherproof connectors for durability.

Technological advancements in thin-film modules have led to the creation of cells on glass substrates through monolithic integration, achieving sunlight conversion rates of 6-12%. Flexible thin-film modules are developed by depositing necessary layers on flexible substrates, such as polyester or polyimide film, and are laminated to protective fluoropolymer sheets. These advancements enable the use of solar technology in varied and innovative applications, from standard agricultural setups to advanced space missions [11]. The flexible nature and evolving efficiency of these modules enhances the adaptability and functionality of solar power systems in diverse environments.

8.5.4 Power Supply

A power supply is a component that provides electrical power to components by converting one type of electrical power to another, or even from a different energy form, such as solar, mechanical, or chemical, into electrical energy. In computers, power supplies typically convert AC to DC current and are located inside the case, often with a fan for cooling [13]. Most computer power supplies feature an input voltage switch, allowing users to select between 110v/115v or 220v/240v, depending on the power supply in their geographic location, making it important for compatibility with local power outlets.

Finally, voltage regulator regulates the voltage to 5V and finally, a blocking diode is used for taking the pulsating waveform.

8.5.5 AC Power (Adapter)

A 12V AC adapter used in WSN (Wireless Sensor Networks) for industrial monitoring is a power supply that converts AC from a standard outlet to a stable 12V DC output, which is required to power the sensor nodes and communication devices in the network. In WSN applications, especially in industrial settings, these adapters ensure that wireless sensors, data collectors, and transceivers have a continuous power source to gather and transmit environmental or system data. The 12V output is ideal for powering multiple devices within the network, while the adapter provides protection from power surges and ensures stable operation in challenging industrial environments, supporting efficient monitoring and control in real time.

8.5.6 Relay

A relay is an electromagnetic switch used to control circuits with a low power signal or manage multiple circuits with one signal. Conversely, de-energizing the relay opens the circuit. Relays are used in applications ranging from telephone exchanges and telegraphy to performing logical operations in computers and controlling high-power devices like motors. They are essential in reducing noise in low voltage applications and minimizing arcing in high voltage applications.

Relays are designed to switch one or more poles, each with normally open (NO), normally closed (NC), or change over (CO) contacts. These contacts control circuits by being energized or de-energized. Relays are crucial for performing safety-critical logic functions, providing time delay functions, controlling high voltage and current circuits with low signals, and acting as protective relays to detect and isolate faults. Overload relays, such as thermal or electronic types, protect motors from overloads by responding to current spikes, ensuring the motor's safety. When selecting a relay, consider factors like protection, regulatory approvals, switching time, current and voltage ratings, contact types, and coil-contact isolation.

8.5.7 Raspberry Pi

The Raspberry Pi is a credit-card sized computer developed in the United Kingdom by the Raspberry Pi Foundation, primarily aimed at teaching basic computer science to students and enthusiasts in computer hardware and programming. It's available in three board configurations, distributed by Newark element 14, RS Components, and Egoman, with Egoman's version marked by red coloring and lacking FCC/CE marks [27]. The hardware features a Broadcom BCM2835 system on a chip (SoC) that includes a 700 MHz ARM1176JZF-S processor and a Video Core IV GPU. Initially shipped with 256MB of RAM, later models like the Model B and B+ were upgraded to 512MB. The Raspberry Pi uses an SD card for booting and storage, with the Model B+ adopting a MicroSD card. It does not include a built-in hard disk or SSD.

The Raspberry Pi Foundation supports various operating systems, including Debian and Arch Linux ARM, providing a range of tools for programming, with Python as the primary language. It also supports other programming languages such as BBC BASIC, C, Java, and Perl, making it a versatile tool for a wide array of projects. The Pi's affordability, compact size, and flexibility have made it popular for educational purposes and various DIY projects, from simple programming tasks to complex hardware integrations.

Description of the components on the Raspberry Pi:

- The Raspberry Pi features a Broadcom BCM2835 System on Chip (SoC) with an ARM1176JZF-S processor, operating at 700 MHz, similar to older smartphone chips.
- The Pi consumes 700mA or 3W of power and can be powered by a Micro USB charger or the GPIO header.
- It lacks onboard storage, relying on an SD card for the operating system and data storage, inserted into its SD card slot.
- GPIO (General-Purpose Input/Output) pins are configurable at runtime and are used for various digital control purposes.
- The Display Serial Interface (DSI) connector allows connection of DSI-compatible LCD screens, aimed at reducing display controller costs.
- A standard 3.5 mm audio jack provides stereo audio output, though USB mics or sound cards are needed for audio input.
- The Raspberry Pi includes five status LEDs indicating activity and power status: OK, ACT, PWR, FDX, LNK, and 10M/100.
- Ethernet ports on Models B and B+ allow network or internet connectivity *via* a standard LAN cable, controlled by the Microchip LAN9512 LAN controller chip.

8.5.8 DHT11 (Temperature/Humidity) Sensor

The DHT11 is a highly affordable digital sensor designed for accurate temperature and humidity measurement, widely utilized in diverse applications such as weather stations, HVAC systems, and home automation projects. Its compact design allows it to fit seamlessly into various setups, making it a popular choice among hobbyists and educators. The DHT11 operates efficiently within a temperature range of 0 to 50°C and a humidity range of 20 to 90%, providing reliable data with a relatively low resolution. Its simplicity in communication with microcontrollers like Arduino and Raspberry Pi *via* a single-wire interface significantly simplifies its integration into different projects, enhancing its appeal for DIY enthusiasts and educational purposes.

In terms of specifications, the DHT11 offers a temperature accuracy of ±2°C and a humidity accuracy of ±5%. It has a sampling rate of 1Hz, meaning it can provide one reading per second. The sensor operates within a voltage range of 3 to 5.5V, ensuring compatibility with most microcontroller platforms. Its low power consumption and small dimensions (15.5mm x 12mm x 5.5mm) further enhance its suitability for a broad range of applications. The DHT11's digital output format simplifies interfacing with various microcontrollers, and its pre-calibrated nature means users can obtain accurate readings without the need for additional calibration. This combination of affordability, ease of use, and reliable performance makes the DHT11 a favorite among those looking to build or enhance temperature and humidity sensing projects.

8.5.9 Dallas Temperature Sensor

The DS18B20's working principle involves measuring temperature and converting the analog-to-digital signal with a resolution ranging from 9-bits to 12-bits, with the default being 12-bit. The sensor operates in a low-power inactive state until a Convert T command is issued, initiating the temperature measurement and A-to-D conversion. The resulting data is stored in a 2-byte register within the sensor, which then returns to its inactive state. If powered by an external source, the sensor can provide read time slots following the Convert T command, sending a "0" while the temperature conversion is in progress and a "1" when it is complete. This functionality makes the DS18B20 highly efficient for applications requiring precise temperature monitoring.

8.5.10 MQ135 Gas Sensor

The MQ-135 gas sensor module is designed to detect a variety of harmful gases, including ammonia (NH3), nitrogen oxides (NOx), alcohol, benzene, smoke, and carbon dioxide (CO2). It is highly sensitive to ammonia, sulfide, and benzene steam, making it useful for applications in air quality monitoring and hazardous gas detection. The module incorporates the MQ-135 air quality detector chip and an LM393 analog comparator chip, facilitating easy integration into projects. It operates on a 5V power supply and provides both digital and analog outputs. The digital output is LOW (0) when no gas is detected and goes HIGH (1) when the gas concentration exceeds the threshold set *via* a potentiometer. The analog output ranges from 0 to 4V, corresponding to the gas concentration levels in the environment. For accurate readings, it is recommended to wait at least 20 seconds after powering the module.

The MQ-135 uses tin dioxide (SnO2) as its sensitive material, which has lower conductivity in clean air and increases with the concentration of target gases. It can monitor various toxic gases within a detection range of 10-10,000 ppm and operates at around 5V±0.1V AC or DC. Notable features of this sensor include a long lifespan, low cost, simple driver circuit, and good sensitivity to toxic gases. It is widely used in industrial gas alarms, portable gas detectors, and domestic gas alarms for monitoring CO2 levels. The sensor is calibrated based on the average atmospheric CO2 concentration of 400.7 ppm.

8.5.11 CPU Fan

The CPU fan is essential for maintaining the optimal operating temperature of a computer's central processing unit (CPU), which generates heat during data processing and calculations. Without effective heat dissipation, the CPU could experience thermal throttling or even hardware failure. The fan actively circulates air over the CPU and surrounding components, keeping temperatures within safe limits and enhancing system performance and longevity. Typically, a CPU fan consists of motorized fan blades in a protective casing, designed to create an efficient airflow pattern that moves hot air away from the CPU while drawing in cooler air. Many CPU fans feature speed control mechanisms, adjusting rotational speeds based on CPU temperature readings to balance cooling efficiency and noise levels during different operational loads.

Manufacturers have responded to noise concerns by designing fans with improved aerodynamics and quieter motors, often employing fluid dynamic bearings (FDB) for smoother operation. CPU fans are typically part of a larger cooling solution that includes a heatsink and additional case fans. The heatsink absorbs heat from the CPU, which the fan then cools by moving air through the system. In high-performance applications, like gaming or data processing, CPU fans may work alongside liquid cooling systems for enhanced thermal management [16]. Advances in technology have led to more efficient and intelligent CPU fans that use temperature sensors and control algorithms to optimize fan speed based on real-time thermal data, improving cooling performance, extending fan lifespan, and reducing energy consumption. As computing demands increase, the importance of effective cooling solutions, including the CPU fan, remains critical to achieving high performance and reliability in modern computing systems.

8.5.12 GSM Module

A GSM module in an industrial monitoring WSN enables wireless data transmission between various sensors and a central control system. These sensors continuously monitor environmental parameters such as temperature, humidity, gas levels, and other critical metrics. The GSM module uses cellular communication to send real-time data, alerts, and updates to a remote monitoring system or directly to the user's mobile device *via* SMS. This ensures that any anomalies or hazardous conditions are quickly identified and addressed, enhancing the safety and efficiency of industrial operations.

In self-powered WSNs, the GSM module is often integrated with energy harvesting technologies, which convert ambient energy sources like solar power, thermal energy, or mechanical vibrations into electrical energy. This integration allows the sensor network to operate independently of external power supplies, reducing maintenance costs and improving sustainability. By combining GSM communication with self-powered sensors, industrial monitoring systems can provide robust, reliable, and efficient monitoring solutions, ensuring continuous data transmission and timely alerts without relying on wired power sources. This capability is particularly valuable in remote or hard-to-reach industrial environments where traditional power infrastructure may be impractical.

8.5.13 Motor Driver (L293D)

The L293D is a highly versatile motor driver integrated circuit (IC) commonly used to control both DC motors and stepper motors. It provides bidirectional control, allowing motors to rotate in both clockwise and counterclockwise directions. Each channel of the L293D can handle up to 600 mA of continuous current, with peak currents up to 1.2 A per channel, making it suitable for small to medium-sized motor control applications. The IC operates over a wide voltage range of 4.5V to 36V, allowing it to be used in various projects. One of the standout features of the L293D is its built-in diodes for back electromotive force (EMF) protection, which safeguards both the IC and the connected components from potential voltage spikes generated by the motors. This built-in protection ensures reliable and safe operation, which is critical in applications such as robotics and automation.

In terms of specifications, the L293D includes four H-bridge drivers, each capable of driving a motor. It has a thermal shutdown feature to prevent overheating, and its logic supply voltage ranges from 4.5V to 7V. The enable pins allow for precise control of the motor operation, enabling or disabling the output stages as needed. The L293D's pin configuration and ease of interfacing with popular microcontrollers such as Arduino and Raspberry Pi make it an accessible choice for both beginners and experienced users. Its ability to control the speed and direction of motors efficiently, combined with its robust performance and built-in safety features, makes the L293D an ideal choice for a wide range of motor control applications, including hobbyist projects, educational purposes, and professional automation systems.

8.5.14 DC Motor

DC motors consist of two main components: a stator, which is the stationary part, and an armature, which rotates. The stator generates a rotating magnetic field that drives the armature's rotation. A simple DC motor uses a stationary set of magnets in the stator and a coil of wire with current running through it to create an electromagnetic field. This field interacts with the magnetic fields of the stationary magnets, generating torque and causing the armature to rotate. The windings of insulated wire in the motor are connected to a commutator, a rotary switch that applies electrical current to the windings in sequence, ensuring a steady rotating force. The motor propels objects by sequentially turning the coils on and off, creating a rotating magnetic field that transforms electrical energy from direct current into mechanical energy.

8.5.15 Push Buttons

A push button switch is an electrical device that turns circuits on or off through a simple mechanism. It can operate with either momentary or latching action depending on the model. Constructed from durable materials such as metal or plastic, push buttons come in various shapes and sizes, making them suitable for numerous applications, including industrial, medical, and everyday uses. In industrial settings, push buttons are often integrated into larger systems and connected mechanically, allowing one button press to trigger another action, such as releasing another button. Examples of common uses include calculator buttons, reset switches, emergency stop buttons in machinery, and arcade game controls.

Push buttons are normally-open tactile switches, meaning they connect the circuit only when pressed and break the connection when released. They are crucial for triggering actions like starting circuits or controlling devices. The switches are prevalent in various applications, from industrial machinery to consumer electronics, due to their reliability and ease of use. Technical specifications typically include a power rating of up to 50mA at 24V DC, an insulation resistance of 100Mohm at 100V, and an operating temperature range of -20 to +70°C. These features ensure that push buttons provide crisp tactile feedback, withstand electrical stresses, and operate effectively in diverse environments.

8.5.16 LCD

A common type of LCD module is the 16x2 display, which shows 16 characters per line across two lines using a matrix of 5x7 pixels for each character. This type of LCD has two registers: Command and Data, for storing instructions and the data to be displayed.

LCDs accept two types of signals: data and control, recognized by the RS pin status. The R/W pin allows data to be read from or written to the display, and the E pin pulse enables the display to read or execute data. It takes approximately 39-43µS to place a character or execute a command, and 1.53ms to 1.64ms to clear the display or return the cursor home. LCDs have two types of RAM: DDRAM, which stores the position of characters to be displayed, and CGRAM, which allows users to create custom characters. The LCD controller reads DDRAM data to display characters, and CGRAM lets users define personalized characters for display. This versatility and programmability make LCDs a popular choice for various applications.

8.5.17 LED

The Light Emitting Diode (LED) is a semiconductor light source with two leads, functioning similarly to a PN junction diode. Discovered by Nick Holonyak in 1962 while working for General Electric, LEDs allow current to flow in the forward direction while blocking it in the reverse direction. LEDs are compact, usually occupying less than 1 mm^2, and are used in various electrical and electronic projects. Constructed as a p-n junction diode, they emit light when forward biased, with electrons and holes recombining and releasing energy as photons. LEDs are highly doped for enhanced performance and use materials such as GaAs and GaP to cover visible spectrums, with structures designed to maximize light emission by preventing photon reabsorption.

LEDs have a multitude of applications, from medical devices and clothing to remote controls, lighting, and indicators. They offer significant advantages, such as higher light output per watt, efficiency in color emission without filters, and durability. LEDs do not change color when dimmed and are ideal for applications requiring frequent on-off cycling. Their solid-state construction makes them resistant to shock and breakage, and they typically have long lifespans, with modern LEDs like the Philips LUXEON K2 rated for around 50,000 hours. LEDs also light up quickly and can be easily integrated onto printed circuit boards, making them suitable for a wide range of uses. Unlike compact fluorescent lights, LEDs do not contain mercury, further enhancing their appeal for various applications.

8.5.18 Buzzer

A buzzer or beeper is an audio signaling device used in various applications like alarms, timers, and user input confirmations. It can be mechanical, electromechanical, or piezoelectric, and is commonly found in computers, printers, telephones, automotive electronics, and electronic toys. The buzzer consists of a 2-pin structure, with a positive pin for power (typically 5V DC) and a negative pin connected to the ground. It operates within a voltage range of 4-8V DC and produces a continuous beep with a resonant frequency of approximately 2300 Hz.

Buzzers are widely used in alarming circuits, communication equipment, and portable electronics due to their compact size and ease of integration into breadboards, perf boards, or PCBs. Simple buzzers emit a continuous beep when powered, while more advanced models generate a beeping pattern. They are typically powered by regulated 5V or 6V DC supplies and are integrated with switching circuits for activation at specific times.

8.6 Software Components

8.6.1 Raspbian OS

Raspbian OS (now called Raspberry Pi OS) is a Debian-based operating system designed specifically for Raspberry Pi single-board computers. It is optimized for the hardware of Raspberry Pi, making it lightweight and efficient.

Key Features:

1. **Debian-Based** – Built on Debian Linux for stability and performance.
2. **Lightweight** – Optimized for Raspberry Pi's low-power hardware.
3. **Desktop Environment** – Uses LXDE (PIXEL desktop) for a smooth GUI experience.
4. **Pre-installed Software** – Includes Python, Scratch, Mathematica, and more.
5. **GPIO Support** – Provides tools for controlling Raspberry Pi's GPIO pins.
6. **Package Management** – Uses apt for installing software.
7. **Networking & SSH** – Supports remote access *via* SSH and VNC.

Variants of Raspberry Pi OS:

1. **Raspberry Pi OS Full** – Includes a GUI and pre-installed applications.
2. **Raspberry Pi OS Lite** – A minimal version without a GUI (command-line only).
3. **Raspberry Pi OS with Recommended Software** – Comes with extra applications.

Common Uses:

- IoT and Embedded Systems Development
- Robotics and Automation
- Home Automation Projects
- Edge Computing and AI
- Education and Learning Linux

8.6.2 Python

Python has emerged as a powerful and versatile programming language for developing, simulating, and analyzing Wireless Sensor Networks (WSNs), especially in industrial monitoring applications. Its simplicity, extensive libraries, and strong community support make it an ideal choice for researchers and engineers working on smart self-powered WSNs.

One of the key advantages of Python is its ability to integrate with hardware platforms such as Raspberry Pi and Arduino, which are commonly used in WSN nodes. Python libraries like RPi.GPIO, pySerial, and Adafruit_BBIO facilitate communication with sensors and actuators, enabling real-time data acquisition and control.

For simulation and modeling of WSNs, Python offers tools such as NS3Py (a Python wrapper for the NS-3 network simulator) and SimPy, a process-based discrete-event simulation framework. These tools allow researchers to model network behavior, evaluate performance metrics, and optimize energy consumption strategies.

Data analysis and visualization are critical components of industrial monitoring. Python excels in this area with libraries such as NumPy, Pandas, and Matplotlib, which enable efficient data processing and graphical representation. Machine learning libraries like scikit-learn and TensorFlow are also used to implement predictive maintenance and anomaly detection algorithms in industrial WSNs.

In addition, Python supports cloud integration through APIs and libraries like requests, boto3, and paho-mqtt, allowing sensor data to be transmitted to cloud platforms for storage, analysis, and remote monitoring.

Overall, Python plays a pivotal role in the development and deployment of smart self-powered WSNs for industrial monitoring, offering a comprehensive ecosystem for hardware interfacing, simulation, data analysis, and cloud connectivity.

8.6.3 VNC Viewer

VNC Viewer is a robust tool that allows users to remotely access and control another computer over a network using the Virtual Network Computing (VNC) protocol. This enables users to interact with the graphical desktop of a remote system as if they were physically present, making it invaluable for IT support, remote work, and managing servers or computers in different locations. The setup process is straightforward: users install a VNC server on the remote machine, configure it, and then connect using the Viewer by entering the server's IP address and authentication credentials.

VNC Viewer supports various platforms, including Windows, macOS, Linux, and mobile devices, ensuring cross-platform compatibility. It transmits screen updates efficiently, even over slower connections, by compressing data between the client and server, which minimizes lag and improves responsiveness. Additionally, it supports multiple display resolutions and color depths, allowing users to adjust settings based on their preferences and bandwidth availability.

Security is a critical aspect of VNC Viewer, which includes features like password protection, encryption settings, and IP address restrictions to safeguard data during transmission. By adhering to best security practices, users can ensure a secure remote connection to their systems. VNC Viewer is a preferred choice for individuals and organizations due to its ease of use, flexibility, and robust features. It enhances productivity and collaboration by enabling technical support, remote training, and server management from virtually anywhere. The advantages of using VNC Viewer include sustainability, reliability, efficiency, automation, scalability, flexibility, real-time access, cost-effectiveness, safety, and predictability. It is widely applied in various fields such as manufacturing, agriculture, healthcare, mining, logistics, automation, smart cities, and transport, demonstrating its versatility and importance in the digital work environment.

8.7 Step-by-Step Procedure

1. Install a solar panel and piezoelectric sensor to generate electrical power, connecting them to a 12V battery for energy storage.
2. Connect the 12V battery to a relay module to switch between battery and AC power as needed.
3. Ensure the relay module is connected to a stable power supply unit for consistent voltage.
4. Connect the Raspberry Pi to the power supply unit to serve as the central processing unit.
5. Attach the DHT 11 sensor to the Raspberry Pi to monitor temperature and humidity levels.
6. Connect the Dallas Temperature sensor to the Raspberry Pi for precise temperature readings.
7. Attach the MQ 135 gas sensor to the Raspberry Pi to detect harmful gases like ammonia, sulfur, and benzene.
8. Install Switch 1 and Switch 2 to the Raspberry Pi for manual control and triggering specific actions.

9. Connect a motor driver to the Raspberry Pi to control the DC motor for mechanical operations.
10. Attach a relay to the Raspberry Pi to manage the operation of a CPU fan for cooling.
11. Connect an LCD display to the Raspberry Pi for real-time data and system status updates.
12. Attach a GSM module to the Raspberry Pi to enable mobile communication and send SMS alerts.
13. Connect a buzzer to the Raspberry Pi for audible alerts in response to specific conditions.
14. Configure the Raspberry Pi to connect to an IoT cloud platform for remote monitoring and data analysis.
15. Power up the system and verify the functionality of each component, ensuring all sensors, actuators, and communication modules operate correctly.

This sequence will help you set up a comprehensive, self-powered wireless sensor network for industrial monitoring, leveraging the Raspberry Pi as the central hub.

8.8 Experimental Setup and Results

- When force or vibrations are applied to the piezoelectric sensor, mechanical energy is converted into electrical energy.
- The generated electrical energy is stored in a battery for future use.
- Any excess or residual electricity produced by the sensor is also captured.
- The stored energy ensures efficient utilization of mechanical movements.
- This process minimizes energy waste and provides a sustainable way to harness energy.
- The solar panel blinks to indicate that sunlight is falling on its surface and energy is being generated.
- It captures light and thermal energy from the sun and converts them into electrical energy through photovoltaic processes.
- The converted electrical energy is then efficiently transferred to a connected battery for storage.

- The stored energy can be utilized later to power various devices or systems.
- This process ensures sustainable and renewable energy generation, reducing dependency on conventional power sources.
- The output configurations as shown in the Figures 8.1 to 8.9.
- When the relay is ON, AC power is activated, allowing electricity to flow from the AC source to the entire system.
- When the relay is OFF, AC power is deactivated, and the system switches to draw power from the battery.
- This setup ensures continuous power supply to the system, regardless of the AC power status.

Abnormal Speed Alert: If the machine speed exceeds 80, the system triggers an alert notification on the LCD, signaling that the speed has surpassed the predefined threshold. This indicates that the machine is operating at an abnormal speed, which may require attention or intervention as shown in the Figures 8.10 and 8.11.

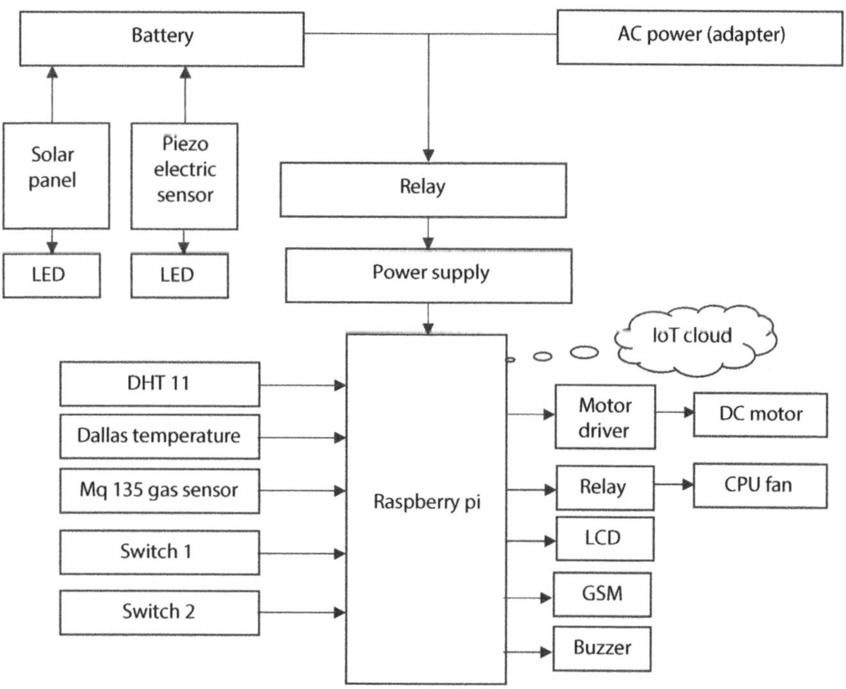

Figure 8.1 Block diagram of proposed system.

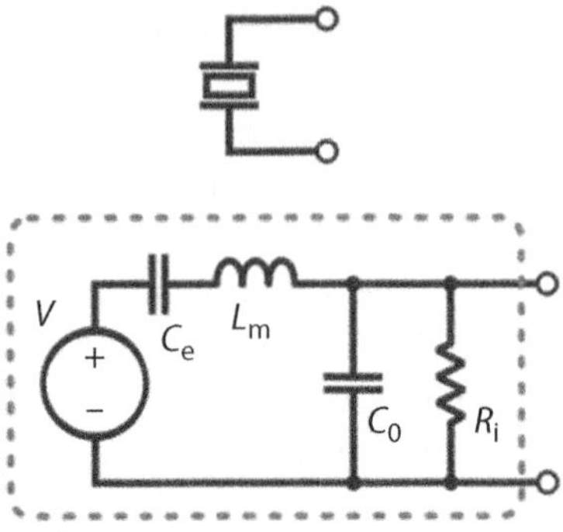

Figure 8.2 Piezoelectric sensor circuit diagram.

Figure 8.3 Power supply circuit diagram.

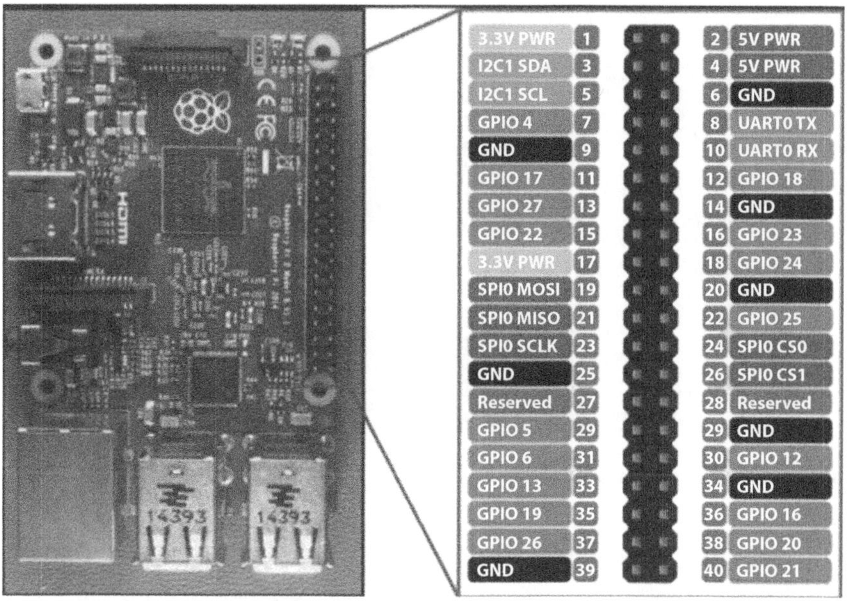

Figure 8.4 Raspberry Pi cross-sectional view.

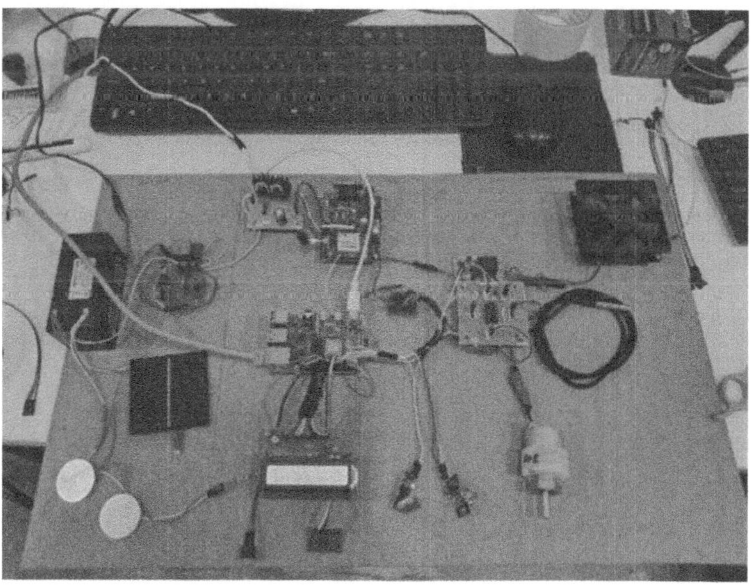

Figure 8.5 Placement of hardware components and their connections.

Figure 8.6 Title displayed on the LCD.

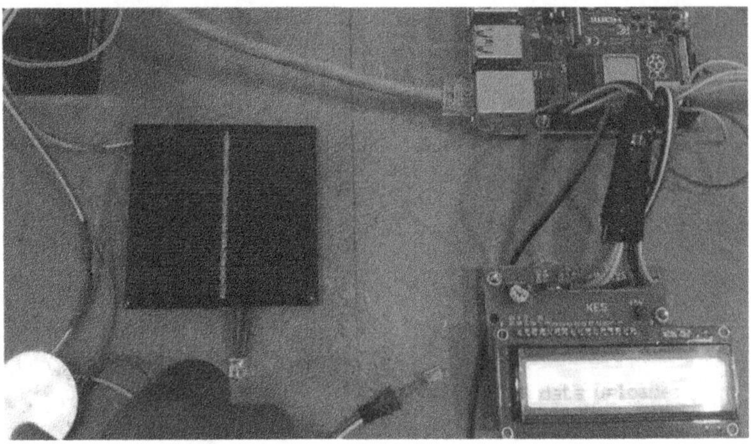

Figure 8.7 Piezoelectric sensor LED blinks.

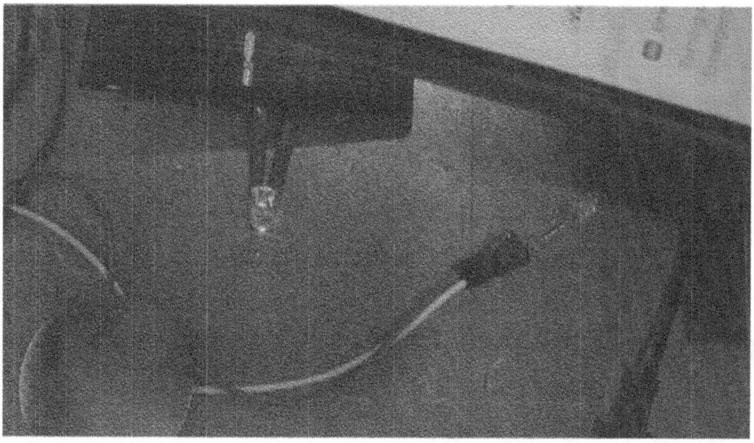

Figure 8.8 Solar panel LED ON.

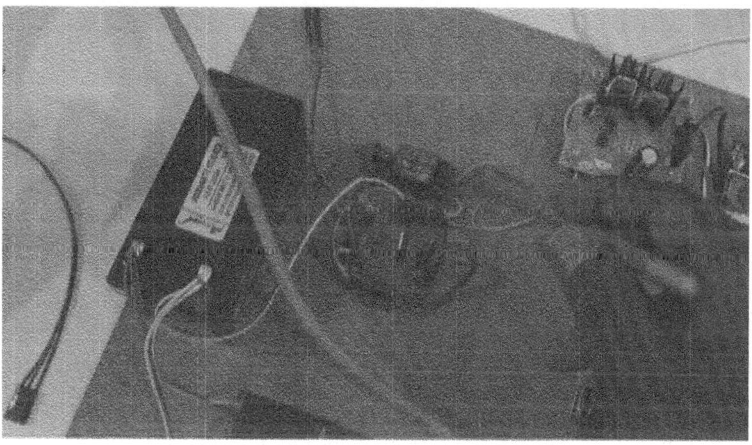

Figure 8.9 Relay shifts the power supply ac adapter and battery.

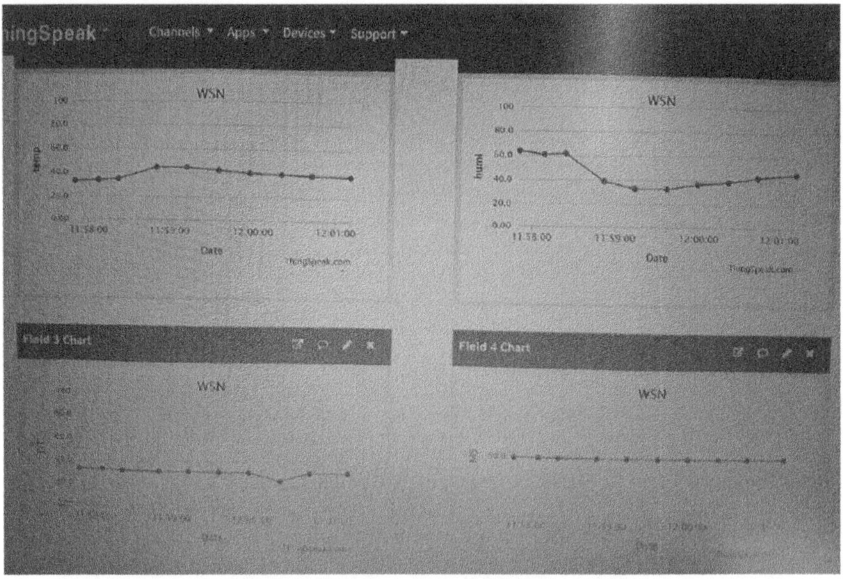

Figure 8.10 ThingSpeak will represent the parameters in graphical format.

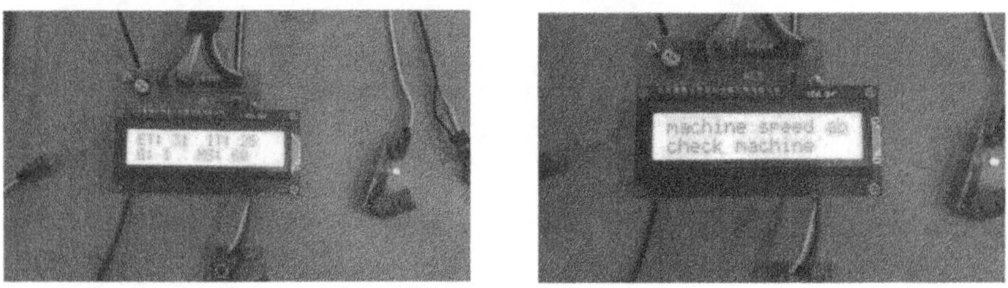

Figure 8.11 Push button to increase or decrease the machine speed.

- **Threshold for Normal Speed:** A machine speed between 60 and 79 is considered normal. This range ensures that the machine operates efficiently without exceeding the optimal performance limits, avoiding potential damage or inefficiency.
- **Speed Monitoring:** The system continuously monitors the machine speed, ensuring that any deviations from the normal range are immediately detected. This proactive monitoring helps in maintaining the machine's health and preventing any issues caused by excessive speed as interpreted in the Figures 8.12 to 8.14.

If the internal temperature exceeds or equals 35°C, it signals that the temperature has reached a potentially harmful level. At this point, the system takes precautionary measures by turning off the machine or DC motor to prevent overheating, which could lead to damage or malfunction. The automatic shutdown helps protect the system's sensitive components, ensuring longevity and safety. On the other hand, if the internal temperature is below 35°C, the machine is considered to be operating in a safe and normal

Figure 8.12 LED blinks on harmful gases detects.

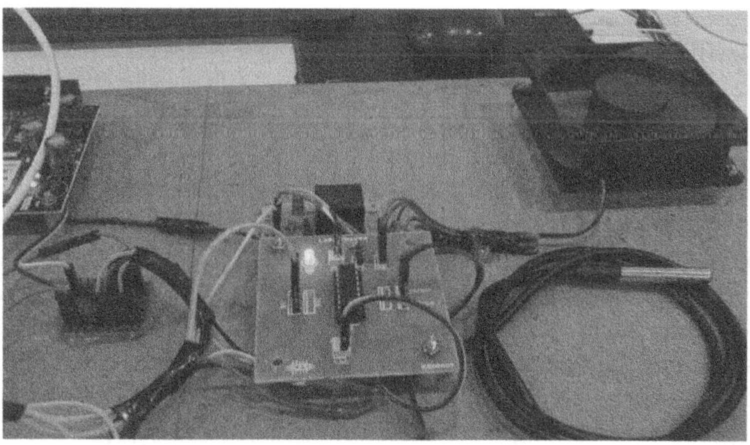

Figure 8.13 An alert message on LCD.

Figure 8.14 DC motor controls *via* temperature.

condition. The DC motor continues to rotate at its average speed, ensuring optimal performance without the risk of thermal stress. This temperature monitoring system is crucial for maintaining the health of the machine, preventing overheating, and ensuring reliable and efficient operation throughout its use. By continuously monitoring and responding to temperature changes, the system guarantees a balance between performance and protection.

8.9 Limitations

Cost: The integration of components such as a Raspberry Pi, DHT 11, Dallas temperature sensor, MQ 135 gas sensor, LCD display, GSM module, solar panel, piezoelectric sensor, battery, and relay makes the project versatile but also expensive. The combined cost of these elements can become a significant limitation for projects operating under strict budget constraints, making it challenging to deploy the system widely or scale it effectively.
Limited Sensor Ranges: The effectiveness of the system is constrained by the operational ranges of its sensors. For example, the DHT 11 is limited in its ability to detect extreme temperature and humidity values, and the MQ 135 gas sensor is only sensitive to specific gas concentrations. These restrictions may hinder the system's ability to provide comprehensive and accurate monitoring in larger industrial settings or environments that demand high precision.
Multiple Power Supplies but Not in Real-Time Implementation: Although the inclusion of solar panels and piezoelectric sensors aims to ensure self-

sufficiency, maintaining a steady power supply in real-time scenarios poses a challenge. Factors such as insufficient sunlight, irregular mechanical vibrations, or variability in power demand can lead to inconsistencies in power generation, impacting the system's reliability and overall operational uptime.

Interference or Signal Loss: Wireless communication methods used for data transmission, such as IoT cloud connectivity and GSM modules, are vulnerable to interference and signal loss. Environmental obstacles, such as physical barriers, and disruptions from other electronic devices can compromise the accuracy of transmitted data, leading to potential gaps in monitoring and a decline in the system's overall reliability.

Mitigation: Overcoming these challenges requires investments in advanced and higher-capacity sensors tailored to industrial requirements, improved power management strategies with reliable energy storage solutions, and robust communication systems equipped with signal boosters or repeaters to minimize interference and ensure consistent performance.

8.9.1 Cost Analysis and Feasibility

The "Smart Self-Powered Wireless Sensor Network (WSN) for Industrial Monitoring" project leverages various components to ensure efficient and reliable monitoring of industrial parameters. The setup includes a Raspberry Pi, multiple sensors like the DHT 11 for temperature and humidity, the Dallas Temperature sensor for precise readings, and the MQ 135 gas sensor for detecting harmful gases. Additionally, an LCD display provides real-time data visualization, while a GSM module facilitates mobile communication. Power management is handled through a solar panel and a piezoelectric sensor, ensuring self-sufficiency. However, several limitations exist. The system's reliance on multiple power supplies means inconsistent power generation due to environmental conditions, such as insufficient sunlight or mechanical vibrations, can affect reliability. The limited sensor ranges restrict the system's effectiveness in broader industrial settings, necessitating more advanced sensors for larger areas or more precise measurements. Wireless communication introduces potential issues with interference and signal loss, compromising the accuracy and reliability of real-time monitoring. Implementing robust communication protocols and using signal repeaters or boosters can help mitigate these challenges.

The project's feasibility is influenced by several factors. Technically, the Raspberry Pi can manage data processing and communication tasks, though its limited processing power may require additional solutions for more demanding applications. Operationally, the system requires regular maintenance, including sensor calibration, software updates, and battery replacement, which can be challenging in remote or resource-limited settings. Security concerns related to IoT cloud connectivity must be addressed through robust measures like encryption and secure authentication to protect sensitive data. Efficient data storage solutions are necessary to prevent data loss or corruption. The setup and maintenance require a certain level of technical expertise, which may not be readily available. Durable, industrial-grade components and protective measures are essential to withstand harsh industrial environments. Despite these limitations, the growing demand for industrial monitoring solutions to improve efficiency, safety, and compliance underscores the market feasibility of the project. By carefully addressing these challenges, the smart self-powered WSN for industrial monitoring can be a valuable asset in various industrial applications.

8.9.2 Graphical Representation and Comparative Analysis

The graphical representation and comparative analysis underline the project's feasibility for industrial applications. The integration of these sensors in a self-powered wireless sensor network ensures low maintenance and high efficiency. The comparison helps industries choose the right sensors based on cost, accuracy, and application needs. Overall, the project emphasizes real-time monitoring, energy efficiency, and data-driven decision-making for enhancing industrial safety and productivity.

In the project, the graphical representation involves plotting sensor data over time to analyze environmental parameters like temperature and air quality. The DHT11 and DALLAS temperature sensors provide temperature readings, which can be plotted on separate line graphs to observe trends, fluctuations, and discrepancies. These graphs allow for comparative evaluation of the two sensors' accuracy and response to environmental changes. For instance, the x-axis represents time, while the y-axis represents temperature values, enabling a clear visualization of variations in readings under similar conditions. This helps identify patterns or anomalies that could be critical in industrial monitoring. Figure 8.15 shows the arrangement of system parameters

Similarly, the MQ-135 gas sensor data can be visualized using a bar or line graph, where the gas concentration levels are plotted over time. This graphical representation highlights periods of high gas concentration, indicating potential hazards. By referencing predefined safety thresholds on the graph, users can quickly identify unsafe conditions and take corrective actions. These graphical insights are essential for industries to ensure operational safety and environmental compliance.

Comparative analysis is a research method used to compare two or more entities to identify their similarities and differences. This approach is

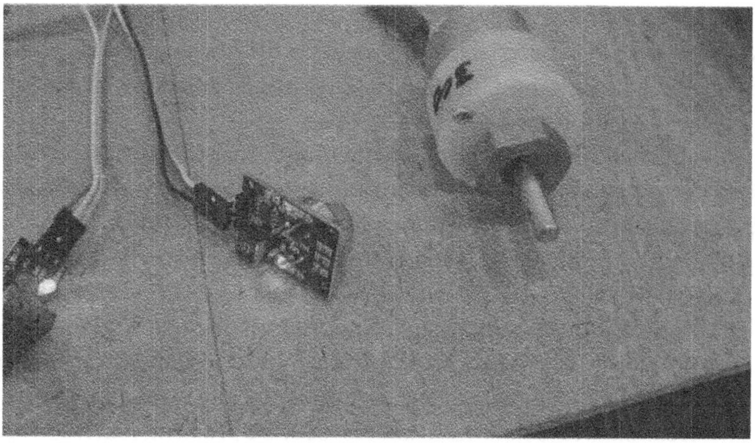

Figure 8.15 Graphical representation of the system parameters.

prevalent in various fields, such as social sciences, business, and technology, as it provides insights that can inform better decision-making. For example, in social sciences, researchers might compare different political systems, educational methods, or economic policies across countries to understand how various factors influence outcomes. By examining these similarities and differences, comparative analysis helps identify best practices and lessons that can be applied to improve existing systems or develop new strategies. This method is valuable for uncovering patterns and correlations that might not be evident when studying a single entity in isolation.

In the business world, comparative analysis is often employed to evaluate the performance of different companies, products, or marketing strategies. By comparing key performance indicators such as revenue, market share, and customer satisfaction, businesses can identify their competitive strengths and weaknesses. This information is crucial for strategic planning, such as deciding which markets to enter, which products to develop or discontinue, and how to allocate resources effectively. For instance, a company might use comparative analysis to benchmark its performance against industry leaders, enabling it to set realistic goals and identify areas for improvement. This approach also helps in understanding consumer preferences and market trends, which are essential for maintaining a competitive edge.

Comparative analysis is a method used to examine the similarities and differences between two or more entities. In social sciences, it might be used to compare different political systems or economic policies across countries. In business, it's helpful for evaluating the performance of companies or products, allowing businesses to identify competitive strengths and weaknesses. In technology, comparative analysis aids in assessing various design options, materials, and processes to choose the best tools for a given project. This method helps in making informed decisions by considering multiple options and their trade-offs.

In technology and engineering, comparative analysis is used to evaluate various design options, materials, and processes. Engineers might compare the performance, cost, and environmental impact of different materials when designing a new product. Similarly, software developers might use comparative analysis to assess different programming languages, development frameworks, or algorithms. By comparing these elements, they can choose the most suitable tools for their projects based on factors such as efficiency, scalability, and ease of use. Comparative analysis in technology also extends to comparing different technological solutions to address specific problems, such as evaluating the effectiveness of various cybersecurity measures or renewable energy sources. This method provides a systematic way to make informed decisions by considering multiple options and their potential trade-offs.

8.9.3 Technical Limitations and Maintenance Challenges

Processing Power: The Raspberry Pi, while versatile and cost-effective, has limited processing capabilities and a restricted number of GPIO pins. This can be a significant limitation when the system is scaled to include multiple sensors and actuators. The increased load could potentially slow down data processing, reduce system responsiveness, and limit the ability to handle concurrent tasks, impacting overall performance and scalability of the WSN.

Sensor Range: The sensors used in this system, such as the DHT 11 for temperature and humidity, and the MQ 135 for gas detection, have specific operational ranges and sensitivities. These ranges might not cover all required parameters in a large industrial setting, leading to potential gaps in monitoring capabilities. For instance, the DHT 11 has a limited temperature and humidity range, which might not be sufficient for environments with extreme conditions.

Power Reliability: The self-powered aspect of the system relies on solar panels and piezoelectric sensors for power generation. However, these power sources can be inconsistent due to environmental factors such as lack of sunlight or insufficient mechanical vibrations. This variability can lead to power shortages, affecting the system's ability to operate continuously and reliably.

Wireless Communication: Wireless data transmission is crucial for the WSN to function, but it is susceptible to interference and signal loss. Industrial environments often have a high density of electronic devices and physical obstacles that can disrupt wireless signals. This interference can lead to data transmission errors, delayed updates, and compromised reliability of real-time monitoring.

Integration Complexity: Combining multiple power supplies (solar, piezoelectric, battery) and ensuring their seamless integration with the Raspberry Pi and other components is complex. This complexity can lead to potential issues in real-time implementation, such as power management inefficiencies and synchronization problems, which can affect overall system reliability and functionality.

Regular Maintenance: To ensure accurate and reliable performance, the system requires regular maintenance, including sensor calibration, software updates, and inspections of power sources and other components. This maintenance can be labor-intensive and time-consuming, particularly in remote or difficult-to-access locations. The need for technical expertise to perform these tasks can further complicate maintenance efforts.

Security and Data Management: Robust security measures are essential to protect sensitive data transmitted over the IoT cloud. Implementing encryption, secure authentication, and regular updates are necessary to prevent unauthorized access and data breaches. Additionally, efficient data storage solutions are required to manage the large volumes of data generated, ensuring it is stored securely and can be retrieved without loss or corruption.

Environmental Durability and Documentation: The components used in the system, especially sensors and power generation units, must be durable enough to withstand harsh industrial environments. This requires using industrial-grade materials and implementing protective measures, such as enclosures and coatings. Comprehensive documentation and support are crucial for users to understand, set up, and maintain the system effectively, ensuring its long-term reliability and efficiency.

8.10 Conclusion

The Smart Self-Powered Wireless Sensor Network for Industrial Monitoring represents a significant advancement in the field of industrial automation. By combining renewable energy sources, intelligent data processing, and a robust sensor framework, the system ensures continuous monitoring and efficient management of critical parameters. The integration of machine learning enables proactive decision-making, enhancing operational safety and efficiency. This project not only demonstrates the feasibility of self-sustaining monitoring systems but also sets the stage for future developments in smart industrial solutions, ultimately contributing to a more sustainable and automated industrial environment.

The project offers an innovative solution for effectively monitoring various industrial parameters. By integrating multiple power sources, including solar panels and piezoelectric sensors, the system ensures continuous and reliable power supply. The Raspberry Pi serves as the central hub, managing data collection from sensors like the DHT 11 for temperature and humidity, the Dallas temperature sensor for precise readings, and the MQ 135 for gas detection. This data is then utilized to control output devices such as motor drivers, CPU fans, LCD displays, GSM modules, and buzzers for alerts. The ability to connect to an IoT cloud platform enables remote access and real-time monitoring, significantly enhancing the functionality and reliability of industrial operations. Despite the technical limitations, such as the Raspberry Pi's processing power constraints and the dependency on environmental conditions for power generation, this system provides a robust framework for improving safety, efficiency, and compliance within industrial environments.

Maintenance and security are crucial aspects of ensuring the system's longevity and effectiveness. Regular calibration of sensors, software updates, and inspections of power sources and other components are essential to maintain accurate and reliable performance. The wireless communication components must be monitored for signal strength and interference, while robust security measures, such as encryption and secure authentication, are necessary to protect sensitive data. The use of durable, industrial-grade components and comprehensive documentation and support further enhance the system's resilience in harsh industrial environments. Overall, the smart self-powered WSN for industrial monitoring represents a significant advancement in leveraging technology to create more resilient and efficient industrial environments. By addressing the outlined challenges through strategic planning and implementation, this system can provide valuable insights and control, making it an indispensable tool for modern industrial applications.

AI disclosure statement: The author acknowledges the use of AI-based tools for minor language editing or grammatical correction. However, all intellectual contributions, analyses, and conclusions presented in this chapter are entirely the author's own.

References

1. Akyildiz, I.F., Su, W., Sankarasubramaniam, Y., Cayirci, E., Wireless sensor networks: A survey. *Comput. Netw.*, 38, 4, 393–422, 2002. https://doi.org/10.1016/S1389-1286(01)00302-4.

2. Ali, A., Iqbal, S., Chen, X., Recent advances in piezoelectric wearable energy harvesting based on human motion: Materials, design, and applications. *Energy Strat.Rev.*, *53*, 2024. https://doi.org/10.1016/j.esr.2024.101422.
3. Anton, S.R. and Sodano, H.A., A review of power harvesting using piezoelectric materials (2003-2006). *Smart Mater. Struct.*, *16*, 3, 2007. https://doi.org/10.1088/0964-1726/16/3/R01.
4. Bakytbekov, A., Nguyen, T.Q., Zhang, G., Strano, M.S., Salama, K.N., Shamim, A., Synergistic multi-source ambient RF and thermal energy harvester for green IoT applications. *Energy Rep.*, *9*, 1875–1885, 2023. https://doi.org/10.1016/j.egyr.2023.01.027.
5. Beeby, S.P., Cao, Z., Almussallam, A., Kinetic, thermoelectric and solar energy harvesting technologies for smart textiles, in: *Multidisciplinary Know-How for Smart-Textiles Developers*, pp. 306–328, 2013, https://doi.org/10.1533/9780857093530.2.306.
6. Beeby, S.P. and O'Donnell, T., Electromagnetic energy harvesting. *Energy Harvesting Technol.*, 129–161, 2009. https://doi.org/10.1007/978-0-387-76464-1_5.
7. Depeursinge, A., Racoceanu, D., Iavindrasana, J., Cohen, G., Platon, A., Poletti, P.-A., Muller, H., Fusing Visual and Clinical Information for Lung Tissue Classification in HRCT Data. *Artif. Intell. Med.*, 2010. ARTMED1118. https://doi.org/10.1016/J.
8. Elvin, N. and Erturk, A., Introduction and methods of mechanical energy harvesting. *Adv. Energy Harv. Methods*, *9781461457053*, 3–14, 2013. https://doi.org/10.1007/978-1-4614-5705-3_1.
9. Guo, W., Zhai, L., Guo, L., Shi, J., Worm propagation control based on spatial correlation in wireless sensor network, in: *Lecture Notes in Computer Science (Including Subseries Lecture Notes in Artificial Intelligence and Lecture Notes in Bioinformatics), 7234 LNCS*, pp. 68–77, 2012, https://doi.org/10.1007/978-3-642-29426-6_10.
10. Haque, M.E. and Baroudi, U., Ambient self-powered cluster-based wireless sensor networks for industry 4.0 applications. *Soft Comput.*, *25*, 3, 1859–1884, 2021. https://doi.org/10.1007/S00500-020-05259-Y.
11. Heaney, P.S. and Bilgen, O., Modeling and experimental characterization of a multi-point loaded piezocomposite beam for actuation and energy harvesting. *J. Intell. Mater. Syst. Struct.*, *30*, 11, 1667–1681, 2019. https://doi.org/10.1177/1045389X19844010.
12. Li, M., Wan, Z., Zou, T., Shen, Z., Li, M., Wang, C., Xiao, X., Artificial intelligence enabled self-powered wireless sensing for smart industry. *Chem. Eng. J.*, *492*, 2024. https://doi.org/10.1016/j.cej.2024.152417.
13. Ling, J.K., Kunwar, R., Li, L., Peng, S., Misnon, I., II, Ab Rahim, M.H., Yang, C.C., Jose, R., Self-rechargeable energizers for sustainability. *EScience*, *2*, 4, 347–364, 2022. https://doi.org/10.1016/j.esci.2022.07.002.
14. Mi, J., Liu, M., Li, Q., Li, X., Zuo, L., Electromagnetic energy harvesting from train induced railway bridge vibrations. *IFAC-PapersOnLine*, *54*, 20, 185–190, 2021. https://doi.org/10.1016/j.ifacol.2021.11.173.

15. Mitcheson, P.D., Yeatman, E.M., Rao, G.K., Holmes, A.S., Green, T.C., Energy harvesting from human and machine motion for wireless electronic devices. *Proc. IEEE*, *96*, 9, 1457–1486, 2008. https://doi.org/10.1109/JPROC.2008.927494.
16. Niu, S., Wang, X., Yi, F., Zhou, Y.S., Wang, Z.L., A universal self-charging system driven by random biomechanical energy for sustainable operation of mobile electronics. *Nat. Commun.*, *6*, 2015. https://doi.org/10.1038/NCOMMS9975.
17. Noras, M.A. and Vo, M.N., Optimization of electromagnetic and vibration energy harvesters utilizing flexible piezoelectric cantilever beams. *Int. J. Energy Res.*, *44*, 4, 2998–3007, 2020. https://doi.org/10.1002/ER.5127.
18. Pavelková, R., Vala, D., Gecová, K., Energy harvesting systems using human body motion. *IFAC-PapersOnLine*, *51*, 6, 36–41, 2018. https://doi.org/10.1016/j.ifacol.2018.07.126.
19. Pradeesh, E.L., Udhayakumar, S., Vasundhara, M.G., Kalavathi, G.K., A review on piezoelectric energy harvesting. *Microsyst. Technol.*, *28*, 8, 1797–1830, 2022. https://doi.org/10.1007/S00542-022-05334-4.
20. Sezer, N. and Koç, M., A comprehensive review on the state-of-the-art of piezoelectric energy harvesting. *Nano Energy*, *80*, 2021. https://doi.org/10.1016/j.nanoen.2020.105567.
21. Shirai, H., Mitamura, H., Arai, N., Moriya, K., Study of Energy Harvesting from Low-Frequency Vibration with Ferromagnetic Powder and Non-magnetic Fluid. *Plasmonics*, *15*, 2, 559–571, 2020. https://doi.org/10.1007/S11468-019-01067-9.
22. Snyder, G.J., Thermoelectric energy harvesting. *Energy Harvesting Technol.*, 325–336, 2009. https://doi.org/10.1007/978-0-387-76464-1_11.
23. Wang, H., Xia, X., Fu, J., Li, J., Chen, C., Dai, Y., Fan, Z., Hu, G., Zi, Y., A general self-powered wireless sensing solution based on triboelectric-discharge effect. *Nano Energy*, *105*, 2023. https://doi.org/10.1016/j.nanoen.2022.107982.
24. Yick, J., Mukherjee, B., Ghosal, D., Wireless sensor network survey. *Comput. Netw.*, *52*, 12, 2292–2330, 2008. https://doi.org/10.1016/j.comnet.2008.04.002.
25. Yu, R., Feng, S., Sun, Q., Xu, H., Jiang, Q., Guo, J., Dai, B., Cui, D., Wang, K., Ambient energy harvesters in wearable electronics: fundamentals, methodologies, and applications. *J. Nanobiotechnol.*, *22*, 1, 2024. https://doi.org/10.1186/S12951-024-02774-0.
26. Zhao, Y.Z., Miao, C., Ma, M., Zhang, J.B., Leung, C., A survey and projection on medium access control protocols for wireless sensor networks. *ACM Comput. Surv.*, *45*, 1, 2012. https://doi.org/10.1145/2379776.2379783.
27. Zhou, C., Yang, Y., Sun, N., Wen, Z., Cheng, P., Xie, X., Shao, H., Shen, Q., Chen, X., Liu, Y., Wang, Z.L., Sun, X., Flexible self-charging power units for portable electronics based on folded carbon paper. *Nano Res.*, *11*, 8, 4313–4322, 2018. https://doi.org/10.1007/S12274-018-2018-8.

9

Recent Advancements in High Electron Mobility Transistor-Based Monolithic Microwave Integrated Circuits: A Review

S.K. Hima Bindhu, Yogesh Kumar Verma* and Kamal Bhatia

School of Electronics and Electrical Engineering, Lovely Professional University, Jalandhar, Punjab, India

Abstract
Monolithic Microwave Integrated Circuits (MMICs) have become a cornerstone in high-frequency applications, ranging from wireless communications (5G and beyond) to radar systems and satellite technologies. MMICs are designed to integrate multiple microwave components (amplifiers, oscillators, mixers, etc.) on a single semiconductor substrate, improving performance, reducing size, and enhancing reliability. Recent advancements in MMIC technology focus on improving material efficiency, integration techniques, power performance, and operating frequency. HEMTs are used in microwave and millimeter-wave power amplifiers in 5G and future 6G communication systems. Their ability to operate at high frequencies with high efficiency is crucial for high-bandwidth data transmission.

Keywords: MMIC, HEMT, 5G, high-frequency, VLSI, semiconductors

9.1 Introduction

The High Electron Mobility Transistor (HEMT) is a critical device in high-frequency, high-power, and low-noise applications. The combination of high electron mobility, superior power handling, and low noise makes HEMTs ideal for a range of industries, including telecommunications, radar, space systems, and power electronics [1–5]. HEMTs will continue to play a pivotal role in enabling the next generation of electronic systems.

**Corresponding author*: yogeshkvermaec@gmail.com

Abhishek Kumar and Suman Lata Tripathi (eds.) Smart Chips for Smart Devices: VLSI Design for Next-Generation IoT Solutions, (179–190) © 2026 Scrivener Publishing LLC

HEMTs are being explored for cryogenic low-noise amplifiers used in quantum computing, where their low noise and high-speed characteristics are valuable for precise signal detection and manipulation at low temperatures. HEMTs are used in millimeter-wave imaging systems for security screening and non-destructive testing, as well as in terahertz imaging for medical diagnostics and industrial inspection. Gallium Nitride (GaN)-based HEMTs are gaining traction in power electronics for use in efficient power converters, inverters, and wireless power transfer systems. HEMTs are used in millimeter-wave imaging systems for security screening and non-destructive testing, as well as in terahertz imaging for medical diagnostics and industrial inspection. HEMTs are being explored for cryogenic low-noise amplifiers used in quantum computing, where their low noise and high-speed characteristics are valuable for precise signal detection and manipulation at low temperatures. HEMTs are used in microwave and millimeter-wave power amplifiers in 5G and future 6G communication systems. Their ability to operate at high frequencies with high efficiency is crucial for high-bandwidth data transmission [6–10]. The high-frequency capabilities of HEMTs make them ideal for radar systems in defense, aerospace, and automotive industries, including synthetic aperture radar (SAR) and automotive radar. HEMTs are extensively used in low-noise amplifiers (LNAs) and high-power amplifiers in satellite communication systems due to their low noise figure and high efficiency [11, 12]. Recent advancements in High Electron Mobility Transistor (HEMT) based Monolithic Microwave Integrated Circuits (MMICs) have led to the development of high-power density GaN HEMT MMICs. These MMICs have excellent high-frequency operation characteristics and power density, making them ideal for communication systems that require high-power and high-efficiency power amplifiers. GaN HEMT on silicon carbide (SiC) substrate-based MMICs have also shown significant progress, with power density as high as 12.2 Watts/mm at X-band. Additionally, GaN HEMT MMICs have been used to create novel reconfigurable fully solid-state microwave power amplifiers [13–20]. These advancements in HEMT-based MMICs have the potential to revolutionize the field of microwave engineering, enabling the creation of more efficient and powerful microwave devices. GaN has emerged as a key material in MMIC technology due to its high breakdown voltage, excellent thermal conductivity, and ability to operate at high power and frequencies. GaN-based MMICs are increasingly used in power amplifiers for 5G, satellite communications, and radar systems. GaAs MMICs have been widely used due to their high electron mobility, making them ideal for high-frequency applications. Though GaN is becoming more dominant, GaAs still remain relevant for applications where power output is less critical but high efficiency is needed. Indium Phosphide (InP)-based MMICs are known for their high electron mobility, making them ideal for extremely high-frequency applications (beyond 100GHz). InP-based MMICs are being used in millimeter-wave and terahertz applications, particularly for next-generation radar and communication systems. One of the most significant advancements is the ability of MMICs to operate at millimeter-wave (30–300GHz) and terahertz (0.1–10 THz) frequencies. These frequencies are crucial for 5G, 6G, and advanced radar systems [21–25]. Recent breakthrough sine beam lithography and material engineering have allowed the design of MMICs that operate at these extremely high frequencies with high efficiency and low power loss [26].

MMICs operating in the sub-terahertz and terahertz range are emerging for next-generation communication systems (6G), which will demand enormous bandwidths and very low latency. One of the key trends in MMIC development is the push for higher efficiency in power amplifiers. GaN MMICs, in particular, have shown improvements in efficiency and power handling capabilities. By improving heat dissipation and utilizing high-efficiency load modulation techniques (Doherty amplifiers), MMICs can handle higher power with reduced thermal management issues. These techniques are used to increase power amplifier efficiency. Envelope tracking adjusts the power supply voltage to match the signal requirements, while DPD helps to linearize the output of amplifiers in MMICs, improving efficiency while maintaining signal quality.

Advances in 3D integration techniques have allowed more complex and compact MMIC designs, stacking multiple layers of circuits vertically to reduce the overall footprint. 3D MMICs are particularly useful for phased-array radar systems and multi-channel communication systems [27–30].

9.2 Recent Advancements in HEMT-Based MMICs

HEMT-based MMICs are being explored for their application in quantum computing. The low-noise and high-frequency performance of HEMTs make them suitable for cryogenic amplification and signal processing in quantum processors. HEMT-based MMICs are integral to automotive radar systems used in autonomous vehicles, where high-frequency operation and low-noise performance are critical for detecting objects and ensuring safe navigation [31–34]. HEMT-based MMICs are crucial in 5G infrastructure, particularly in mm-Wavebands for base stations and user equipment. With the development of 6G technologies, operating in higher frequency ranges (above 100 GHz), HEMTs are expected to play an even larger role due to their high-frequency capabilities and power efficiency. New techniques in epitaxial layer growth have led to reduced defect densities in GaN and InP HEMTs, improving the overall reliability and performance of MMICs in high-power applications. GaNHEMT-based MMICs are now more reliable in extreme environments, such as in military, aerospace, and space applications [35, 36]. These devices are now engineered to withstand high radiation and temperature extremes, making them ideal for space communications and defense systems. New techniques in epitaxial layer growth have led to reduced defect densities in GaN and InP HEMTs, improving the overall reliability and performance of MMICs in high-power applications. GaNHEMT-based MMICs are now more reliable in extreme environments, such as in military, aerospace, and space applications. These devices are now engineered to withstand high radiation and temperature extremes, making them ideal for space communications and defense systems. Figure 9.1 represents the device structure.

Recent advancements include heterogeneous integration of different semiconductor materials, such as combining GaN on Si or GaN on SiC to optimize cost, performance, and thermal properties. This has resulted in high-performance MMICs suitable for mass production. HEMT-based MMICs are increasingly being integrated with digital, analog, and RF components to create fully functional System-on-Chip (SoC) designs.

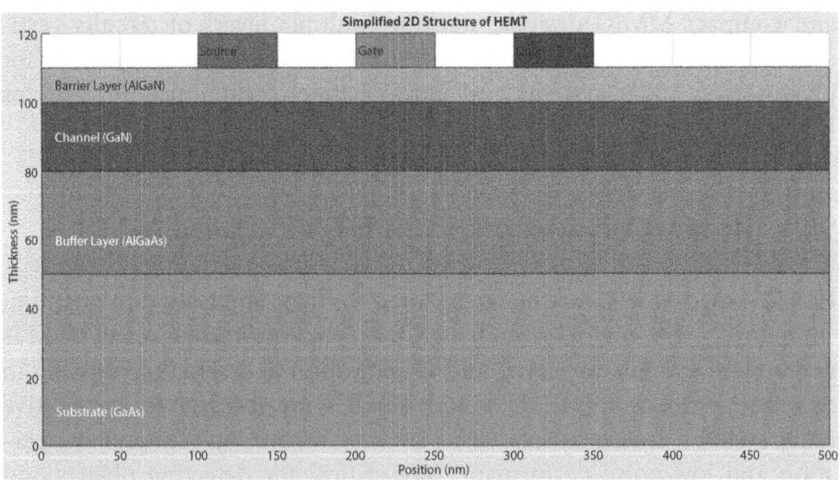

Figure 9.1 Device structure.

In aviation, HEMT-based MMICs are utilized for high-frequency communication systems that support air-to-ground and air-to-air communication, radar, and navigation systems. GaN HEMT MMICs are also being used in wireless power transfer systems, where efficient power amplification and high-frequency operation are crucial for transferring energy wirelessly over short distances. HEMT-based MMICs, particularly those using GaN, are being applied in power electronics for high-efficiency power conversion systems [37–41].

The footprint of the overall system improves power efficiency, especially in portable and high-frequency systems. Three-dimensional (3D) integration techniques have allowed for the vertical stacking of components in MMICs, reducing the overall size while enhancing performance. This is particularly useful for phased-array antennas, where space and weight are critical, and for multi-functional MMICs in advanced radar and communication systems. To enhance heat dissipation, advanced packaging technologies like flip-chip designs diamond heat spreaders have been integrated into HEMT-based MMICs. These techniques allow for better thermal management and improved reliability, especially in high-power applications. One of the major challenges for GaN-based HEMT MMICs has been heat dissipation due to high power densities. Recent advancements include the use of high thermal conductivity substrates like Silicon Carbide (SiC), which dissipates heat more efficiently than traditional materials. These substrates allow HEMTs to operate at higher powers without performance degradation.

Innovations in passivation techniques, such as the use of SiN (Silicon Nitride) layers, have minimized surface states that contribute to noise, leading to better overall signal quality and reduced phase noise in HEMT-based MMICs. GaN and InP HEMT-based MMICs are advancing in the design of low-noise amplifiers, essential for radar, satellite communication, and high-sensitivity radio frequency (RF) systems. These LNAs offer high gain and extremely low noise figures, enabling more precise signal reception at high frequencies. Integration of envelope tracking and DPD in HEMT-based MMICs has enhanced amplifier efficiency and linearity. Envelope tracking dynamically adjusts the supply voltage based on the signal's amplitude, while DPD helps correct signal distortion, improving performance in power amplifier applications. The Doherty power amplifier design, combined with GaNHEMTs, has been a key innovation, enabling higher power efficiency, especially in wireless infrastructure applications. This design allows the amplifier to maintain high efficiency over a wide dynamic range of output power. GaNHEMT-based power amplifiers have demonstrated high output power, better linearity, and greater efficiency, making them ideal for 5G base stations, satellite communication, and radar systems. The scaling down of gate lengths to sub-50nm has been a significant advancement, enabling HEMTs to achieve higher frequencies with greater efficiency. This miniaturization allows for faster switching speeds and higher operational frequencies, supporting ultra-high frequency MMIC designs. HEMTs have advanced to operate efficiently in the millimeter-wave (30GHz–300GHz) and terahertz (0.1–10THz) frequency ranges. GaN and InP HEMTs are crucial for next-generation communication systems like 5G, 6G, and high-resolution imaging radar. These frequencies are critical for applications requiring large bandwidth, such as high-speed data transmission and automotive radar.

Advancements in materials, such as using Aluminium Gallium Nitride (AlGaN) and Indium Aluminium Nitride (InAlN) as barrier layers, have improved the performance of GaN-based HEMTs. These materials enhance the carrier concentration and mobility, leading to better power handling, higher gain, and lower noise figures. Indium Phosphide (InP) HEMTs continue to be used in applications that require extremely high-frequency operation (up to terahertz frequencies) due to their superior electron mobility compared to GaN. InP-based MMICs are critical for applications in millimeter-wave (mm-Wave) and terahertz (THz) communication systems. GaN-based HEMTs have become the dominant technology for high-power, high-frequency MMICs due to their high breakdown voltage, wide bandgap, and ability to operate at high power densities. GaN HEMTs are now being extensively used in power amplifiers, offering higher efficiency and improved thermal management. GaN-based HEMTs have become the dominant technology for high-power, high-frequency MMICs due to their high breakdown voltage, wide bandgap, and ability to operate at high power densities. GaNHEMTs are now being extensively used in power amplifiers, offering higher efficiency and improved thermal management.

High Electron Mobility Transistor (HEMT)-based Monolithic Microwave Integrated Circuits (MMICs) have seen substantial advancements, driven by the growing demand for high-performance systems in applications like 5G, satellite communications, radar, and power electronics. HEMTs, particularly those based on Gallium Nitride (GaN) and Gallium Arsenide (GaAs), are integral to MMIC designs because of their high electron mobility, low noise, and excellent high-frequency performance.

9.3 Applications of Recent Advancements in HEMT MMIC

HEMT-based MMICs are used in IoT devices that require reliable communication in harsh environments, such as industrial IoT systems deployed in remote or high-temperature areas (oil rigs, factories). The high efficiency and frequency performance of HEMTs ensure that IoT devices maintain connectivity with minimal energy consumption. HEMT-based MMICs are essential in satellite transponders used for signal relay between ground stations and satellites. The high power and efficiency of GaN-based MMICs are critical for ensuring reliable communication in space. In aviation, HEMT-based MMICs are utilized for high-frequency communication systems that support air-to-ground and air-to-air communication, radar, and navigation systems. GaN HEMT MMICs are also being used in wireless power transfer systems, where efficient power amplification and high-frequency operation are crucial for transferring energy wirelessly over short distances. HEMT-based MMICs, particularly those using GaN, are being applied in power electronics for high-efficiency power conversion systems. These systems are used in data centers, renewable energy installations, and electric vehicles (EVs), where GaNHEMTs provide superior performance over silicon-based devices in terms of power handling, efficiency, and switching speed. HEMT-based MMICs are also being explored for use in quantum sensing systems, where ultra-sensitive measurements of physical properties (magnetic fields, temperature) are essential for emerging technologies like quantum imaging and quantum radar. In scientific research and industrial applications, terahertz spectroscopy powered by HEMT-based MMICs is used for chemical analysis, pharmaceutical quality control, and gas detection. The high sensitivity and resolution of THz MMICs make them invaluable in these fields. HEMT-based MMICs operating in the terahertzrange (0.1–10 THz) are crucial for future wireless communication systems that require ultra-high data rates. These systems are being researched for applications such as ultra-fast wireless data transfer, short-range communication, and high-resolution sensing. GaNHEMTs are being explored for use in high-power microwave (HPM) and laser systems in DEWs, where high energy and precise targeting are essential for neutralizing threats like drones or missiles. HEMTMMICs are used in high-power RF transmitters for defense applications, such as communication systems for military air craft and naval ships, where reliability and performance in harsh environments are crucial. GaN HEMT-based MMICs are used in electronic warfare systems, where high-power amplifiers and oscillators are needed for jamming enemy radar and communication signals. The high-frequency operation and power efficiency of HEMTs make them ideal for these applications, offering superior performance in compact, rugged designs. Terahertz-based MMICs, utilizing HEMT technology, are being developed for medical imaging applications such as cancer detection, where the non-ionizing radiation at THz frequencies can safely penetrate biological tissues, offering potential for early and accurate diagnosis. Millimeter-wave and THz imaging systems powered by HEMT MMICs are increasingly being used in industrial applications for non-destructive testing of materials, ensuring quality control in manufacturing without damaging the products. HEMT-based MMICs are critical in millimeter-wave imaging systems used in security scanners at airports and other high-security areas. These systems operate at frequencies ranging from 30 GHz to 300 GHz, providing non-invasive imaging to detect concealed objects. HEMT-based MMICs are used in SAR for Earth images and remote sensing, where high-frequency operation and high-power output are necessary for improving resolution and accuracy. High-power GaN-based MMICs are used in phased-array radar systems for military applications, including airborne surveillance, missile detection, and target tracking. The high efficiency and power handling of HEMT-based MMICs allow for enhanced radar performance, offering greater range and sensitivity while reducing system size and cooling requirements. HEMT-based MMICs are widely used in advanced driver-assistance systems (ADAS) for automotive radar. These systems, particularly in the 77GHz range, rely on HEMT technology to provide high-resolution imaging for vehicle navigation, collision avoidance, and autonomous driving. The ability of GaNHEMTs to operate at high frequencies with low noise and high output power is essential for detecting objects at greater distances with better precision. HEMT-based MMICs have helped reduce the size and weight of RF communication systems in satellites, improving the efficiency and longevity of space missions. These compact, high-frequency circuits are ideal for next-generation small satellite constellations used in global internet coverage and Earth observation systems. Recent advancements in GaNHEMTs have enabled high-power, low-noise amplifiers for satellite communications, particularly in the Ka-band (26.5–40 GHz) and Q-band (33–50GHz). These MMICs provide the necessary power amplification for long-distance communication between satellites and ground stations, as well as inter-satellite links. Future6G networks are expected to operate at even higher frequencies, approaching terahertz (THz) bands. HEMT-based MMICs, particularly those based on InP (Indium Phosphide) and GaN, are well-suited for this due to their high-frequency performance and ability to handle large bandwidths, crucial for ultra-fast data transfer and ultra-low latency communications. HEMT-based MMICs, particularly those using GaN (Gallium Nitride), are integral to 5G base stations and network infrastructure due to their ability to handle high power and operate at millimeter-wave frequencies (above 24 GHz). GaNHEMTs provide superior power efficiency, enabling more compact, high-frequency power amplifiers for 5G deployment. Recent advancements in High Electron Mobility Transistor (HEMT)-based Monolithic Microwave Integrated Circuits (MMIC) have led to significant improvements in various high-frequency, high-power, and low-noise applications. HEMT-based MMICs are now pivotal in several industries, including telecommunications, aerospace, automotive, and defense.

9.4 Conclusion

Recent advancements in HEMT-based MMIC technology have significantly improved performance in high-frequency, high-power, and low-noise applications. Innovations in materials like GaN, AlGaN, and InP, along with improvements in thermal management, miniaturization, and integration techniques, have expanded the scope of HEMT-based MMICs in emerging technologies like 5G/6G, radar, satellite communications, and quantum computing. As demand for higher power and frequency capabilities grows, HEMT-based MMICs are poised to play a crucial role in the next generation of microwave and millimeter-wave technologies.

AI disclosure statement: The author acknowledges the use of AI-based tools for minor language editing or grammatical correction. However, all intellectual contributions, analyses, and conclusions presented in this chapter are entirely the author's own.

References

1. Ambacher, O., Smart, J., Shealy, J.R., Weimann, N.G., Chu, K., Murphy, M., Schaff, W.J., Eastman, L.F., Two-dimensional electron gases induced by spontaneous and piezoelectric polarization charges in N- and Ga-face AlGaN/GaN heterostructures. *J. Appl. Phys.*, 85, 6, 3222–3233, 1999.
2. Ghosh, S., Dasgupta, A., Khandelwal, S., Agnihotri, S., Chauhan, Y.S., Surface-potential based compact modeling of gate current in AlGaN/GaNHEMTs. *IEEE Trans. Electron Devices*, 62, 2, 443–448, 2015.

3. Ahsan, S.A., Ghosh, S., Sharma, K., Dasgupta, A., Khandelwal, S., Chauhan, Y.S., Capacitance modeling in dual field-plate power GaN HEMT for accurate switching behavior. *IEEE Trans. Electron Devices*, 63, 2, 565–572, 2016.
4. Ganeriwala, M.D., Yadav, C., Mohapatra, N.R., Khandelwal, S., Hu, C., Chauhan, Y.S., Modeling of Charge and Quantum Capacitance in Low Effective Mass III-V FinFETs. *IEEE J. Electron Devices Soc.*, 4, 6, 396–401, 2016.
5. Yadav, C., Duarte, J.P., Khandelwal, S., Agarwal, A., Hu, C., Chauhan, Y.S., Capacitance Modeling in III–V FinFETs. *IEEE Trans. Electron Devices*, 62, 11, 3892–3897, 2015.
6. Yadav, C., Kushwaha, P., Khandelwal, S., Duarte, J.P., Chauhan, Y.S., Hu, C., Modeling of GaN-based normally-off FinFET. *IEEE Electron Device Lett.*, 35, 6, 612–614, 2014.
7. Khandelwal, S. and Fjeldly, T.A., Analysis of drain-current nonlinearity using surface-potential-based model in GaAspHEMTs. *IEEE Trans. Microw. Theory Tech.*, 61, 9, 3265–3270, 2013.
8. Khandelwal, S., Chauhan, Y.S., Fjeldly, T.A., Analytical modeling of surface-potential and intrinsic charges in AlGaN/GaN HEMT devices. *IEEE Trans. Electron Devices*, 59, 10, 2856–2860, 2012.
9. Ye, J.D., Pannirselvam, S., Lim, S.T., Bi, J.F., Sun, X.W., Lo, G.Q., Teo, K.L., Two dimensional electron gas in Zn-polar ZnMgO/ZnO heterostructure grown by metal-organic vapor phase epitaxy. *Appl. Phys. Lett.*, 97, 11, 2010–2012, 2010.
10. Khan, M.A., Singh, R., Mukherjee, S., Kranti, A., Buffer Layer Engineering for High (≥ 1013 cm-2) 2-DEG Density in ZnO-Based Heterostructures. *IEEE Trans. Electron Devices*, 64, 3, 1015–1019, 2017.
11. Khandelwal, S., Goyal, N., Fjeldly, T.A., A physics-based analytical model for2DEG charge density in AlGaN/GaN HEMT devices. *IEEE Trans. Electron Devices*, 58, 10, 3622–3625, 2011.
12. Tampo, H., Shibata, H., Maejima, K., Yamada, A., Matsubara, K., Fons, P., Kashiwaya, S., Niki, S., Chiba, Y., Wakamatsu, T., Kanie, H., Polarization-induced two-dimensional electron gases in ZnMgO/ZnO heterostructures. *Appl. Phys. Lett.*, 93, 20, 11–14, 2008.
13. Kuhn, K.J., Giles, M.D., Becher, D., Kolar, P., Kornfeld, A., Kotlyar, R., Ma, S.T., Maheshwari, A., Mudanai, S., Process technology variation. *IEEE Trans. Electron Devices*, 58, 8, 2197–2208, 2011.
14. Shur, M.S., Gaska, R., Khan, A., Simin, G., Wide band gap electronic devices, in: *Proceedings of the Fourth IEEE International Caracas Conference on Devices, Circuits and Systems*, Oranjestad, Aruba, Netherlands, pp. D051–D051, 2002, (Cat. No. 02TH8611).
15. Anderson, R.L., Germanium-gallium arsenide heterojunctions. *IBM J. Res. Dev.*, 4, 3, 283–287, 1960.

16. Dingle, R., Störmer, H.L., Gossard, A.C., Wiegmann, W., Electron mobilities inmodulation-doped semiconductor heterojunction superlattices. *Appl. Phys. Lett.*, 33, 7, 665–667, 1978.
17. Witkowski, L.C., Drummond, T.J., Stanchak, C.M., Morkoc, H., High mobilities in AlxGa1-xAs-GaAs heterojunctions. *Appl. Phys. Lett.*, 3711, 1033–1035, 1980.
18. Wang, W.I., Wood, C.E.C., Eastman, L.F., Extremely high electron mobilities inmodulation-doped GaAs-AlxGa1-xAs heterojunction superlattices. *Electron. Lett.*, 17, 1, 36–37, 1981.
19. Delagebeaudeuf, D. and Linh, N.T., Metal-(n)AlGaAs-GaAs two-dimensional electron gas FET. *IEEE Trans. Electron Devices*, 29, 6, 955–960, 1982.
20. Ambacher, O., Foutz, B., Smart, J., Shealy, J.R., Weimann, N.G., Chu, K., Murphy, M., Sierakowski, A.J., Schaff, W.J., Eastman, L.F., Two dimensional electron gases induced by spontaneous and piezoelectric polarization in undoped and doped AlGaN/GaN heterostructures. *J. Appl. Phys.*, 87, 1, 334–344, 2000.
21. Shur, M.S., GaN based transistors for high power applications. *Solid State Electron.*, 42, 12, 2131–2138, 1998.
22. Look, D.C., Recent advances in ZnO materials and devices. *Mater. Sci. Eng., B*, 80, 1–3, 383–387, 2001.
23. Özgür, U., Alivov, Y.I., Liu, C., Teke, A., Reshchikov, M.A., Dogan, S., Avrutin, V., Cho, S.-J., Morkoc, H., A comprehensive review of ZnO materials and devices. *Appl. Phys.*, 98, 4, 041301, 2005.
24. Meyer, B.K., Alves, H., Hofmann, D.M., Kriegseis, W., Forster, D., Bertram, F., Christen, J., Hoffmann, A., Straßburg, M., Dworzak, M., Haboeck, U., Rodina, A.V., Bound exciton and donor-acceptor pair recombinations in ZnO. *Phys. Status Solidi Basic Res.*, 241, 2, 231–260, 2004.
25. Singh, R., Khan, M.A., Mukherjee, S., Kranti, A., Analytical Model for 2DEG Density in Graded MgZnO/ZnO Heterostructures with Cap Layer. *IEEE Trans. Electron Devices*, 64, 9, 3661–3667, 2017.
26. Singh, R., Khan, M.A., Mukherjee, S., Kranti, A., Role of Surface States and Interface Charges in 2DEG in Sputtered ZnO Heterostructures. *IEEE Trans. Electron Devices*, 65, 7, 2850–2854, 2018.
27. Yadav, C., Ganeriwala, M.D., Mohapatra, N.R., Agarwal, A., Chauhan, Y.S., Compact Modeling of Gate Capacitance in III-V Channel Quadruple-Gate FETs. *IEEE Trans. Nanotechnol.*, 16, 4, 703–710, 2017.
28. Ferain, I., Colinge, C.A., Colinge, J.-P., Multi gate transistors as the future of classical metal-oxide-semiconductor field-effect transistors. *Nature*, 479, 7373, 310–316, 2011.
29. He, L., Chiang, T.K., Liou, J.J., Zheng, W., Liu, Z., A new analytical subthreshold potential/current model for quadruple-gate junctionless MOSFETs. *IEEE Trans. Electron Devices*, 61, 6, 1972–1978, 2014.

30. Lee, C.-T., Chen, W.-S., Lee, H.-Y., Quadruple Gate-Embedded T Structured GaN-Based Metal-Oxide-Semiconductor High-Electron Mobility Transistors. *IEEE J. Electron Devices Soc.*, 1, 6, 63–67, 2018.
31. Sharma, D. and Vishvakarma, S.K., Analyses of DC and analog/RF performances forshort channel quadruple-gate gate-all-around MOSFET. *Microelectron. J.*, 46, 8, 731–739, 2015.
32. Li, M. and Wang, Y., 2-D Analytical Model for Current–Voltage Characteristics and Transconductance. *IEEE Trans. Electron Devices*, 55, 1, 261–267, 2008.
33. A.U. Manual, *3-D device simulator*, Silvaco Inc., St. Clara, CA, USA, 2012.
34. Mimura, T., Hiyamizu, S., Fujii, T., Nanbu, K., A new field-effect transistor with selectively doped GaAs/n-AlxGa1-xAs heterojunctions. *Jpn. J. Appl. Phys.*, 19, 5, L225, 1980.
35. Wu, Y.-F., *et al.*, Very high breakdown voltage and large transconductance realized on GaN heterojunction field effect transistors. *Appl. Phys. Lett.*, 69, 10, 1438–1440, 1996.
36. Khan, M.A., *et al.*, AlGaN/GaN metal oxide semiconductor heterostructure field effect transistor. *IEEE Electron Device Lett.*, 21, 2, 63–65, 2000.
37. Solomon, P.M. and Morkoc, H., Modulation-doped GaAs/AlGaAs heterojunction field-effect transistors (MODFET's), ultrahigh-speed device for supercomputers. *IEEE Trans. Electron Devices*, 31, 8, 1015–1027, 1984.
38. Lee, K., Shur, M.S., Drummond, T.J., Morkoc, H., Parasitic MESFET in (Al, Ga)As/GaAs modulation doped FET's and MODFET characterization. *IEEE Trans. Electron Devices*, 31, 1, 29–35, 1984.
39. Morkoc, H. and Solomon, P.M., The hemt: A superfast transistor: An experimental GaAs AlGoAs device switches in picoseconds and generates little heat. This is just what super computers need. *IEEE Spectr.*, 21, 28–35, 1984.
40. Esaki, L. and Tsu, R., Superlattice and negative conductivity in semiconductors, in: *IBM Res. Note RC*, p. 2418, 1969.
41. Hiyamizu, S., Mimura, T., Fujii, T., Nanb, K., High mobility of two-dimensional electrons at the GaAs/n-AlGaAs heterojunction interface. *Appl. Phys. Lett.*, 37, 9, 805–807, 1980.

10

Ferroelectric Gated Field Effect Transistors

Ankush Mondal*, **Amandeep Singh, Sweta Chander and Sanjeet K. Sinha**

School of Electronics and Electrical Engineering, Lovely Professional University, Punjab, India

Abstract

Advancements in Semiconductor FET (Field Effect Transistor) design have enabled the designers to push for lower technology nodes while improving the said device performance with the advantage of better modeling methods as well as architectural design principles which allow for scaling down of technology node to sub-nanometer regime. The once neglected effects that arise with the shortening of dimensions of the said devices have now become the focus of concern, especially with the current technology pushing the design to the very limits of sub-nanometer range. Hence, additional measures should be taken to reduce such issues while maintaining or even improving the said performance of the devices. One of the methods is to exploit the negative capacitance property shown by a special class of dielectric materials—known as ferroelectrics—that can mitigate such issues while allowing for performance gains by the same architecture compared to those without them. This manuscript gives insight into the special report of ferroelectrics from a thermodynamic point of view. Furthermore, including such materials with the existing framework of semiconductor manufacturing is also discussed, along with an example of simple MOSFET. The difference in performance is analyzed through various parameters such as potential gradient and current density. Finally, some groundbreaking work in the field of ferroelectrics and some newly developed methods that employ these materials are also discussed.

***Keywords*:** Ferroelectrics, negative capacitance, field effect transistors, gate stacking, electric polarization

**Corresponding author*: ankushmondal399@gmail.com

Abhishek Kumar and Suman Lata Tripathi (eds.) Smart Chips for Smart Devices: VLSI Design for Next-Generation IoT Solutions, (191–214) © 2026 Scrivener Publishing LLC

10.1 Introduction

Field effect transistors (FETs) have been a staple for semiconductor technology. FETs control the current flow between two terminals, the source and drain, through a channel modulated by an electric field applied at the gate terminal. This makes them highly efficient and scalable, which is why they are key component of integrated circuits (ICs), especially in Metal-Oxide-Semiconductor FETs (MOSFETs), which dominate the microelectronics industry. As technology advances, there is a constant push toward miniaturizing devices, leading to lower technology nodes measured in nanometers (nm). It is imperative that FETs be made smaller to meet the requirements of contemporary applications like mobile devices, high-speed computing, and the Internet of Things (IoT). Doing so improves device performance, increases transistor density, and lowers power consumption.

Recent advancements in CMOS technology have made the transistor size to nanometer regime that offers improved performance and less power dissipation. However, as the size of the CMOS transistor decreases over time, other short-channel effects arise that challenge the feasibility of operating the FETs in low technology nodes and lower operating voltages. Some of the common problems faced by CMOS transistors in lower gate lengths include Drain-Induced Barrier Lowering (DIBL), Enhanced Leakage current due to lowering of these parathion between the source and the drain region, advanced heating issues, quantum effects and so on. Furthermore, the constant lowering of operating voltages due to lowering power consumption necessities has made the traditional MOS structures unable to operate reliably due to their inherent voltage requirements.

To compensate for such drawbacks, Tunnel Field effect transistors were introduced as an alternative Device configuration for FETs, and while they have mitigated a considerable number of problems that were being faced, they face their own sets of issues, prominent of which are the Low ON current offered as compared to the MOS counterparts and Ambipolar conduction in opposite polarity which compromises the reliability of the said device [1]. Various methods are being employed to compensate for such problems, from developing Device geometries to changing the fabrication material. One of these methods includes including a ferroelectric material atop the Gate oxide to facilitate voltage boosting *via* the inherent charge-holding properties of the ferroelectric material due to its inherent electric polarization. This chapter deals with the properties of ferroelectric materials *via* the Landau theory of phase transitions for solids. Furthermore, the application of a ferroelectric gate stack on a FET is discussed, along with the supposed fabrication of such devices. Finally, some considerations regarding the application of such extra material layer and recent advancements in the field of such material are also discussed.

10.2 Ferroelectric Materials

Ferroelectric materials, a subset of Dielectric materials, are often marked by their intense polarization in the presence of an external electric field. Figure 10.1 shows the hysteresis curve of a general ferroelectric material $BatiO_3$ at 20°C against an applied electric field.

The main principle behind the working of a ferroelectric can be explained by the Landau-Devonshire theory of phase transitions, described in detail in the section below.

10.2.1 Landau-Devonshire Theory of Phase Transition

The Landau-Devonshire theory is a foundational framework for describing ferroelectric materials' phase transitions and thermodynamic properties. This phenomenological theory expresses free energy as a function of spontaneous polarization and sheds light on the behavior of ferroelectrics. It is based on Landau's theory of phase transitions. The theory accounts for the polarization states of ferroelectric materials and predicts their response to external factors like temperature and electric fields. By expanding the free energy in terms of polarization and incorporating higher-order terms,

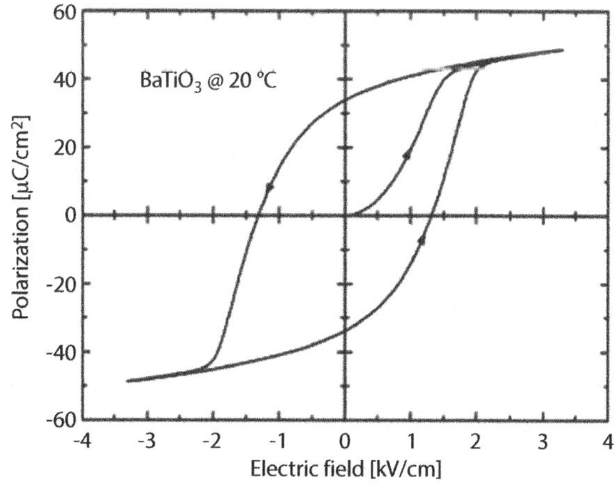

Figure 10.1 Ferroelectric hysteresis loop for $BatiO_3$ at 20°C [2].

the Landau-Devonshire theory successfully explains key phenomena such as spontaneous polarization, hysteresis, and domain formation. This theory plays a critical role in understanding the complex behaviors of ferroelectrics, particularly near phase transitions, and serves as a basis for developing advanced materials and applications.

The Landau-Devonshire theory is used to model the thermodynamic behavior of ferroelectrics, mainly focusing on how the polarization of the material evolves near phase transitions [3]. The theory expands the free energy **F** of a ferroelectric system as a power series in terms of the polarization **P**, assuming that close to the phase transition, the system's behavior can be described by a few key parameters. The free energy is expressed as follows

$$F(P)=F_0+ \alpha P^2+ \beta P^4+ \gamma P^6+ \dots \tag{10.1}$$

where F_0 is a reference free energy (independent of P) P is polarization of given material terms α, β and γ are temperature dependent coefficients. Equation (10.1) thus shows the relation between the electric polarization of any ferroelectric material and its thermodynamic energy. The nature and behavior of a ferroelectric material depends upon the constants mentioned above α, β and γ. These constants are called Landau free energy coefficients for the given ferroelectric. These coefficients decide when the phase transition occurs from its paraelectric phase to its ferroelectric phase. Typically, the phase transition for the given material is decided by the second-order coefficient α. There are two main case scenarios with the material being either in the paraelectric phase or ferroelectric phase, which is being dictated by the Landau coefficients as follows:

- ***Paraelectric phase*** (at high temperatures), the second order coefficient $\alpha>0$, indicating minimum Landau free energy at $P = 0$ and no spontaneous polarization.
- As the temperature decreases to the critical limit, the value of α decreases until it becomes less than 0($\alpha <0$), and that is when the material shows spontaneous polarization, as free energy F shows minima at values $P \neq 0$. The material shows **ferroelectric behavior**.

The critical temperature where this transition takes place is generally the Curie temperature (T_c). Near the phase transition, the dependence of α is generally linear and can be expressed as

$$\alpha=\alpha_0(T-T_c) \tag{10.2}$$

where α_0 is a constant determined by the specific ferroelectric material, and T is the current temperature [3]. Equation (10.2) gives the relation between the first order coefficients α, which is important to know the symmetry of the free energy about the absolute minima. For example, Figure 10.2 shows the free energy landscape concerning electric polarization for the compound $BaTiO_3$ at 20°C.

From Figure 10.2, it is clear that despite the initial Landau parameter α, the minima of the Free energy function is decided by the higher order parameters β and γ. Now, the equilibrium Polarization is found by minimizing the free energy (**F**) with respect to Polarization (**P**) as

$$\frac{dF}{dt} = 0 \tag{10.3}$$

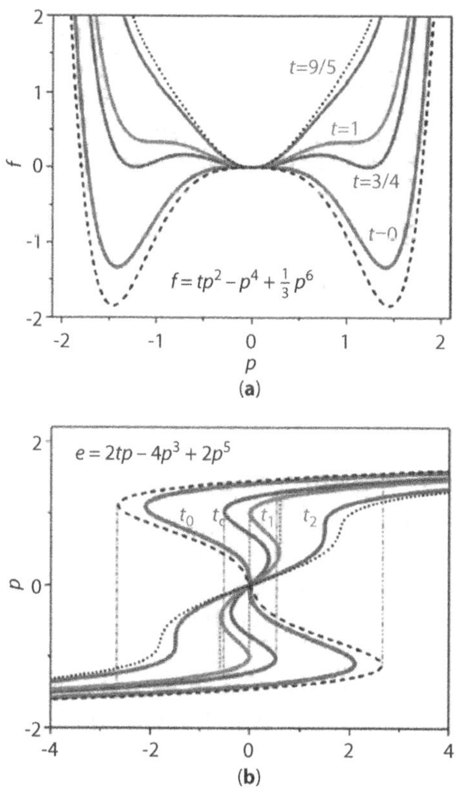

Figure 10.2 Landau free energy and polarization with respect to applied external electric field for $BaTiO_3$ at 20°C [4].

Only the P^2 term is essential for a second-order phase transition. However, the widely used convention for ferroelectric behavior is expressed as 6th order transition for more accurate Polarization behavior. Thus, the higher order terms like P^4 and P^6 also become essential. These higher-order terms may cause the material to show discontinuous jumps in spontaneous polarization. In the presence of an external Electric Field, another term is added to the Landau free energy expression, and the total expression becomes:

$$F=F_0+\alpha P^2+\beta P^4+\gamma P^6-EP \tag{10.4}$$

Where the term ***–EP*** accounts for the interaction of the applied external electric field and spontaneous polarization of the material. Thus, Equation (10.4) gives us the complete picture of the change in polarization of any ferroelectric material with respect to its inherent remnant polarization, the applied external field and its thermodynamic properties. The hysteresis behavior observed in ferroelectrics, as shown in the polarization-electric field (P-E) curve, canals derived from the Landau-Devonshire theory. The existence of multiple minima in the free energy results in polarization switching between stable and metastable states, leading to the characteristic hysteresis loop. The coercive field E_c represents the external electric field required to reverse the polarization.

10.2.2 Landau-Khalatnikov Equation and Dynamic Phase Transition

However, on its own, the Landau-Devonshire theory only provides a static description of ferroelectric materials, focusing on the thermodynamic equilibrium properties such as polarization and phase transitions. However, to understand the dynamic response, especially the time dependent evolution of polarization under external fields, the Landau-Khalatnikov equation is introduced. This equation describes the non-equilibrium dynamics of polarization and is essential for modeling the switching behavior of ferroelectric domains, particularly in transient regimes like hysteresis loop formation and polarization switching.

The Landau-Khalatnikov equation describes the evolution of time-dependent Electric Polarization P(t). It relates the change in Polarization value with respect to the gradient of the Landau free energy with respect to the said Polarization. The equation is given in the

$$\tau \frac{dP}{dt} = -d\,\hat{G}\,/\,dP \tag{10.5}$$

where τ is a kinetic coefficient. This equation expresses the fact that the regression dP/dt of a given polarization fluctuation towards equilibrium is the faster, the larger the **'thermodynamic force'**, $-\partial\,\hat{G}(P,E)/\partial P$, that is applied [2]. Now, incorporating the free energy expression from the Landau-Devonshire theory, the equation becomes

$$\tau \frac{dP}{dt} = -(2\alpha + 4\beta P^3 + 6\gamma P^5 - E) \tag{10.6}$$

which becomes,

$$\frac{dP}{dt} = -\gamma(2\alpha + 4\beta P^3 + 6\gamma P^5 - E) \tag{10.7}$$

Where the γ outside of the free energy expression denotes the damping constant of the Ferroelectric system under the time-variant Electric Field. The Landau-Khalatnikov equation also helps explain the formation of the hysteresis loop in ferroelectric materials. As an external electric field **E** is applied and reversed, the polarization does not follow the same path, leading to a characteristic loop. This hysteresis behavior is a direct consequence of the time dependent response of the polarization, which lags behind the applied field due to the energy barriers between table polarization states [4].

10.3 Ferroelectric Stack on Gate for Voltage Amplification

Now that we know the behavior of ferroelectric material, the next question arises is how we employ the polarization property of the said material in FET devices to improve their performance, but before we delve into that, it is essential to know the main problems the said devices face in short technology nodes.

10.3.1 Common Short-Channel Effects at Lower Technology Nodes

Traditional FET architectures face increasing challenges as the industry pushes toward ever smaller technology nodes (sub-7 nm and beyond). At these lower nodes, transistors encounter issues such as short-channel effects, increased leakage currents, reduced gate control, and heightened variability in manufacturing processes. These factors threaten the continued gains in performance, energy efficiency, and scaling, necessitating innovations in both transistor design and the materials used in their construction. Below are some of the prominent problems faced in the sub-manometer region of FET devices.

10.3.1.1 Drain-Induced Barrier Lowering (DIBL)

In the long-channel FETs, the drain has little influence on the barrier at the source end. Nevertheless, the drain's electric field might affect the barrier at the source end as the channel length decreases, lowering the source-channel potential barrier [5]. This lowering of the barrier means that carriers can be injected from the source to the channel even when the gate voltage is lower than the threshold voltage. As a result, the device can turn on partially, even with a lower gate voltage. Possible impacts include reducing the Threshold Voltage (V_{th}) of the MOS transistor, thus increasing the OFF-state leakage current.

10.3.1.2 Threshold Voltage Roll-Off

In short-channel devices, the threshold voltage depends on the channel length. As the channel length decreases, the gate control over the channel diminishes, making the threshold voltage drop [6]. The device's threshold voltage becomes lower for shorter channel lengths. The threshold voltage is no longer a constant for the device but varies with channel dimensions. The variation in threshold voltage complicates the design of circuits, especially when uniform threshold voltages are necessary for multiple transistors. Furthermore, lowering the threshold voltage can increase leakage current and reduce FET reliability.

10.3.1.3 Velocity Saturation

In long-channel devices, carrier velocity increases linearly with the electric field. However, at high electric fields, typical in short-channel MOSFETs, carrier velocity saturates due to scattering processes, especially for electrons. This saturation of carrier velocity limits the current drive of the FET, regardless of further increases in the electric field [7]. The transconductance (gain of the FET) decreases because the current is no longer proportional to the voltage. Moreover, since the carrier velocity saturates, the FETs' maximum current (switching speed) is constrained.

10.3.1.4 Channel Length Modulation

In an ideal long-channel MOSFET, the channel length is fixed, and the current is independent of the drain voltage in the saturation region. On the other hand, with short-channel devices, the pinch-off zone extends from the drain towards the source, reducing the effective channel length as the drain voltage rises. Even though the device allegedly runs in saturation, the decrease in channel length increases drain current in the saturation area [8]. This leads to an on flat I-V curve in the saturation region, affecting the predictability of the device's performance. The output resistance decreases, making the device less effective as a voltage-controlled current source.

10.3.1.5 Hot Carrier Effect (HCE)

The electric field near the drain becomes very large when the channel length is small. As a result, carriers (electrons or holes) gain high kinetic energy (becoming "hot carriers"). Some of these hot carriers can be introduced into the gate oxide or induced to induce impact ionization to create electron-hole pairs. Hot carriers can make the oxide stuck with charges or degrade the gate oxide [9]. Over time, the injection of hot carriers into the gate oxide causes a gradual degradation in device performance, shifting the threshold voltage and reducing mobility. The damage to the oxide can also lead to increased gate leakage current, reducing the overall reliability of the FET device.

10.3.1.6 Subthreshold Slope Degradation

The subthreshold region is where the FET is supposed to be off, and the current ideally decays exponentially as the gate voltage falls below the threshold. Due to weak gate control over the channel, this transition becomes less sharp in short-channel devices [10]. The subthreshold slope (steepness of the current-vs-gate voltage curve in the subthreshold region) becomes worse (less steep). The degradation in the subthreshold slope causes more current to flow in the off-state, leading to higher leakage power. A less sharp transition means slower switching times between on and off states.

10.3.1.7 Gate-Induced Drain Leakage (GIDL)

High electric fields close to the drain in short-channel devices can cause band-to-band tunneling of electrons from the valence band to the conduction band, especially when the gate is negatively biased with respect to the drain. Even in situations where the device is meant to be off, this effect results in leakage currents between the gate and the drain [11]. The leakage current leads to higher static power dissipation. Over time, excessive leakage can degrade the reliability of circuits.

10.3.1.8 Punch Through

As the channel length becomes comparable to the depletion regions extending from the drain and the source, these depletion regions can merge. This can cause the drain to influence the source directly, by passing the gate's control [12]. Regardless of the gate voltage, the current flows directly from the drain to the source. Once the gate loses its ability to control the channel current, the device cannot be turned off. Punch through leads to very high leakage currents, drastically reducing the effectiveness of the FETs as a switching device in digital circuits.

10.3.2 Ferroelectric Stacking

Due to the problems mentioned above faced by the FET devices, traditional approaches don't work in short channel lengths. Hence, there is a need for alternative means of application of FETs to overcome such issues [13]. One such method is to use ferroelectric materials to provide voltage amplification for the device.

There are other benefits of including ferroelectric materials. Ferroelectric materials enhance the electric field control over the channel due to their higher permittivity. This stronger control reduces the influence of the drain on the source-channel barrier, thus minimizing DIBL. Ferroelectric materials exhibit a negative capacitance effect, which can stabilize the threshold

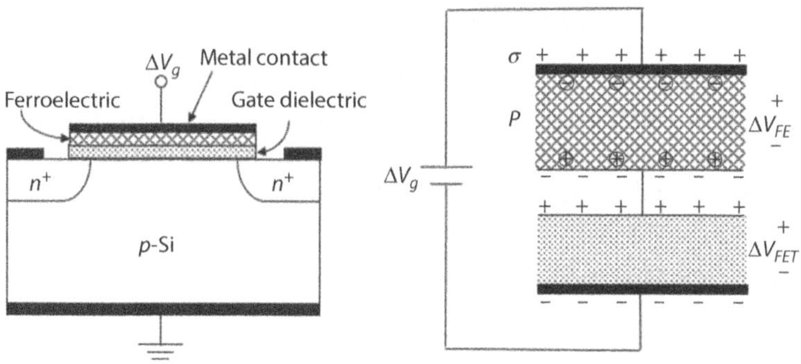

Figure 10.3 Ferroelectric gate stacking on a simple MOSFET [14].

voltage and reduce its dependence on channel length. This results in less variation of threshold voltage at shorter channel lengths. Enhanced gate control by ferroelectric materials leads to a more uniform charge distribution in the channel, reducing channel length modulation effects by keeping the channel length more stable. Ferroelectric materials, especially in negative capacitance FeFETs, can achieve a subthreshold slope lower than 60 mV/decade. This leads to sharper switching between the on and off states, improving the subthreshold slope and reducing leakage. The strong gate control offered by ferroelectrics can suppress the high electric fields responsible for band-to-band tunneling, thereby reducing GIDL and associated leakage currents. The most common method for the application of ferroelectric is by providing it as an external stack over the gate oxide of the FETs. Figure 10.3 shows the implementation process of such a stack of ferroelectric material in a standard MOSFET.

As seen in the above Figure 10.3, the ferroelectric layer is now introduced between the gate oxide and the gate metal. The ferroelectric material now exhibits spontaneous polarization when the device's gate terminal is powered on, which allows its internal dipoles to align when an external electric field is applied. Once the field is removed, the dipoles remain aligned; giving rise to a non-volatile memory effect [15]. This improves the electrostatic control over the channel and achieves additional functionality like memory storage. The ferroelectric layer introduces several significant physical effects, as described below.

10.3.2.1 Negative Capacitance Effect

Negative capacitance is a property of ferroelectric materials that increases the gate stack's total capacitance. This boosts gate control over the channel and allows for a steeper subthreshold slope, improving the device's switching speed and reducing power consumption [16].

10.3.2.2 Charge Accumulation

The ferroelectric layer aligns its dipoles in response to a voltage supplied to the gate, thus creating an electric field that aids in modulating the charge in the underlying channel through the gate oxide. This reduces the required gate voltage for switching the device on or off, effectively lowering the threshold voltage (V_{th}) without increasing leakage current [17]. The process of implementation of such a layer involves the following processes:

Step-1: *Substrate Preparation*: As with conventional MOSFET fabrication, the process starts with a silicon substrate. Depending on whether the FeFET is PMOS or NMOS, the source and drain regions are doped by ion implantation to produce n-type or p-type regions.

Step-2: *Gate Oxide Deposition*: On the silicon surface, a thin coating of high-k material such as HfO_2or gate oxide $SiO_2$is deposited. This acts as a barrier between the channel and the gate.

Step-3: *Ferroelectric Layer Deposition*: A thin ferroelectric material layer is placed on top of the gate oxide. Usually, methods like chemical vapor deposition (CVD) or atomic layer deposition (ALD) are used for this. This ferroelectric layer's thickness is precisely regulated to guarantee adequate polarization characteristics and correct dipole alignment.

Step-4: *Gate Electrode Deposition*: A metal gate (tungsten, titanium nitride, or other materials) is deposited on top of the ferroelectric layer once it has been deposited. This creates the gate electrode, which will use the combined effects of the oxide and ferroelectric layers to control the channel.

Step-5: *Source and Drain Contacts*: Metal deposition and etching techniques create contact contacts to the source and drain areas. After that, the gadget is linked to the external interface. One important element affecting a FET device's performance having an additional ferroelectric layer is its thickness. Complete polarization switching is ensured by a suitably thick layer, which is necessary for effective charge storage and device operation [18]. Furthermore, a thicker layer can enhance device

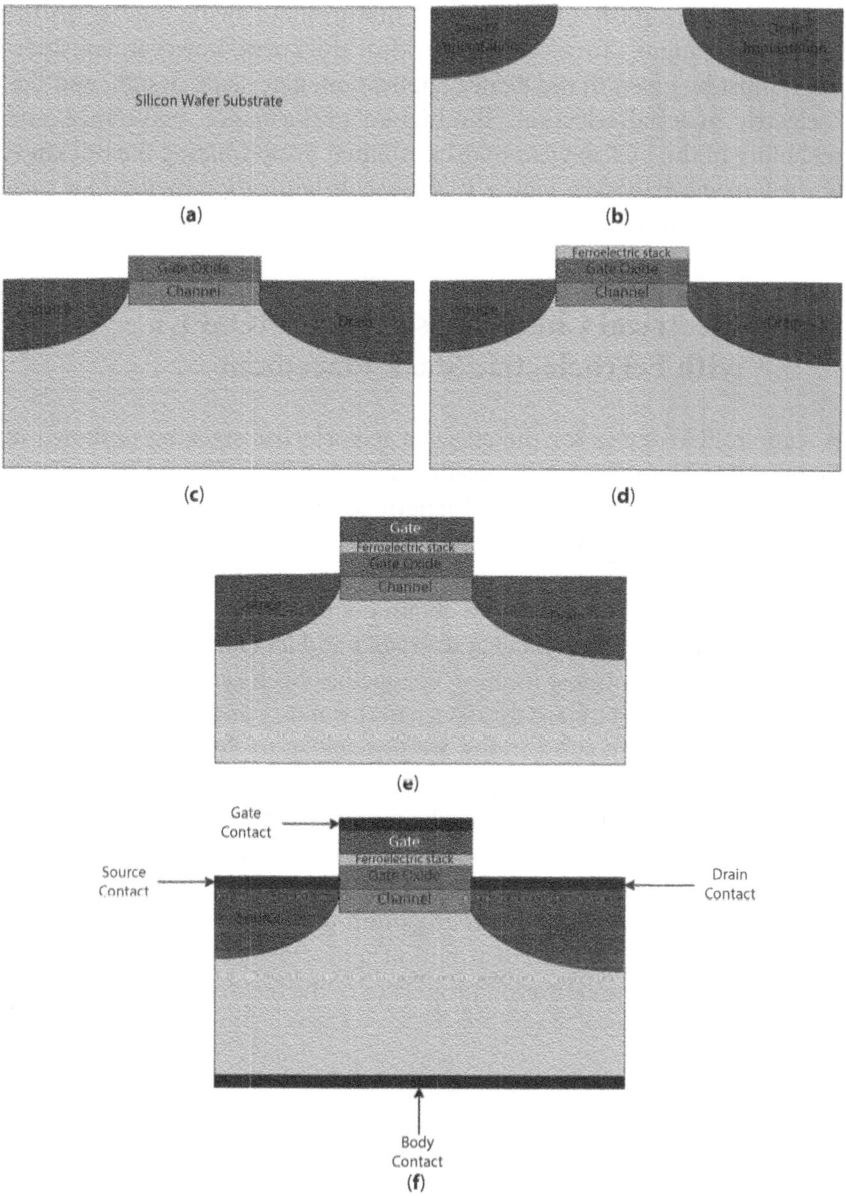

Figure 10.4 (a) Base wafer for FET substrate; (b) source and drain implantation; (c) gate oxide placement; (d) ferroelectric stack placement; (e) gate metal placement; (f) Ohmic contacts placement.

properties like threshold voltage and subthreshold swing and lessen the impacts of fatigue. To ascertain the ideal thickness, however, consideration must be given to the device application, gate stack design, and ferroelectric material selection. The highest performance, endurance, and reliability of the FET devices may be attained by optimizing the thickness of the ferroelectric layer. Figure 10.4 presents fabrication method for such a FET architecture.

10.4 Electrical Characterisation of MOSFET with Ferroelectric Stack Placement

A sample FET device for the effect of ferroelectric stacking is shown in Figure 10.5. Here, we have considered a MOSFET (Metal Oxide Semiconductor FET) with a channel length of 60 nm. Both the traditional MOSFET and the one with the Ferroelectric Gate stack (Fe-MOSFET) have identical properties regarding their building material and doping profiles. The devices are then simulated to obtain their performance characteristics mainly the Input. (Transfer) characteristics and its Output characteristics. In both cases, the utmost limit of Voltage has been set to 1.0 V, which is widely used as the standard for device performance analysis.

In the subsequent subsections, we provide an in-depth analysis of the impact of the ferroelectric stack on the overall performance of the device. This discussion will be multifaceted, covering several critical aspects that contribute to understanding how the ferroelectric material influences the device's behavior under various conditions. Specifically, we will explore the device's key characteristics, including input-output current versus voltage (I-V) behavior, electric field distribution across the device, current density variations, and the polarization properties of the ferroelectric material.

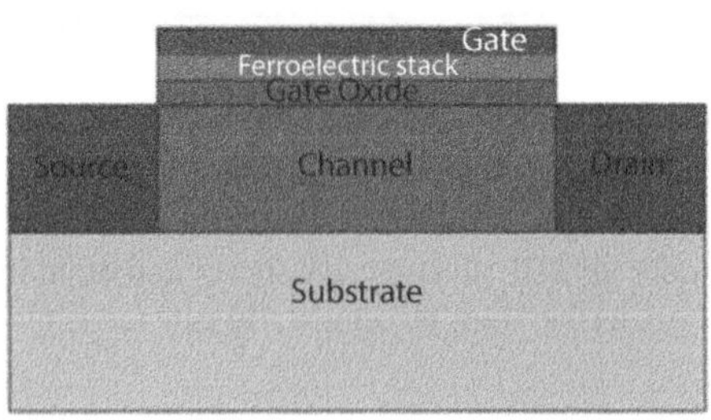

Figure 10.5 Simulated MOSFET with Ferroelectric Gate Stack (Fe-MOSFET).

We begin by examining the performance of the device, with a focus on its current-voltage (I-V) characteristics. Following the analysis of the I-V characteristics, we delve into the spatial distribution of the electric field within the device. The electric field profile will be evaluated along the length of the device for different configurations, with and without the ferroelectric stack. Next, we investigate the variation in charge carrier density within the device. This section will focus on how the introduction of the ferroelectric material affects the distribution and movement of charge carriers, which are essential for the conduction process. The interaction between the ferroelectric polarization and the charge carriers will be carefully analyzed, illustrating how the ferroelectric stack can either enhance or impede the flow of carriers through the device. Finally, we turn our attention to the ferroelectric polarization itself, which is a critical parameter in determining the performance of devices incorporating ferroelectric materials.

10.4.1 Characteristics

Figure 10.6 give us the transfer and output characteristics of the simulated MOSFETs in the voltage range of 0 to 1.0 V. The figure shows that using a ferroelectric stacked MOSFET, as opposed to a standard MOSFET, results in a significant improvement in the decrease of the OFF-state current in terms of transfer characteristics. This reduction in OFF-state current is critical, especially for low-power applications, because it directly impacts leakage current, resulting in lower power consumption and increased efficiency. A critical factor in accomplishing this reduction is the addition of

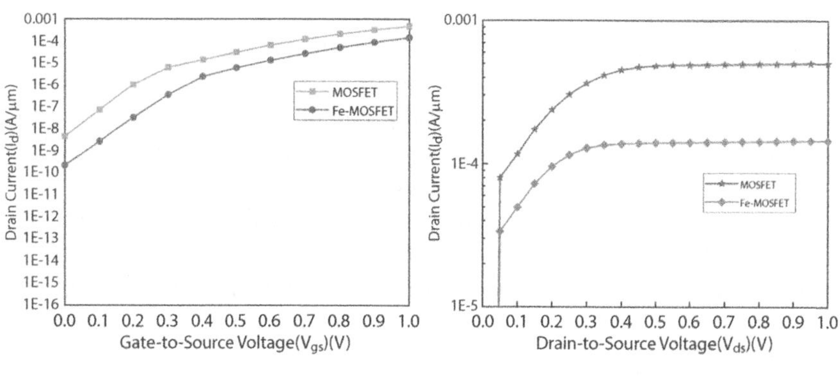

Figure 10.6 Transfer characteristics and output characteristics for simple MOSFET and gate ferroelectric stacked MOSFET (Fe-MOSFET).

the ferroelectric material to the MOSFET stack. This is primarily because of its unique characteristic of offering negative capacitance, which will be further discussed [19].

However, this enhancement in OFF-state current comes at the cost of a slight compromise in the device's output characteristics. In particular, the Fe-MOSFET has a somewhat lower ON-state current than the conventional kind. The interaction between the ferroelectric material and the charge carriers in the device is responsible for this modest decrease in ON-state current. So, while the negative capacitance lowers the threshold voltage, aiding in faster switching and reducing power dissipation during the OFF-state, it can also slightly hinder the full movement of charge carriers during the ON-state, leading to the observed lower ON current. This behavior indicates a trade-off between the benefits of reduced leakage current and the slight degradation in ON-state performance, which needs to be considered in specific application contexts. Despite the lower ON current, the ferroelectric MOSFET still offers considerable advantages for energy-efficient devices. Reducing OFF-state leakage and overall power consumption is more critical than maximizing ON-state performance. In summary, introducing the ferroelectric stack in MOSFETs demonstrates its potential to enhance performance, but with an inherent balance between improving OFF-state leakage and slightly impacting ON-state current conduction.

10.4.2 Electric Field and Electric Potential Profiles

The electric field in the channel of a MOSFET is crucial for establishing the potential gradient that drives charge carriers from source to drain, enabling current flow when the device is ON. The strength and uniformity of this field directly affect how efficiently charge carriers move, influencing the device's current output and overall performance. Introducing a ferroelectric material into the MOSFET structure significantly enhances the uniformity of the electric field across the device length. This more consistent field ensures a steady potential gradient, improving charge carrier conduction and minimizing regions where the field might weaken, which otherwise would hinder how current could flow [20].

Figures 10.7 and 10.8 compares the electric field profile and potential gradient between a traditional MOSFET and the Fe-MOSFET.

From the Figures 10.7 to 10.9, it is evident that the Fe-MOSFET exhibits a significantly improved electric field profile and a more uniform potential gradient compared to its traditional counterpart. This enhanced field uniformity results in better charge carrier movement through the channel,

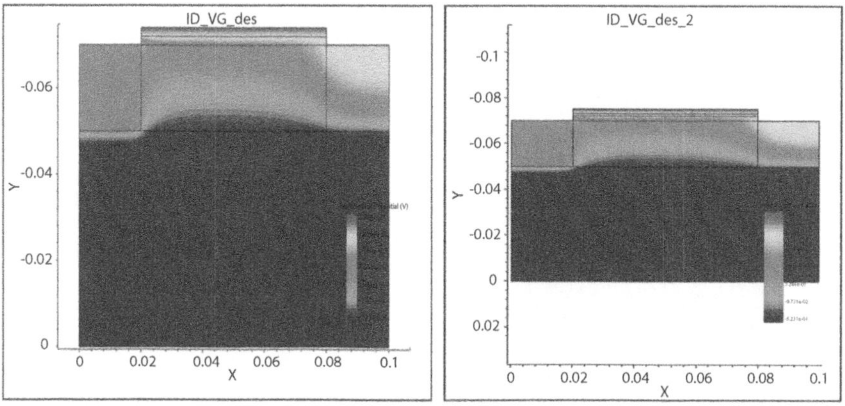

Figure 10.7 Electric field profile for simple MOSFET and Fe-MOSFET.

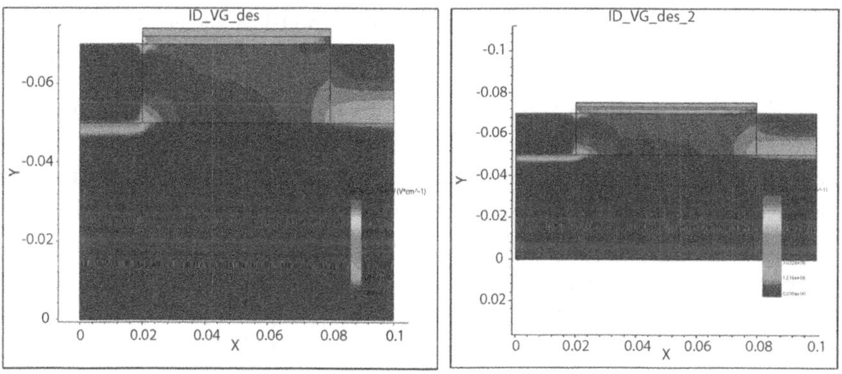

Figure 10.8 Potential gradient profile for simple MOSFET and Fe-MOSFET.

contributing to Fe-MOSFET's superior performance. The smoother and more consistent electric field across the device helps reduce carrier scattering and inefficiencies common in conventional MOSFETs. Additionally, this optimized potential gradient plays a key role in minimizing leakage current, particularly in the sub-threshold region, where controlling unwanted current flow is crucial for energy-efficient operation.

10.4.3 Current Density Profiles

The current density profile serves as a visual indicator of how effectively charge carriers move across the device, providing insights into the overall conduction performance. A well-distributed current density throughout

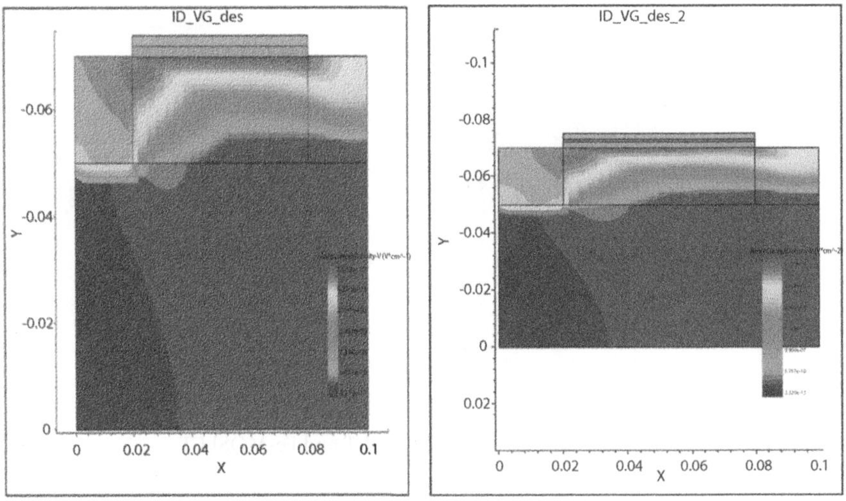

Figure 10.9 Electron current density profile for simple MOSFET and Fe-MOSFET.

the device's dimensions signifies efficient carrier transport, which in turn leads to better conduction in the FET [21]. In contrast, non-uniform current density may suggest area so flow carrier mobility efficient conduction pathways. Figure 10.9 illustrates the current density profiles for both the traditional MOSFET and the Fe-MOSFET, simulated under similar conditions. By comparing these profiles, we can observe how the introduction of the Ferroelectric material in the Fe-MOSFET improves the uniformity of current distribution, resulting in enhanced carrier flow and overall device performance.

It is clearly visible from Figure 10.9 the Fe-MOSFET exhibits a more uniform and well distributed current density compared to its traditional counterpart. This improvement is primarily attributed to the consistent electric field provided by the Ferroelectric material, which ensures a more even spread of charge carriers across the channel region. The Ferroelectric layer contributes to stabilizing the electric field throughout the device, minimizing fluctuations that could disrupt the flow of carriers. As a result, charge carriers can move more efficiently and uniformly, reducing areas of low carrier concentration and enhancing the overall conduction within the FET. This leads to a higher and more consistent current density, thereby improving the performance and efficiency of the FET.

10.4.4 Ferroelectric Polarization of the Fe-MOSFET

Until now, we have considered the various electric profiles that show the difference between the two implementations of the FET; now, we must consider the leading cause of such a performance increase in ferroelectric stacked MOSFET. As we remember from Section 2, the main working of ferroelectric could be explained through its polarization. Hence, it becomes crucial to know how the polarization occurs across the ferroelectric stack to understand its impact [22]. Figure 10.10 below gives the visualization of the ferroelectric polarization for Fe-MOSFET.

As can be inferred from Figure 10.10, there is an uneven distribution of polarization across the length of the ferroelectric material. Although this imbalance is non-detrimental in this specific case, it is crucial to understand that for fully harnessing the benefits of negative capacitance, uniform polarization across the ferroelectric material is essential [23]. A consistent polarization ensures that the negative capacitance effect is optimally utilized, enhancing the device's performance by stabilizing the electric field and improving charge carrier flow.

However, in this instance, despite the uneven polarization, there is still a uniform gradient across the ferroelectric material, which continues to

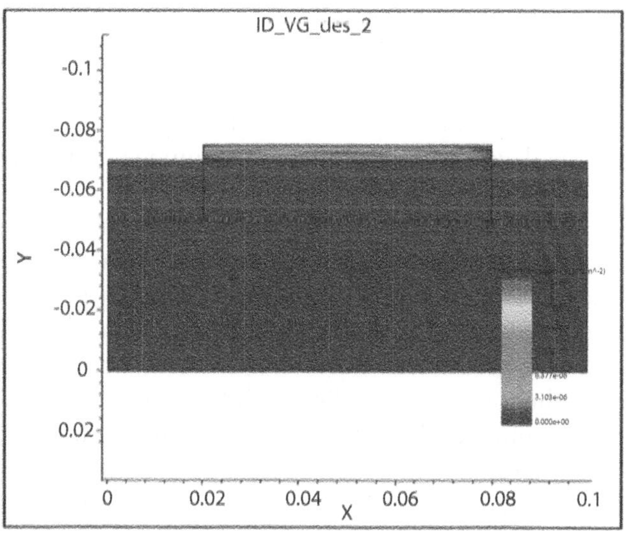

Figure 10.10 Ferroelectric polarization profile in FE-MOSFET.

contribute to the desired ferroelectric effect. This gradient plays a key role in stabilizing the electric field and improving the current density in the Fe-MOSFET, even with slight polarization variations.

As device dimensions shrink into the sub-10nm range, the choice of ferroelectric material and its thickness become increasingly critical factors that directly influence the effectiveness and reliability of the ferroelectric layer [24]. The polarization properties, material quality, and thickness control will all affect how efficiently the negative capacitance effect can be utilized, impacting the overall performance of the MOSFET. Therefore, further considerations and research are necessary in the field of ferroelectrics to optimize these parameters. Some of these challenges and potential solutions are discussed in the subsequent section, which explores advancements in ferroelectric materials and their applications in nanoscale devices.

10.5 Further Considerations and Improvements

As discussed in Section 10.3, adding a ferroelectric gate stack compensates for the short-channel effects of traditional FET. However, with the rise of the need for faster switching and ultra-low power consumption, there arises a new set of challenges; for example, ferroelectric materials must be compatible with standard CMOS processes, requiring careful tuning of deposition methods and material properties. Over time, the repeated switching of polarization in ferroelectric materials can cause degradation (fatigue), which may limit the long-term reliability of the device. As FETs continue to scale down in size, the ferroelectric layer must be thin enough to fit within the device while maintaining its unique properties. Various techniques are being developed in addition to ferroelectric engineering to help combat the shortcomings mentioned earlier. Hafnium polymers [25] provide strong ferroelectric properties as compared to naturally found ferroelectric materials like $BaTiO_3$. Additionally, different plane geometries of the said polymers are being studied to provide a better phase transition that leads to better ferroelectric behavior even at extremely low voltages.

Another method is the material design of the underlying FET devices with alternative compound materials with the same molecular weight as the ferroelectric layer to maximize the performance offered by the polarization effect of the ferroelectric material [26]. Another method is by employing efficient FET device architectures to accomplish the impact of ferroelectric voltage gain even with significantly shorter channel lengths (sub-5nm) [24]. Furthermore, with advancements in nanotechnology, complex polymers are being studied with properties that allow them to show a stable higher-order Landau transition that enables faster and more stable polarization [27]. Alternatively, ongoing work in advanced device structures such as FinFETs and nanowire sheet FETs also offers exceptional performance with the inclusion of ferroelectric material [28].

Acknowledgment

This work is funded and supported by the Science and Engineering Research Board (SERB), Department of Science & Technology, Government of India, CRG/2020/006229, dated 05/04/2021.

AI disclosure statement: The author confirms that no artificial intelligence (AI)-generated content or AI-assisted writing tools were used in the preparation of this chapter (including summaries and the abstract). All text, analysis, and conclusions are the result of the author's original work.

References

1. Hannah Blessy, P., Shenbagavalli, A., Arun Samuel, T.S., A Comprehensive Review on the Single Gate, Double Gate, Tri-Gate, and Heterojunction Tunnel FET for Future Generation Devices. *Silicon*, 15, 5, 2385–2405, 2023.
2. Tae Kwon, S., Landau-Khalatnikov simulations for ferroelectric switching in ferroelectric random access memory application. *J. Korean Phys. Soc*, 46, 1, 5, 2005.
3. Lye-Hock, O., Osman, J., Tilley, D.R., Landau theory of second-order phase transitions in ferroelectric films. *Phys. Rev. B*, 63, 14, 144109, 2001.
4. Lu, X., Li, H., Cao, W., Landau expansion parameters for $BaTiO_3$. *J. Appl. Phys.*, 114, 224106, 2013.
5. Singh, A., Sinha, S.K., Chander, S., Device Structure Modifications in Conventional Tunnel Field Effect Transistor (TFET) for Low-power Applications. *Nanoelectron. Devices Design Materials Appl. (Part I)*, 133, 2023.
6. Shih, H.C., Chen, Y.M., Lien, C., An analytical threshold voltage roll-off equation for MOSFET by using effective-doping model. *Solid State Electron.*, 49, 5, 808–812, 2005.
7. Hwang, S., Im, H., Song, M., Ishida, K., Hiramoto, T., Sakurai, T., Velocity saturation effects in a short-channel Si-MOSFET and its small signal characteristics. *J. Korean Phys. Soc*, 55, 2, 581–584, 2009.
8. Veeraraghavan, S. and Fossum, J.G., Short-channel effects in SOI MOSFETs. *IEEE Trans. Electron Devices*, 36, 3, 522–528, 1989.

9. Takeda, E., Yang, C.Y., Yang, C.Y.W., Miura-Hamada, A., Hot-carrier effects in MOS devices, 1995.
10. Zou, C., Overcoming sub-threshold swing degradation in sub-10nm technologies, in: *Asia Conference on Electronic Technology (ACET 2024)*, vol. 13211, SPIE, pp. 95–100.
11. Chan, T.Y., Chen, J., Ko, P.K., Hu, C., The impact of gate-induced drain leakage current on MOSFET scaling, in: *1987 International Electron Devices Meeting*, IEEE, pp. 718–721, 1987, December.
12. Barnes, J.J., Shimohigashi, K., Dutton, R.W., Short-channel MOSFETs in the punch-through current mode. *IEEE Trans. Electron Devices*, 26, 4, 446–453, 1979.
13. Rusu, A., Salvatore, G.A., Jimenez, D., Ionescu, A.M., Metal-ferroelectric-metal-oxide-semiconductor field effect transistor with sub-60mV/decade subthreshold swing and internal voltage amplification, in: *2010 International Electron Devices Meeting*, IEEE, pp. 16–3, 2010, December.
14. Blinc, R. and Zeks, B., *Soft Modes in Ferroelectrics and Antiferroelectrics*, p. 21, 1974.
15. Kwok, N., Steven, H., Alexei, G., Transient Nature of Negative Capacitance in Ferroelectric Field-Effect Transistors. *Solid State Commun.*, 265, 2017.
16. Tu, L., Wang, X., Wang, J., Meng, X., Chu, J., Ferroelectric negative capacitance field effect transistor. *Adv. Electron. Mater.*, 4, 11, 1800231, 2018.
17. Ko, E., Shin, J., Shin, C., Steep switching devices for low power applications: negative differential capacitance/resistance field effect transistors. *Nano Convergence*, 5, 1–9, 2018.
18. Prakash, O., Gupta, A., Pahwa, G., Chauhan, Y.S., Amrouch, H., On the critical role of ferroelectric thickness for negative capacitance transistor optimization, in: *2021 5th IEEE Electron Devices Technology & Manufacturing Conference (EDTM)*, IEEE, pp. 1–3, 2021, April.
19. Singh, A. and Sinha, S.K., Performance analysis of device characteristics in negative capacitance field effect transistor, in: *2021 Second International Conference on Electronics and Sustainable Communication Systems (ICESC)*, IEEE, pp. 236–239, 2021, August.
20. Lee, K., Lee, J., Kim, S., *et al.*, Analysis on fully depleted negative capacitance field-effect transistor (NCFET) based on electrostatic potential difference, in: *2019 Electron Devices Technology and Manufacturing Conference (EDTM)*, IEEE, pp. 422–424, 2019, March.
21. Cam, T., Wang, J.K., Wong, M., *et al.*, Sustained benefits of NCFETs under extreme scaling to the end of the IRDS. *IEEE Trans. Electron Devices*, 67, 9, 3843–3851, 2020.
22. Darmis, N., Alam, A.Z., Hashim, M.M., Ferroelectric behaviour and NCFETs-TCAD simulation. *Asian J. Electr. Electron. Eng.*, 1, 1, 30–41, 2021.
23. Wang, H., Yang, M., Huang, Q., *et al.*, New insights into the physical origin of negative capacitance and hysteresis in NCFETs, in: *2018 IEEE International Electron Devices Meeting (IEDM)*, IEEE, pp. 31.1.1–31.1.4, 2018, December.

24. Singh, A., Sinha, S.K., Chander, S., Effect of negative capacitance on heterojunction tunnel field effect transistor, in: *2021 5th International Conference on Electronics, Communication and Aerospace Technology (ICECA)*, IEEE, pp. 297–303, 2021, December.
25. Taekyong, K., del Alamo, J.A., Antoniadis, D.A., Dynamics of $HfZrO_2$ ferroelectric structures: Experiments and models, in: *2020 IEEE International Electron Devices Meeting (IEDM)*, IEEE, 2020.
26. Kwon, D., *et al.*, Near threshold capacitance matching in a negative capacitance FET with 1 nm effective oxide thickness gate stack. *IEEE Electron Device Lett.*, 41, 1, 179–182, 2019.
27. Jain, A. and Alam, M.A., Prospects of hysteresis-free abrupt switching (0mV/decade) in Landau switches. *IEEE Trans. Electron Devices*, 60, 12, 4269–4276, 2013.
28. Hiroyuki, O., *et al.*, Structural advantages of silicon-on-insulator FETs over FinFETs in steep subthreshold-swing operation in ferroelectric-gate FETs. *Jpn. J. Appl. Phys.*, 56, 4S, 2017.

11

Design and Analysis of a Boolean Function Using CMOS-Based Logic Styles

Vanshika Nanda[1] **and Suman Lata Tripathi**[2*]

[1]*School of Electronic and Electrical Engineering, Lovely Professional University, Punjab, India*
[2]*Department of E & TC, Symbiosis Institute of Technology, Symbiosis International (Deemed University), Pune, India*

Abstract
CMOS-based ICs are frequently used for design and implementation of logic functions, registers and memory elements. Mostly the implementations are based on CMOS logic style with an equal number of PMOS and NMOS due to its perfect logic transition, less delay and low power consumption. Delay and power consumptions are dependent on transistor aspect ratio (W/L). A selective aspects ratio for different logic blocks helps in delay and power optimisation process. Pseudo-NMOS, pass transistor logic, CMOS transmission gate and dynamic logic style are different logic styles that are similar to CMOS logic showing variation in their performance at the same technology node. This chapter presents a comprehensive analysis of logic styles for the implementation of the Boolean function. The logic styles considered include complementary CMOS, pass-transistor, and domino logic. The performance analysis metrics include power consumption, delay, and area utilization, to provide insights into their suitability for various design requirements. The Cadence Virtuoso is used for the analysis of different logic style at 45nm technology node.

Keywords: CMOS logic, transistor sizing, VLSI design, digital circuit implementation, power dissipation, propagation delay, switching activity, performance analysis

**Corresponding author*: tri.suman78@gmail.com

Abhishek Kumar and Suman Lata Tripathi (eds.) Smart Chips for Smart Devices: VLSI Design for Next-Generation IoT Solutions, (215–232) © 2026 Scrivener Publishing LLC

11.1 Introduction

Complementary Metal-Oxide-Semiconductor (CMOS)-based ICs are popular in several computational blocks performing certain task, arithmetic and logic operations along with registers and memory blocks [1–6]. The design of digital ICs often involves selecting an appropriate logic style to meet specific design goals evaluated majorly on the basis of their output logic level, power and delay performance [7–11]. For fixed technology node, the circuit performance is optimised through transistor sizing, aspect ratio, choice of logic style and transistor threshold levels. To reduce circuit delay, the transistor with high W/L ratio is preferred while for low power low W/L ratio transistors are suitable [12–15]. This chapter describes the implementation of a Boolean function using three different logic styles, i.e., CMOS logic, Pseudo-NMOS logic style and Dynamic logic style. This chapter aims to compare these logic styles with respect to Transient response, DC response, and Power responses and further analysis on factors including average Power and Delay. Figure 11.1 shows logic block of Boolean function f=(a·b+c)′.

This paper will also do an analysis of variation displayed by the above parameters when the W/L ratio is varied in the transistors used, i.e., from (W/L) 120nm to 2(W/L) 240nm. We have used 45-nanometer technology.

11.2 CMOS Logic Style

Complementary Metal-Oxide-Semiconductor (CMOS) technology is a widely used semiconductor technology that plays a fundamental role in modern electronic devices, including microprocessors, memory chips, and integrated circuits. It is known for its low power consumption, high noise immunity, and compatibility with both digital and analog circuitry [16–22]. Here are some key aspects of CMOS technology:

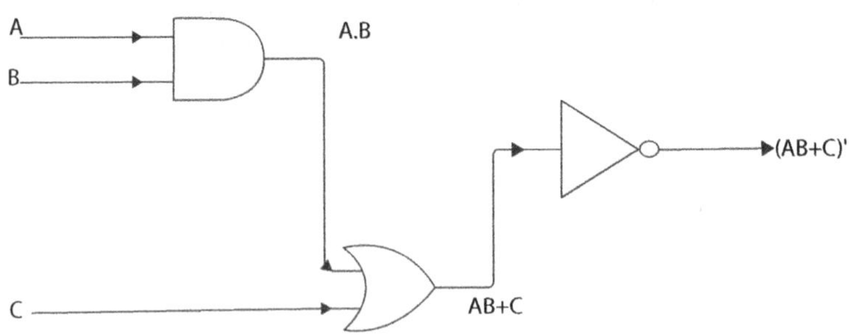

Figure 11.1 Boolean function f=(a·b+c)′.

Basic Structure: CMOS technology is based on the use of complementary pairs of metal-oxide-semiconductor field-effect transistors (MOSFETs). These transistors are composed of four layers: metal (gate), oxide (insulator), semiconductor (channel), and metal or polysilicon (source/drain). Figure 11.2 depicts the CMOS logic basic structure.

Low Power Consumption: CMOS circuits are known for their low static power consumption. When a CMOS gate is stable (high or low), it consumes very little power. This property makes CMOS ideal for battery-powered devices and energy-efficient electronics.

Digital and Analog: CMOS technology is versatile and can be used to implement both digital and analog circuits. Digital CMOS logic gates are the building blocks of digital integrated circuits, while CMOS analog circuits are used in applications such as amplifiers, filters, and analog-to-digital converters.

High Noise Immunity: CMOS logic gates offer high noise immunity because they rely on voltage levels rather than current flow to represent logic values. This makes them less susceptible to noise and interference.

Scalability: CMOS technology has seen continuous scaling down of transistor sizes, allowing for the integration of more transistors on a single chip. This has led to increased computational power and functionality in electronic devices while reducing their size and power consumption.

Manufacturing Process: CMOS fabrication involves a series of photolithography, etching, and doping steps to create transistors, interconnects, and other components on a silicon wafer. Advanced manufacturing techniques, such as silicon-on-insulator (SOI) and FinFET (Fin Field-Effect Transistor), have been developed to enhance CMOS performance.

Figure 11.3 shows the schematic of the given function using CMOS logic style.

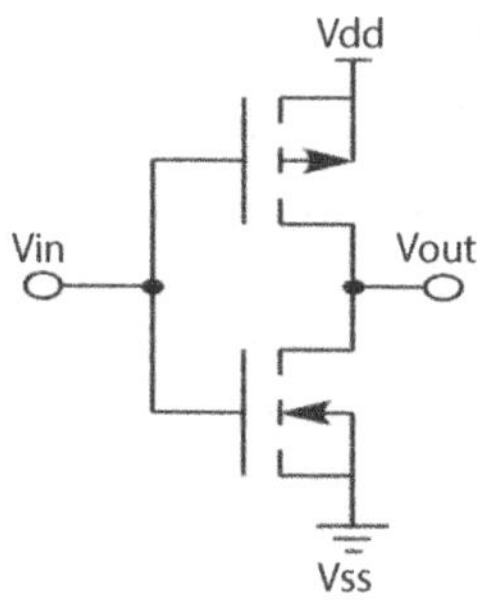

Figure 11.2 CMOS logic basic structure.

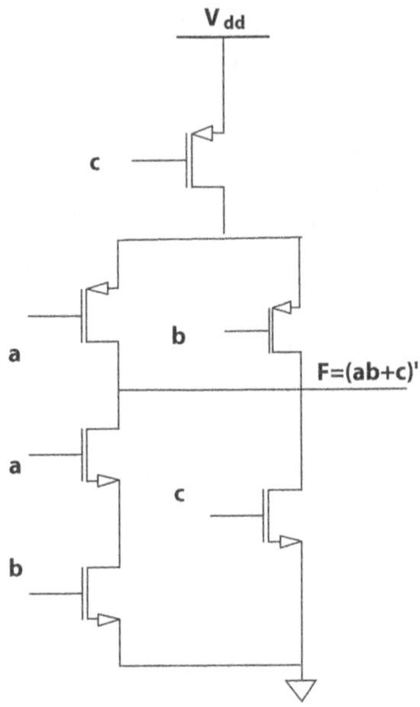

Figure 11.3 Schematic of f=(a·b+c)′ using CMOS logic.

Applications: CMOS technology is used in a wide range of applications, including microprocessors, memory chips (such as DRAM and SRAM), digital signal processors (DSPs), image sensors (CMOS cameras), microcontrollers, and many other integrated circuits found in computers, smartphones, cameras, and various consumer electronics.

Overall, CMOS technology has been a driving force behind the miniaturization and energy efficiency of electronic devices, enabling the development of more powerful and portable technology.

11.2.1 Pseudo Logic Style

Pseudo-NMOS logic is a type of digital logic family that was used in the early days of integrated circuit design, primarily in the 1970s and 1980s. It's an alternative to static CMOS logic and is characterized by its use of a pull-up network consisting of a single PMOS transistor and a pull-down network of NMOS transistors to implement logic gates. Here are some key points about Pseudo-NMOS logic:

Single Pull-Up PMOS Transistor: In Pseudo-NMOS logic, each logic gate has a single PMOS transistor acting as the pull-up device. This PMOS

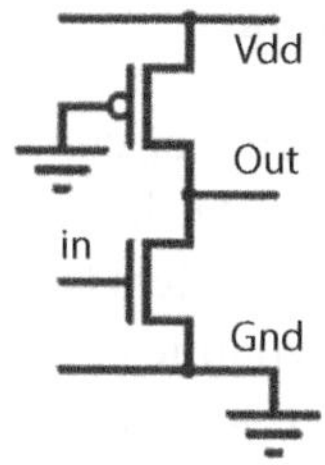

Figure 11.4 Pseudo-NMOS logic basic structure.

transistor is always "ON" because its gate is connected to ground (logic low). When the NMOS network is active, it creates a path to ground, causing the output node to discharge to logic low. Figure 11.4 shows Pseudo-NMOS logic.

N+1 Transistor Logic: The total number of transistors used in a Pseudo-NMOS gate is N+1, where N is the number of inputs to the gate. This means that for an N-input gate, you would have N NMOS transistors in the pull-down network and 1 PMOS transistor in the pull-up network.

Ratioed Logic: Pseudo-NMOS logic is a ratioed logic family, meaning that the dimensions (width and length) of the transistors are carefully chosen to achieve the desired logic functionality. The width of the PMOS transistor is typically wider than that of the NMOS transistors to ensure that it can provide sufficient current when the output is high.

Speed vs. Power Trade-off: Pseudo-NMOS logic offers faster switching speeds compared to static CMOS logic, especially in large fan-in NOR gates. However, this speed improvement comes at the cost of increased power consumption. When the NMOS network is active, a substantial current flow from the supply voltage (Vdd) to the ground, leading to dynamic power dissipation.

Applications: Pseudo-NMOS logic was commonly used in early digital integrated circuits, particularly for implementing logic gates, flip-flops, and other basic digital functions. However, it has largely been replaced by static CMOS logic in modern semiconductor technology due to its lower power consumption in most applications.

In summary, Pseudo-NMOS logic was an early logic family that provided fast switching speeds but at the expense of higher power consumption. It played a significant role in the history of integrated circuit design but has largely been supplanted by more power-efficient logic families like static CMOS in modern semiconductor technology.

11.2.2 Dynamic Logic Style

Dynamic logic is a family of digital circuit design techniques used in integrated circuits to implement digital logic gates and other sequential and combinational logic functions. It differs from other logic families like static CMOS logic in that it uses a dynamic approach to store and transfer logic values. Dynamic logic is known for its potential for high-speed operation and compact circuitry. Here are the key aspects of dynamic logic:

Charge Storage: In dynamic logic, information is stored as an electrical charge on capacitors rather than as voltage levels on nodes. The presence or absence of charge represents the logical states (1 or 0).
Clocking: Dynamic logic relies heavily on clock signals. A clock signal controls the precharge and evaluation phases of the logic operation. The precharge phase initializes the capacitors to a known state, while the evaluation phase performs the actual logic computation.
Pass Transistors: Dynamic logic gates typically employ pass-transistor logic, where a clocked transistor selectively connects capacitors to the input signals during the evaluation phase. The result is temporarily stored as a charge on the capacitors.
Dynamic Nodes: The output of dynamic logic gates is often a dynamic node, which means the voltage on the node is not stable but varies with time. This dynamic nature allows for high-speed operation but necessitates careful clock synchronization.
Pipelining: Dynamic logic is well-suited for pipelined designs, where multiple stages of logic are connected sequentially, each driven by a clock signal. This allows for high throughput and reduced pipeline latency.
Power Consumption: Dynamic logic can consume more power than other logic families because it relies on charging and discharging capacitors during each clock cycle. However, it can be power-efficient when used judiciously in applications that require high performance.
Applications: Dynamic logic is often used in high-performance and high-frequency digital circuits, such as microprocessors, digital signal processors (DSPs), and custom digital ASICs where speed is a critical factor.
Variants: Various variants of dynamic logic exist, including domino logic, self-resetting logic, and NORA (Negative-OR-AND) logic, each with its own advantages and trade-offs.

Figure 11.5 shows the schematic of the given function using pseudo NMOS logic style. Figure 11.6 shows the circuit diagram of 2-input NAND gate using dynamic logic. One of the main advantages of dynamic logic is its potential for high-speed operation due to parallelism and pipelining. However, it also presents challenges related to clock distribution, charge leakage, and power consumption. Modern semiconductor technology has developed numerous strategies to address these challenges, making dynamic logic a valuable tool in the design of high-performance digital circuits. Figure 11.7 shows Boolean function using dynamic CMOS in simulation. Figure 11.7 shows the circuit diagram of the given function using dynamic logic.

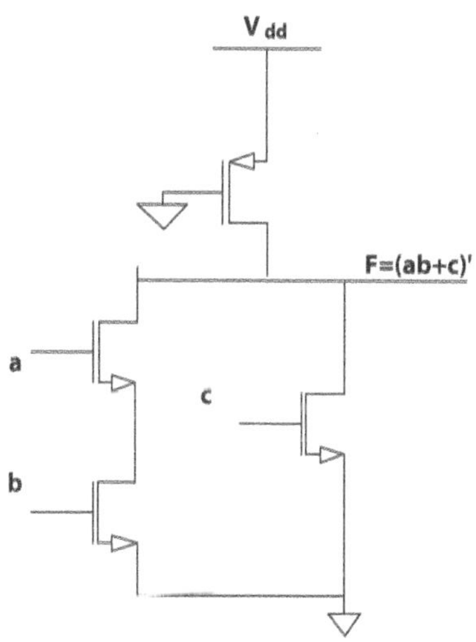

Figure 11.5 Schematic of f=(a·b+c)′ using pseudo-NMOS logic.

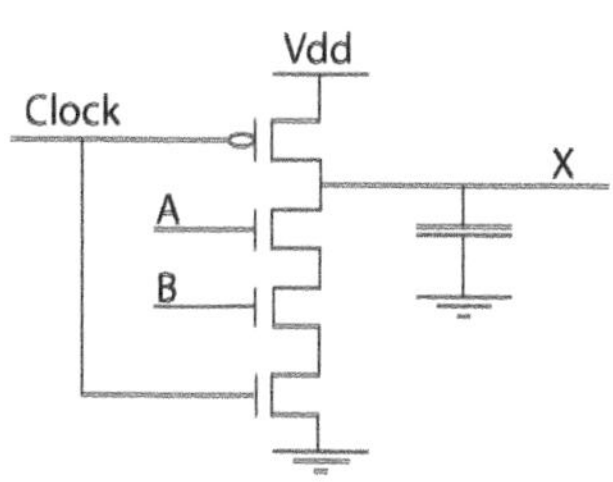

Figure 11.6 Dynamic logic basic structure.

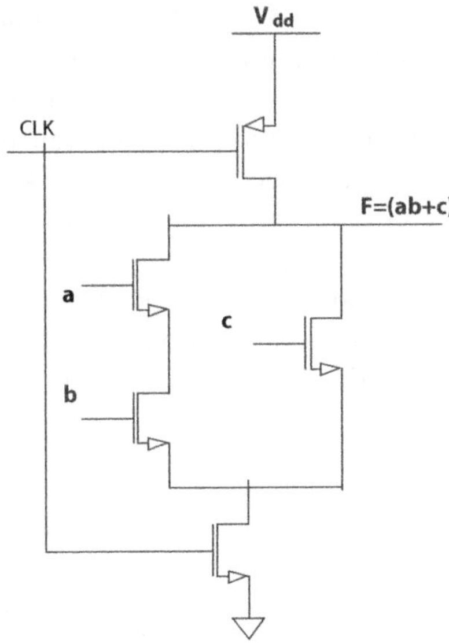

Figure 11.7 Boolean function using dynamic CMOS in simulation.

11.3 Performance Parameters

11.3.1 Leakage Current

The current when transistor is in OFF state.

$$I_{reverse} = I_s(e^{\frac{V}{V_T}} - 1)$$

The current I_s is the reverse saturation current dependent on the fabrication process and the PN-junction area. V_T is thermal voltage.

11.3.2 On Current

It is the current when transistor is in ON state.

11.3.3 Propagation Delay

Defined as time taken in propagating logic from input to output node. It is average of rise and fall delay. *Propagation delay (tp)* of a gate defines how quickly it responds to a change at its input(s). It expresses *the delay experienced by a signal when passing through a gate.* It is measured between the 50% transition points of the input and output waveforms

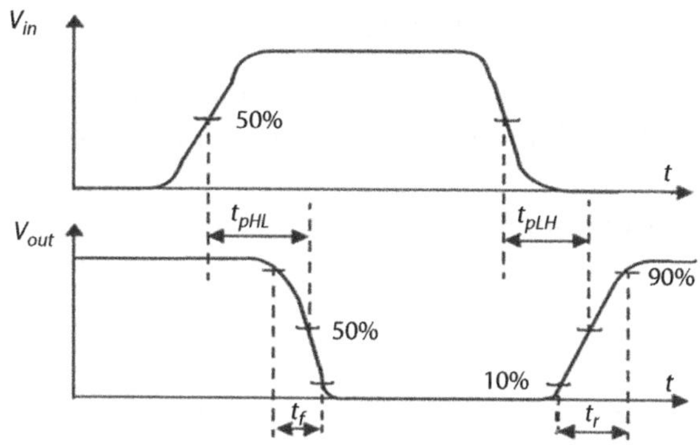

Figure 11.8 Delay schema of CMOS inverter.

t_{pLH}= response time of gate from low to high transition
t_{pHL}= response time of gate from high to low transition

Hence, the propagation delay $t_p = \frac{t_{pLH} + t_{pHL}}{2}$

Figure 11.8 shows the input and output function with time used in delay calculation.

11.3.4 Power Consumption

This is the total power consumed while charging and discharging capacitances in transistor structures when logic circuits switch states. Figure 11.9 shows the different types of power consumption.

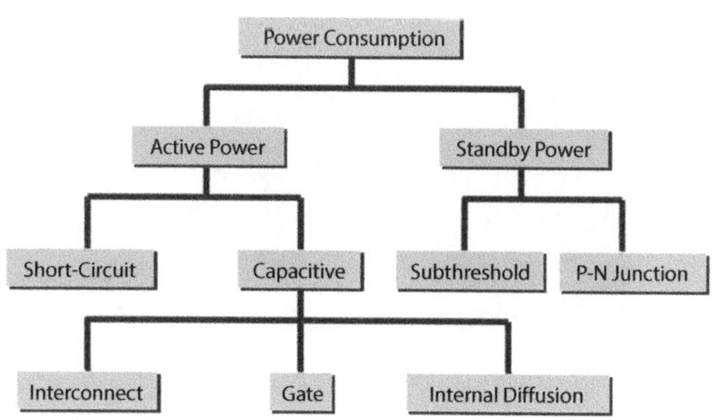

Figure 11.9 Power dissipation.

11.3.4.1 Static Power Consumption

This is the product of operating voltage and leakage current. Even when transistors are OFF, some current leaks through gates, which is dissipated as heat.

Total Power Consumption: $P_{total} = P_{static} + P_{dynamic} + P_{short\ circuit}$

$$P_{total} = V_{dd} I_{static} + \alpha V_{dd}{}^{2} C_L f_{clk} + t_{sc} V_{dd} I_{peak} f_{clk}$$

Here, α=activity factor; C_L *= load capacitance;* f_{clk} *= clock frequency;* t_{sc} *= switching time*

11.3.4.2 Power Delay Product

The power-delay product (PDP) is a fundamental parameter which is often used for measuring the quality and the performance of a CMOS process and gate design. PDP=P.Td

Figure 11.10 shows the waveform of the input and output signals.

The initial value of the (W/L) ratio was 120; now we have doubled the value of (W/L) to 240. Now, we will observe the Transient, DC, and Power response with the (W/L) ratio of the transistor as 240.

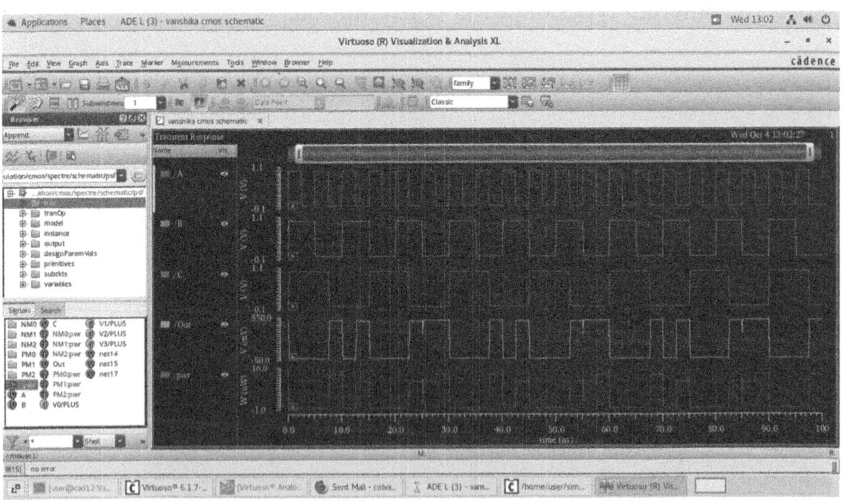

Figure 11.10 Transient response of the given Boolean function using CMOS logic.

Figure 11.11, Figure 11.12 and Figure 11.13 show the waveforms of the input and output signals with 2x(W/L) ratio for CMOS logic, pseudo-NMOS NMOS and dynamic logic circuits, respectively.

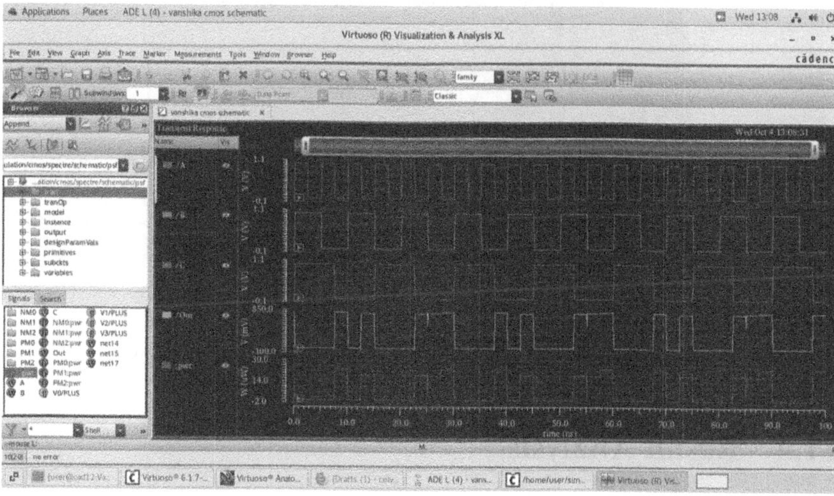

Figure 11.11 Transient response of Boolean function using CMOS logic.

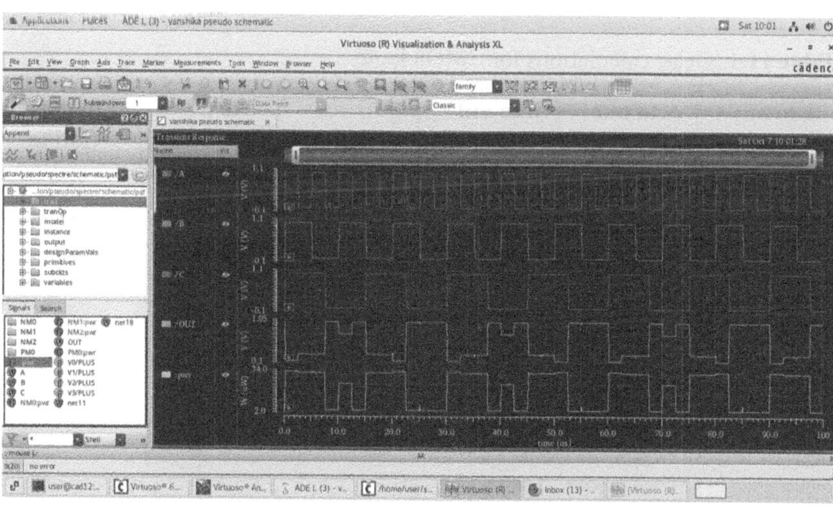

Figure 11.12 Transient response of Boolean function using pseudo-NMOS.

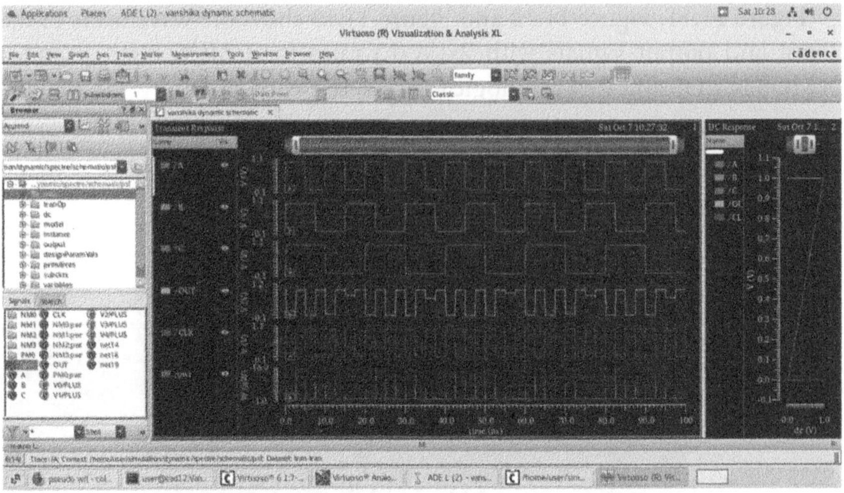

Figure 11.13 Transient response of Boolean function using dynamic CMOS.

11.4 Result Analysis

Table 11.1 shows the delay calculations for CMOS logic circuit. Table 11.2 shows the value of average power consumption in CMOS logic circuit for the specific Boolean function.

Table 11.1 Delay calculations for CMOS logic.

A to out (rise to fall)	A to out (fall to rise)	B to out (rise to fall)	B to out (fall to rise)	C to out (rise to fall)	C to out (fall to rise)
42.29 E-12	5.088 E-9	17.29 E-12	2.563 E-9	-7.710 E-12	37.96 E-12

Table 11.2 Average power consumption.

Average power consumption
4.685 $\times 10^{-6}$

Tables 11.3 and 11.4 shows the delay calculation and average power consumption with 2x(W/L) ratio for the same circuit using CMOS logic style. Table 11.5 and Table 11.6 show the delay and power calculations of a similar circuit using pseudo-logic style. Table 11.7 and Table 11.8 show the delay and power calculations for dynamic logic style. An overall comparison of power consumption for all logic styles is shown in Table 11.9.

Table 11.3 Delay calculations for CMOS logic for 2(W/L) ratio.

A to out (rise to fall)	A to out (fall to rise)	B to out (rise to fall)	B to out (fall to rise)	C to out (rise to fall)	C to out (fall to rise)
47.79 E-12	5.089 E-9	17.79 E-12	2.564 E-9	-7.212 E-12	38.83 E-12

Table 11.4 Average power CMOS logic for 2*(W/L) ratio.

Average power consumption
8.339 x10^{-6}

Table 11.5 Pseudo logic style delay calculations.

W/L ratio	A to out (rise to fall)	A to out (fall to rise)	B to out (rise to fall)	B to out (fall to rise)	C to out (rise to fall)	C to out (fall to rise)
W/L	82.12 E-12	4.995 E-9	61.12 E-12	2.470 E-9	36.12 E-12	-55.05 E-12
2W/L	84.64 E-12	4.996 E-9	59.64 E-12	2.471 E-9	34.64 E-12	-54.07 E-12

Table 11.6 Pseudo logic style average power.

W/L ratio	Average power consumption
W/L	18.54 x10^{-6}
2W/L	34.37 x10^{-6}

Table 11.7 Dynamic CMOS logic style delay calculations.

W/L ratio	A to out (rise to fall)	A to out (fall to rise)	B to out (rise to fall)	B to out (fall to rise)	C to out (rise to fall)	C to out (fall to rise)
W/L	66.24 E-12	-2.504 E-9	41.24 E-12	-5.029 E-9	16.24 E-12	-7.554 E-9
2W/L	67.42 E-12	-2.504 E-9	42.42 E-12	-5.029 E-9	17.42 E-12	-7.554 E-9

Table 11.8 Dynamic CMOS logic style average power.

W/L ratio	Average power consumption
W/L	57.10 x10^{-9}
2W/L	109.0 x10^{-9}

Table 11.9 Pseudo logic style comparative power performance.

Increase in consumption of average power (uW) with the increase in W/L ratio	
CMOS Logic style	1.77993597
Pseudo Logic style	1.85382956
Dynamic logic style	1.9089317

From the above table, we can say that in Dynamic logic maximum change in power consumption is observed. Hence, the choice between CMOS, pseudo, and dynamic logic styles depends on factors like power consumption, delay, area constraints, and noise immunity. CMOS is a robust and versatile choice suitable for many applications, while pseudo and dynamic logic styles can offer advantages in specific situations where optimizing for speed or area is more critical. Engineers must carefully evaluate the trade-offs and requirements of their particular design to make an informed choice.

11.5 Advantages and Disadvantages

Using CMOS logic style, pseudo logic style, and dynamic logic style to implement a Boolean expression each has its advantages and disadvantages.

The choice of which style to use depends on the specific requirements of the application and the trade-offs that are acceptable. Here's a brief conclusion for each of these logic styles:

Table 11.10, Table 11.11 and Table 11.12 show the advantages and disadvantages of CMOS, pseuodo-NMOS and dynamic logic circuits.

Table 11.10 Comparative table of CMOS logic.

Advantages	Disadvantages
Low static power consumption: CMOS logic consumes minimal power when not switching.	High dynamic power consumption: CMOS logic consumes more power when transitioning between logic levels.
Excellent noise margins: It provides a high noise immunity, making it robust against noise and voltage variations.	Larger transistor count: Implementing complex Boolean expressions can require more transistors, leading to larger chip area.
High fan-out capability: CMOS gates can drive multiple loads without significant degradation of signal quality.	Slower operation: CMOS gates typically have higher propagation delays compared to other logic styles.

Table 11.11 Comparative table of pseudo logic style.

Advantage	Disadvantages
Reduced transistor count: Pseudo logic can sometimes implement Boolean expressions with fewer transistors than CMOS.	Higher static power consumption: It may not offer as low static power consumption as CMOS.
Faster operation: Pseudo logic can have lower propagation delays due to simplified transistor configurations.	Limited noise margin: Pseudo logic can be more susceptible to noise and voltage fluctuations compared to CMOS.
Simplified Design: Pseudo logic can offer a simpler and more intuitive design approach for specific logic functions, making it easier to understand and implement certain logic operations efficiently.	Limited fan-out: Pseudo logic gates may not be as good at driving multiple loads.

Table 11.12 Comparative table of dynamic logic style.

Advantage	**Disadvantage**
• Low transistor count: Dynamic logic can often implement Boolean expressions with a smaller number of transistors than static logic.	• High dynamic power consumption: Dynamic logic consumes a significant amount of power during transitions.
• Fast operation: Dynamic logic circuits can operate faster than static logic circuits because they have less parasitic capacitance and resistance. This is because dynamic logic circuits have fewer transistors and wires.	• Limited noise margin: It can be sensitive to noise and voltage variations, requiring careful design.
• Low power consumption: Dynamic logic circuits have lower static power consumption than static logic circuits because they do not have any direct paths between the power supply and ground.	• Complex clocking and signal restoration circuits: Dynamic logic often requires additional circuitry to manage clock signals and restore logic levels.

11.6 Conclusion

This chapter describes the comparative performance for a Boolean function with CMOS, Pseudo-NMOS and dynamic CMOS logic style. With the increase in (W/L) ratio in CMOS technology shows that a negligible increase of delay is observed and power consumption is increased, i.e., the average power consumption of the circuit is increased by a factor of 1.77993597, i.e., from 4.685×10^{-6} to 8.339×10^{-6}. In the case of Pseudo logic style, there is a variable response of delay with increase in (W/L) ratio average power consumption is increased by a factor of 1.85382956, i.e., from 18.54×10^{-6} to 34.37×10^{-6}. A negligible increase in delay can be observed with an increase in (W/L) ratio. The average power consumption is increased by a factor of 1.9089, i.e., from 57.10×10^{-9} to 109.0×10^{-9} respectively.

AI disclosure statement: The author confirms that no artificial intelligence (AI)-generated content or AI-assisted writing tools were used in the preparation of this chapter (including summaries and the abstract). All text, analysis, and conclusions are the result of the author's original work.

References

1. Kang, S.-M. and Leblebici, Y., *CMOS Digital Integrated Circuits*, Tata McGraw-Hill.
2. Mendiratta, N. and Tripathi, S.L., 18nm n-channel and p-channel Dopingless Asymmetrical Junctionless DG-MOSFET: Low Power CMOS Based Digital and Memory Applications. *Silicon*, 14, 6435–6446.
3. Gupta, P., Tripathi, S.L., Mishra, S., A novel Low leakage current and Power Consideration of 2 to 4 decoder. *IJAER*, 7, 2012. Online ISSN 1087-1090.
4. Pindoo, I.A., Dhariwal, S., Sharma, R., Lata, S., Speed Enhancement in the Performance of Two Phase Clocked Adiabatic Static CMOS Logic Circuits, in: *2018 International Conference on Intelligent Circuits and Systems (ICICS)*, pp. 149–154, 2018, doi: 10.1109/ICICS.2018.00040.
5. Kumar Singh, A. and Lata Tripathi, S., *Power Analysis to Ensure Secure CMOS Architecture*, Nova Publishers, Feb. 2020, 978-1-53616-556-2.
6. Tripathi, S.L., Kumar, A., Rao, K.S., Mudimela, P.R., *Nanodevices for Integrated Circuit Design*, Wiley-Scrivener Publishing LLC, USA, October 2023, DOI:10.1002/9781394186396.
7. Kagaris, D. and Haniotakis, T., A Methodology for Transistor-Efficient Supergate Design. *IEEE Trans. VLSI Systems New York USA*, 15, 4, 488–492, Apr. 2007.
8. Da Rosa Junior, L.S., Automatic Generation and Evaluation of Transistor Networks in Different Logic Styles, in: *Tese (Programa de Pós-Graduação em Microeletrônica) – Instituto de Informática*, p. 147 f, UFRGS, Porto Alegre, 2008.
9. Xue, J., Al-Khalili, D., Rozon, C.N., Technology mapping in library-free logic synthesis, in: *Proceedings of SPIE*, 2005.
10. Weste, N.H.E. and Harris, D., *CMOS VLSI Design: A Circuits and Systems Perspective*, 3rd, Pearson/Addison Wesley, Boston, 2005.
11. Thorp, T.J., Yee, G.S., Sechen, C.M., Design and synthesis of dynamic circuits. *IEEE Trans. VLSI Systems New Y.*, 11, 1, 141–149, Feb. 2003.
12. Lata Tripathi, S., Ahmad Alvi, P., Subramaniam, U. (Eds.), *Electrical and Electronic Devices, Circuits and Materials: Technological challenges and Solutions*, Wiley-Scrivener Publishing, USA, 2021, ISBN: 1119750369, 978119750369, DOI:10.1002/9781119755104 (Scopus).
13. Lata Tripathi, S., Padmanabhan, S., Kumar, D., Raja, P. (Eds.), *Design and Development of Energy Efficient Systems*, Wiley-Scrivener Publishing, USA, 2021, ISBN: 1119761638, 9781119761631 DOI:10.1002/9781119761785 (Scopus).
14. Lata Tripathi, S. and Dwivedi, S. (Eds.), *Electronic Devices and Circuit Design: Challenges and Applications in the Internet of Things*, 1st, Apple Academic Press, Boca Raton, 2022.

15. Tripathi, S.L., Saxena, S., Mohapatra, S.K. (Eds.), *Advanced VLSI Design and Testability Issues*, 1st, CRC Press, Boca Raton, 2020, https://doi.org/10.1201/9781003083436.
16. Lata Tripathi, S. and Kumar Sinha, S. (Eds.), *Recent Advancement in Electronic Devices, Circuit and Materials*, Nova Science Publishers, New York, 2020.
17. Lata Tripathi, S., Saxena, S., Kumar Sinha, S., Singh Patel, G., *Digital VLSI Design Problems and Solution with Verilog*, John Wiley & Sons, Ltd, USA, 2021, DOI:10.1002/9781119778097 (Scopus).
18. Lata Tripathi, S., Kumar, A., Pathak, J., *Programming and GUI Fundamentals: TCL-TK for Electronic Design Automation (EDA)*, pp. 2021–22, Wiley-IEEE, USA.
19. Lata Tripathi Kolla Prakash, S., Balas, V., Mohapatra, S., Nayak, J. (Eds.), *Electronic Device and Circuit Design Challenges for Biomedical Applications*, Elsevier, ISBN: 9780323851725, https://doi.org/10.1016/C2020-0-02289-9 (Scopus).
20. Tripathi, S.L., Alvi, P.A., Subramaniam, U., *Electrical and Electronic Devices, Circuits and Materials: Design and Applications*, 1st, CRC Press, Boca Raton, 2021, https://doi.org/10.1201/9781003097723.
21. Arora, S. and Lata Tripath, S., A comprehensive analysis of ultra-low power GNRFET based 20T hybrid full adder for computing applications. *Phys. Scr.*, 99, 8, 085008, 2024. DOI 10.1088/1402-4896/ad5c0b.
22. Kumar, A., Lata Tripathi, S., Srinivasa Rao, K., *Machine Learning for VLSI Chip Design*, Wiley-Scrivener Publishing LLC, USA, 2023, DOI:10.1002/9781119910497.

12

SAFETRACK: A Next-Generation Vehicle Tracking and Safety System Using Arduino and GPS

Manoj Kumar Yadav[1]*, Dinesh Singh[2] and Jhulan Kumar[1]

[1]*School of Electronics and Electrical Engineering, Lovely Professional University, Phagwara, India*
[2]*Department of Electronics & Communication Engineering, Dr. B. R. Ambedkar National Institute of Technology, Jalandhar, Punjab, India*

Abstract
Road accidents rates, especially for two-wheelers, are very high nowadays. Timely medical aid can help in saving lives. There are various systems and technologies being developed to detect accidents and alert nearby medical centers to provide immediate medical aid. These systems use accelerometers and heartbeat sensors to detect accidents. By utilizing the accelerometer and heartbeat sensor, the system can detect the severity of the accident and automatically send alerts to nearby medical centers and friends. An accelerometer can detect the sudden change in the axes of a vehicle, such as a sudden impact or collision, and trigger an alert. The GSM module can then send an alert message to the user's mobile phone, including the location of the accident.

Keywords: Accident detection, Arduino Nano, GPS, GSM, accelerometer sensor

12.1 Introduction

The high demand for automobiles has led to an increase in traffic hazards and road accidents, putting people's lives at risk. This is compounded by the lack of adequate emergency facilities available. A report published by the Ministry of Road Transport and Highways that on average, six accidents

**Corresponding author*: mymanoj347@gmail.com

Abhishek Kumar and Suman Lata Tripathi (eds.) Smart Chips for Smart Devices: VLSI Design for Next-Generation IoT Solutions, (233–248) © 2026 Scrivener Publishing LLC

occur for every 10,000 vehicles in a year. In 2022, this was a 3.53% growth in accidents due to a rapid increase in the number of vehicles on the road. However, technology can help improve safety, such as automatic accident detection and messaging systems that use accelerometers to detect accidents and enhance security. An accident detection and alert system can be used to send a message to registered mobile phones, hospitals, and police stations in less time when such accidents happen in a particular location in a city [1–3]. Arduino is a very popular and easy access open-source electronics platform which is used to sense any input signal from a sensor and according to programme upload to the control operation of the actuator. It is often used in projects that involve transferring messages between different devices in a system. An accelerometer sensor can be used to detect an accident in a vehicle, and the information can be transferred to a registered number through a GSM module [4–7]. The system can then notify nearby medical centres and registered mobile numbers about the location of the accident using GPS and GSM modules. The GPS system can be used to determine the location of an accident. By integrating GPS and GSM modules, the proposed system can detect when an accident has occurred and notify nearby medical centers and registered mobile numbers about the accident's location. This system will guide emergency responders to reach on accident site quickly and make assistance available to victims [8–10]. Through a tracking system that covers the geographic coordinates over the area, the accident's position can be sent. When a vibration or impact is detected, the system can use the GPS module to determine the location of the accident and send notifications to emergency services and registered mobile numbers [11–15]. This can help ensure that help arrives quickly and that those involved in the accident receive timely medical assistance.

12.1.1 Problem Statement

One of the primary causes of the increase in traffic congestion, which in turn causes an increase in the number of accidents, is the growing number of vehicles on the road. The loss of life and injuries resulting from accidents can be reduced by ensuring prompt medical attention to the victims. The current manual system for reporting accidents and calling for emergency services can have its limitations, as it relies on the availability and responsiveness of nearby individuals.

12.1.2 Proposed System

This paper's major goal is to create a mobile application that uses a smartphone's accelerometer and GPS to detect collisions. This will be achieved through the use of a sensor fusion-based algorithm that can detect sudden external disturbances in speed, which may indicate a collision. The accelerometer sensor in a mobile phone can detect sudden changes in acceleration or movement, which can indicate a collision or other type of accident. If no emergency is actually present, they can cancel the request for help within a specified time limit, such as 10 seconds. However, if the user clicks the "Call Help" button or leaves the alert message ignored for longer than the predetermined amount of time—for example, 10 seconds—the program will automatically send an assistance request to emergency services, the user's chosen contacts, and family members.

12.1.3 Existing System

This concept was first presented at the beginning of the current era of mobile phones. The evolution of GPS-based security applications started from the integration of GPS sensors in mobile phones. With the integration of GPS sensors in mobile phones, it became possible to develop location-based security applications that could use the GPS data to track the location of the mobile phone and the user. However, there is a drawback to spending more money on additional gear [5–7]. But this drawback has been significantly reduced in the previous 10 years due to the rapid advancement of mobile phones. The present application of the paper may be limited to a few countries or regions, but it has the potential to be expanded and implemented in other areas as well. By providing information to both relatives and friends, as well as emergency services, the efficiency and effectiveness of the application can be greatly increased.

12.2 System Architecture

The system as shown in Figures 12.1 to 12.3 is an accident detection and emergency response system that uses GPS technology to locate the user in case of an emergency. The accident detection module detects an external disturbance and triggers the Location Detection Module to find the current location of the user using GPS data.

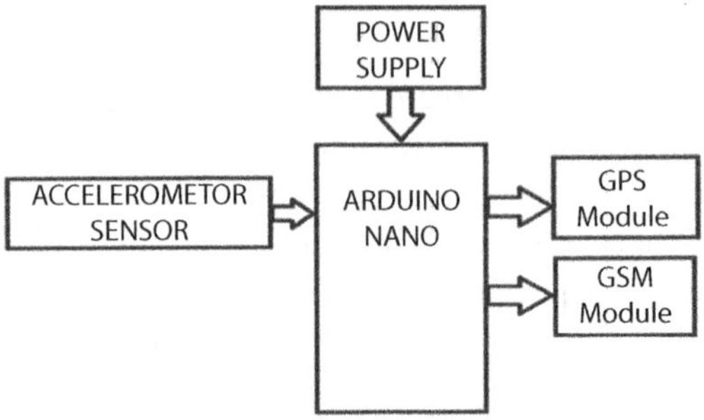

Figure 12.1 Block diagram.

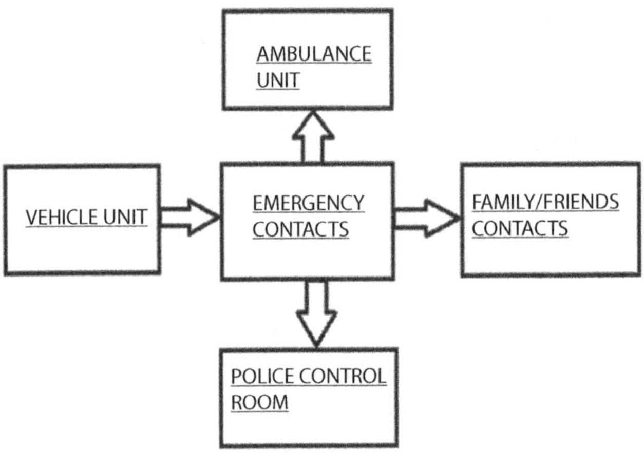

Figure 12.2 Simplifying vehicle unit servicing.

Once the location data is obtained, it is sent to the emergency services to request help. This system is designed to provide quick and efficient emergency response and help save lives in case of accidents or other emergencies. The system is a vehicle tracking and emergency response system that uses an accelerometer to keep track of the vehicle's position. The accelerometer continuously informs the microcontroller about the vehicle's coordinates. If the vehicle is found to be in a random location, the GPS location tracker is activated to track the vehicle's exact location. Once the GPS location is determined, the system uses a GSM SIM module to inform the emergency number about the location of the vehicle. The system provides the emergency number with the latitude, longitude, and Google Maps position of the vehicle to enable a quick and efficient response to any emergency situation. This system can help improve the safety and security of vehicles and their passengers.

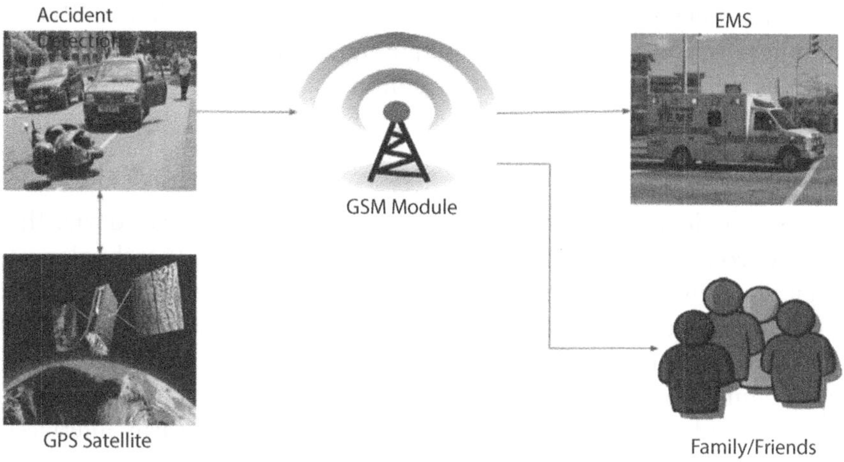

Figure 12.3 Hardware implementation of safety system.

12.2.1 Arduino

The Arduino Nano as shown in Figure 12.4 features the same pinout and connectivity choices as the Arduino UNO and is powered by the same microcontroller, the ATmega328P or (older version) ATmega168. The Atmel ATmega328P microcontroller, which has a clock speed of 16 MHz and 32KB of flash memory, forms the foundation of the Nano board.

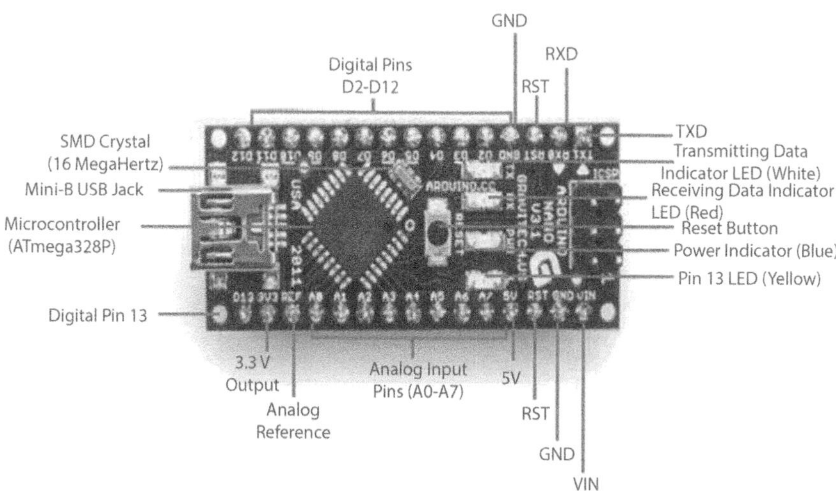

Figure 12.4 Arduino Nano.

It includes eight analog input pins, six pulse-width modulation (PWM) output pins, and 14 digital input/output pins. It also features a USB interface for serial connection and programming.

12.2.2 GSM Module

A GSM modem or GSM module, as shown in Figure 12.5, is a device that enables communication with a mobile network using GSM technology. It typically consists of a hardware module that includes a cellular radio, a SIM card slot as well as software that allows it to communicate with other devices using various protocols. The SIM800L can be easily interfaced with an Arduino board for various applications that require mobile communication and internet connectivity. The SIM800L can be easily interfaced with an Arduino board for various applications that require mobile communication and internet connectivity.

12.2.3 GPS Module

The Global Positioning System, or GPS for short, is a satellite-based navigation system that may be used to find location and time information worldwide. GPS technology has many applications, from navigation and mapping to aviation and emergency services. GPS technology has a wide range of applications across many industries and fields. Here the GPS used is NEO-6M. The NEO-6M is a popular GPS module that has been widely used in various applications which has been shown in Figure 12.6. Due to its great sensitivity, the NEO-6M GPS module can precisely locate itself even in difficult-to-reach environments and receive weak GPS signals. Because of its extremely low power consumption, the module is perfect

Figure 12.5 SIM800L.

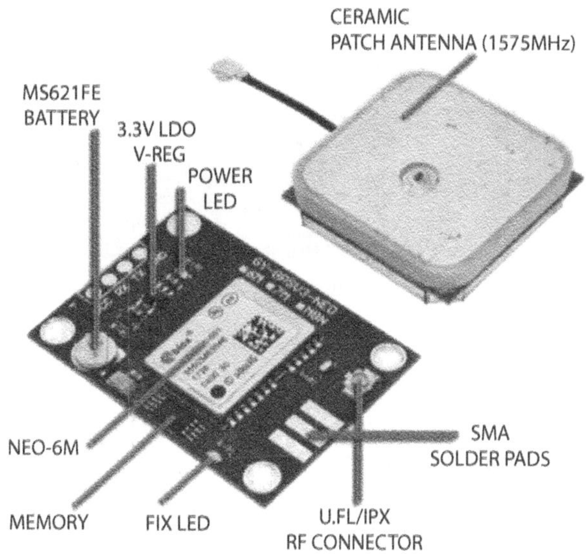

Figure 12.6 NEO-6M.

for battery-powered applications, using GPS and GSM modules in tandem to track and send a vehicle's location. The GPS module will receive signals from satellites and determine the coordinates of the vehicle's location. The coordinates can then be sent to an Arduino board which processes the data and sends it to a GSM module for transmission to a designated contact or server. The GSM module uses cellular network signals to send the location data as an SMS or through a data connection.

12.2.4 LM2596 Step Converter

A step-down voltage regulator module, the LM2596, shown in Figure 12.7, changes a greater input voltage to a lower output value. It is commonly used in electronic circuits and projects where a stable and adjustable DC voltage supply is required. The LM2596 module can produce an output voltage between 1.25V and 35V with a maximum output current of 3A. Typically, it operates with an input voltage range of 3.2V to 40V. The output voltage can be adjusted by turning a potentiometer on the module. However, the tracking device may require a lower voltage supply to power its various components such as the GPS module, GSM module, and microcontroller.

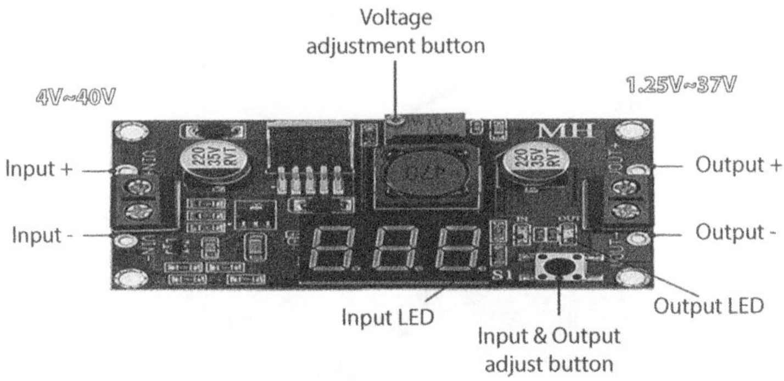

Figure 12.7 LM2596.

12.2.5 ADXL-345 Sensor

ADXL-345, as shown in Figure 12.8, is a small, low-power, 3-axis accelerometer sensor produced by Analog Devices Inc. It is a highly integrated device that can measure acceleration up to ±16g on each of the three axes. The sensor outputs digital data in 16-bit twos complement format, which can be accessed *via* an I2C digital interface or an SPI digital interface (3- or 4-wire). The ADXL-345 is compatible for mobile device use due to its small size, low power consumption, and high accuracy. The sensor can measure both static and dynamic acceleration, making it useful for a wide range of uses, comply tilt sensing, motion detection, and shock sensing. Its high sensitivity of 3.9 mg/LSB enables particular measurement of leaning deviations less than 1.0°, making it ideal for uses that need high accuracy.

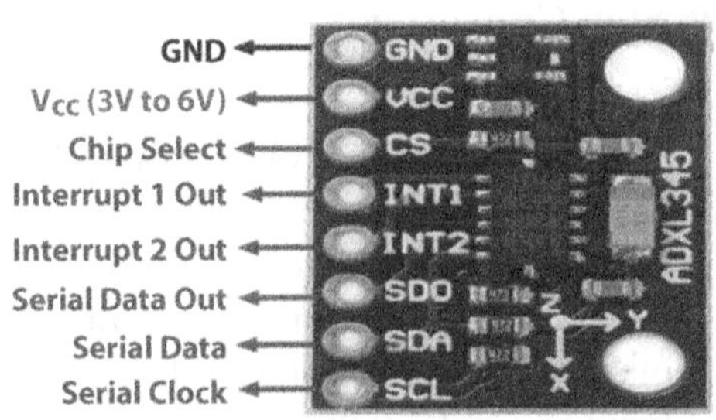

Figure 12.8 ADXL-345.

12.2.6 Flow Diagram

Figure 12.9 shows in a flowchart the basic working flow of an accident detection system.

1. The system starts when the sensors are activated.
2. The sensors detect whether an accident has occurred or not.
3. If an accident is detected, the GPS system is activated to find the exact location of the vehicle.
4. Once the location is determined, the emergency services are notified of the accident and its location.
5. The system ends once the emergency services have been notified.

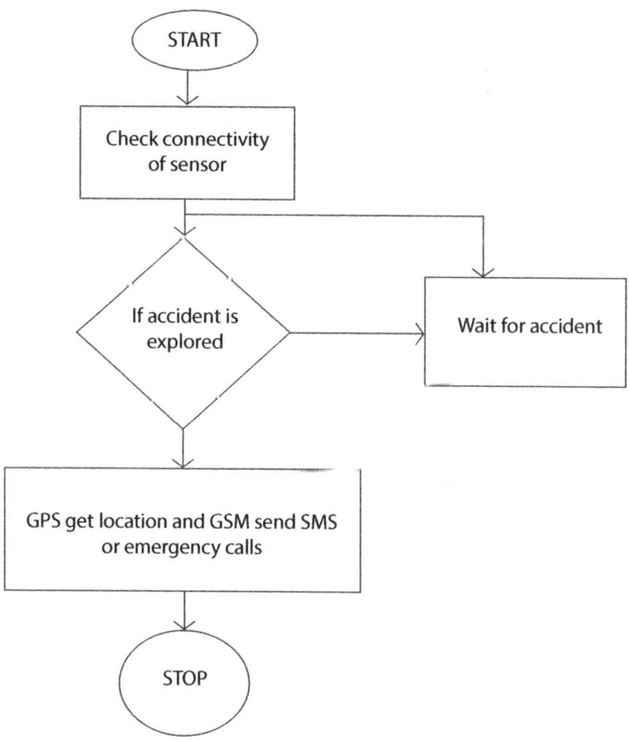

Figure 12.9 Flow chart.

12.3 Methodology

The accident alert system is designed to detect accidents and notify emergency contacts. A circuit diagram and PCB diagram of the system has been shown in Figure 12.10 and Figure 12.11 and its working model has been shown in Figure 12.12. The system uses various components to detect accidents, including the Neo 6M GPS, GSM 800L, Arduino Nano, LM2596 step converter, ADXL 335, and 12V 2A power supply. The Neo 6M GPS is a popular module used to determine the location of the accident. The

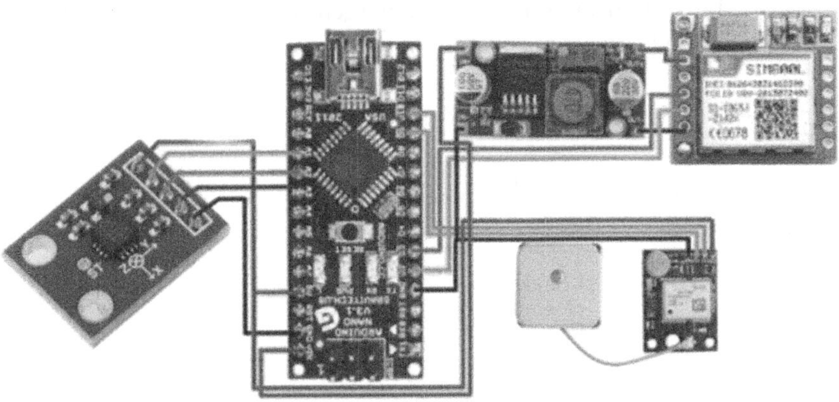

Figure 12.10 Circuit diagram.

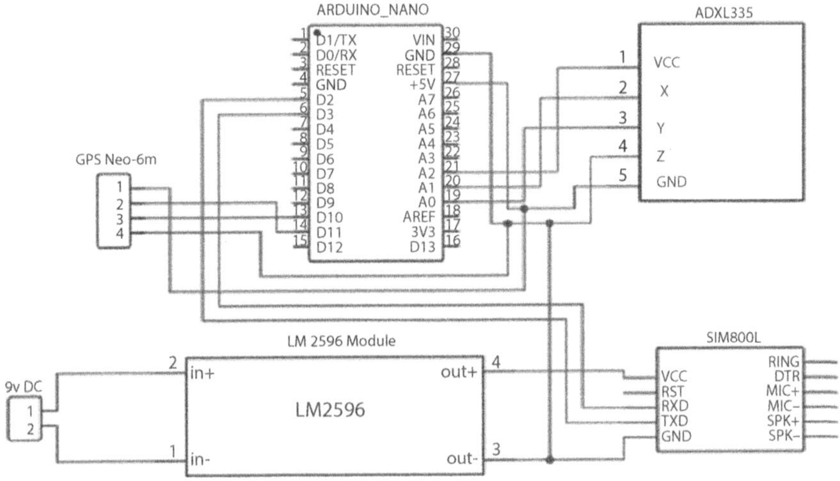

Figure 12.11 PCB diagram.

Figure 12.12 Working model.

module provides accurate GPS positioning data, which is essential to alert emergency services and family members about the accident. The GPS module works by receiving signals from GPS satellites and using triangulation to calculate the location of the device.

In the event of an emergency, the GSM 800L module is utilized to call or send SMS messages to programmed emergency contacts. The GSM module functions by establishing a network connection and using it to send and receive phone calls and messages. The module is a well-liked option for microcontroller-based applications because it is inexpensive and simple to utilize. The system's central processor unit is the Arduino Nano microcontroller. It is in charge of analyzing the sensor data and sending out notifications in the event that an accident is discovered. Programming the Arduino Nano is simple, and it can be used for a lot of things. The Arduino IDE is a free software development environment that is open-source that may be used to program microcontrollers. The 12V/2A power supply's voltage is controlled by the LM2596 step converter. The step converter makes sure the system has a steady and dependable power supply, which is necessary for the system to operate correctly. When an input voltage is delivered into the system, the step converter transforms it into a controlled output voltage. The ADXL 335 accelerometer is used to detect sudden changes in acceleration or impact force that may indicate an accident has occurred. The accelerometer is a popular and low-cost sensor that can be used in a variety of applications. The accelerometer works by measuring the acceleration of the device in three axes, x, y, and z. The data from the accelerometer is processed by the Arduino Nano to determine if an accident has occurred.

The system works by acquiring data from various sensors, including the Neo 6M GPS and ADXL 335 accelerometer. The GPS module provides accurate location data, while the accelerometer detects sudden changes in acceleration or impact force that may indicate an accident has occurred. The data from these sensors is processed by the Arduino Nano, which compares the accelerometer data to predefined thresholds or uses machine learning algorithms to detect patterns in the data. If an accident is detected, the system generates an alert using the GSM 800L module. The GSM module is used to send SMS messages or make phone calls to preconfigured emergency contacts, such as emergency services or family members. 12V and 2A power supply is provided to the system by which it is regulated by the LM2596 step converter to ensure that the system receives a stable and reliable power supply. The system may also include a user interface, such as a physical button or switch, which can be used to activate or deactivate the system.

12.4 Results and Discussion

The overall result of the project is an application that helps people who are involved in accidents and unable to ask for help. The application uses the sensors available at low cost, such as the Neo 6M GPS and ADXL 335 accelerometer, along with the Arduino Nano microcontroller and GSM 800L module, to detect accidents and notify emergency services and family

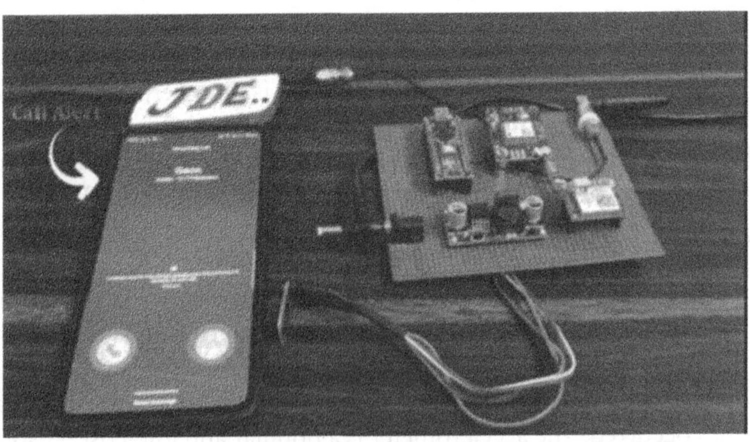

Figure 12.13 ADXL-335 sensor detected Rafe movement and sends call to a particular mobile number.

members about the same. When this program is used, a help request is automatically submitted together with the accident's location, assisting emergency personnel in providing assistance as soon as is feasible. This can be crucial in situations where time is of the essence and immediate medical attention is required. The chapter demonstrates technology can be used to improve safety and save lives. By using low-cost sensors and micro-controllers, the project shows that it is possible to create effective solutions that are accessible to a wider audience. With further development and improvements, such applications can be used in a variety of settings to improve safety and provide timely assistance in the event of an accident

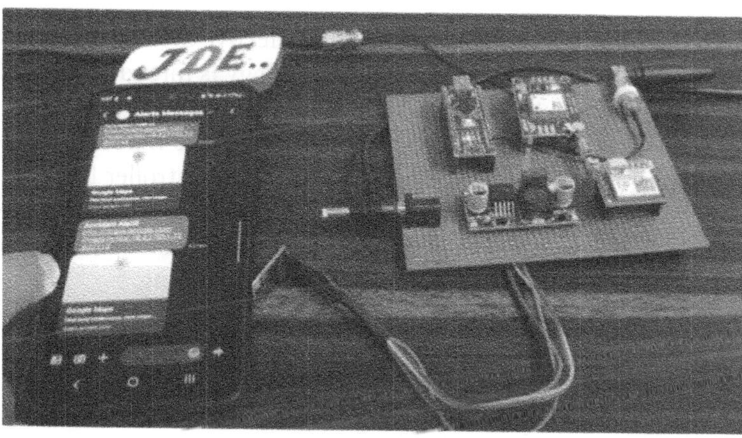

Figure 12.14 ADXL-335 sensor text message sent to a specific mobile number.

Figure 12.15 Real-time victims' locations on Google Maps.

or emergency. In Figure 12.13 the ADXL-335 sensor is showing detection of Rafe movement. Then GSM gets to call a particular mobile number.

In the Figure 12.13, ADXL-335 Sensor identified Rafe movement. Then GSM, if there is any sort of trouble or network issue or to disconnect call for any reason, sends a text message to a specific mobile number (Figure 12.14). The link will activate and open when you click on the text message. Utilizing GPS, victims' locations can be tracked in real time on Google Maps as shown in Figure 12.15.

12.5 Conclusion

The accident alert system is a reliable and effective system for detecting accidents and alerting authorities. The technology is intended to use GPS to precisely locate an accident and use an accelerometer to detect it. The system then sends an alert message to a stated phone number using the GSM module. The use of an Arduino Nano microcontroller board and a step-down voltage converter ensures that the system is compact and efficient. Overall, this system can help improve the safety of drivers and passengers on the road and prevent accidents from escalating.

AI disclosure statement: The author confirms that no artificial intelligence (AI)-generated content or AI-assisted writing tools were used in the preparation of this chapter (including summaries and the abstract). All text, analysis, and conclusions are the result of the author's original work.

References

1. Varma, S.K.C., Poornesh, Varma, T., Harsha, Automatic vehicle accident detection and messaging system using GPS and GSM modems. *Int. J. Sci. Eng. Res.*, 4, 8, August 2013.
2. Bari, A.S., Falalu, M.A., Umar, M.A., Sulaiman, Y.Y., Gamble, A.M., Baballe., M.A., Accident Detection and Alerting Systems: A Review. *Global J. Res. Eng. Comput. Sci.*, 2, 4, 24–29, 2022.
3. Doshi, A., Shah, B., Kamdar, J., Accident_Detection_And_Alert_System. *Int. J. All Res. Educ. Sci. Methods*, 9, 11, November 2021.
4. Rajendra Prasad, M. and Aswani, P., An automated traffic accident detection and alarm device. *Int. J. Technol. Explor. Learn. (IJTEL)*, 1, 1, August 2012.
5. Moumen, I., Rafalia, N., Abouchabaka, J., Aoufi, M., Real-time GPS Tracking System for IoT-Enabled Connected Vehicles, in: *E3S Web of Conferences*, 2023.

6. Kumar Yadav, M. and Kumar Gupta, S., FeAl/MgO/FeAl MTJ with enhanced TMR and low resistance area product for MRAM: A first principle study, in: *Micro and Nanostructures*, vol. 165, p. 207192, Elsevier, 2022.
7. Kumar Yadav, M. and Kumar Gupta, S., First Principle Study of Spin Tunneling Current under Field Effect in Magnetic Tunnel Junction for Possible Application in STT-RAM. *IEEE Trans. Electron Devices*, 69, 9, 4894–4899, September 2022.
8. Trebuňa, P., Mizerák, M., Pekarčíková, M., Kliment, M., Matiscsák, M., Tracking of Trucks Using the GPS System for the Purpose of Logistics Analysis, in: *Design, Simulation, Manufacturing: The Innovation Exchange*, pp. 162–171, Springer, 2023.
9. Idriss, M., Jaafar, A., Najat, R., Adaptive traffic lights based on traffic flow prediction using machine learning models. *Int. J. Electr. Comput. Eng. (IJECE)*, 2023.
10. Kumar Yadav, M., Kumar Gupta, S., Rai, S., C. Pandey, A., Al embedded MgO barrier MTJ: A first principle study for application in fast and compact STT-MRAMs. *Supperlattice Microstrucutre* (Elsevier), IF-2.685 103, 314, March 2017.
11. Poonam, Srivastava, A.K., Yadav, M.K., Kripal, R., Singh, V., Bok Lee, D., Lee, J.-H., Synthesis and Dielectric of characterization of Polycarbonate/Multi-wall carbon nanotube nanocomposite. *Arabian J. Chem.*, 12, 440–446, March 2019.
12. Kumar Yadav, M. and Kumar Gupta, S., A comparative first principles study of quantum well states in MgO barrier MTJs for STT-RAMs. *Microelectron. J.* (Elsevier), IF-1.685, 105, 104909, 14 September 2020.
13. Wirawan, H. and Suharjito, S., Multi Depot Vehicle Routing Problem and Geographical Information System Integration: Retail Stores Case Study, in: *E3S Web of Conferences*, vol. 388, EDP Sciences, 2023.
14. Zhian, L. and Han, H., A bus management system based on ZigBee and GSM/GPRS, in: *2010 International Conference on Computer Application and System Modeling (ICCASM 2010)*, vol. 7, IEEE, pp. V7-210-V7-213, 2010.
15. Vigneshwaran, S., Nithya, B., Raghul, K., Nivas, B., Kishore, V., Design of bus tracking and fuel monitoring system, in: *2020 6th International Conference on Advanced Computing and Communication Systems (ICACCS)*, IEEE, pp. 348–351.

13

The Role of VLSI in Next-Generation IoT Systems

S. Deepa*, S. Pavan Kumar, V. Arun and J. Lalidesh

Karpagam College of Engineering, Coimbatore, India

Abstract

The exponential growth of the Internet of Things (IoT) has been creating tons of real-time data, which makes it need equally efficient and scalable processing solutions. Very Large Scale Integration (VLSI) technology plays a pertinent role in improving the performance, power efficiency, and reliability of IoT data processing. The VLSI technology helps in integrating millions to billions of transistors onto a single chip to pack the design into a compact architecture that allows a high-speed computing system for IoT applications. This chapter discusses VLSI technology in IoT networks, especially how it contributes to edge computing, power-aware design, and hardware acceleration. Several important VLSI architectures are covered, such as neuromorphic computing, FPGA, ASIC, and System-on-Chip (SoC). The chapter also examines some of the security risks associated with VLSI-based Internet of Things systems, including cryptographic implementations, side-channel attacks, and hardware trojans. We also talked about certain hardware accelerators like GPUs, DSPs, and TPUs that enhance real-time AI-driven processing of IoT tasks. Promising technologies like 3D Integrated Circuits (3D ICs) and quantum computing are believed to transform the future of data processing in Internet of Things systems by boosting energy efficiency, decreasing latency, and improving data throughput. By exploiting advances in VLSI, researchers and engineers can develop IoT solutions that support the requirements for real-time analytics, intelligent decision-making, and secure communication. In this chapter, we look into the changing regime of VLSI changing IoT networks, emphasizing the major designs and future trajectories guiding the coming generations.

**Corresponding author*: deepa.s@kce.ac.in
S. Pavan Kumar: pavankumar26540@gmail.com
V. Arun: arunsmart701@gmail.com
J. Lalidesh: lalidheshj24@gmail.com

Abhishek Kumar and Suman Lata Tripathi (eds.) Smart Chips for Smart Devices: VLSI Design for Next-Generation IoT Solutions, (249–268) © 2026 Scrivener Publishing LLC

Keywords: IoT, VLSI, data processing, edge computing, hardware acceleration, ASIC, intelligent decision-making

13.1 Introduction

The fast growth of the Internet of Things has created a huge amount of data that needs effective processing systems to handle it properly. Several types of connected devices control IoT networks through sensors, actuators, and edge nodes that send real-time data continuously. To make these applications work right now, we need constant data processing. Through VLSI technology, chipmakers solve big data processing issues faster by using less power and combining multiple systems onto one device. VLSI gives rise to tiny computers that pack thousands or even millions of transistors on one small chip, thereby making powerful components like the CPU, DSP, and hardware accelerators possible [1–3]. VLSI technology gives IoT devices extreme performance at a minimum power. Current computer systems based on conventional methods face excessive delays and power consumption and therefore are unable to meet the demands of fast and immediate applications. The custom-made chips (ASICs and FPGAs) speed up the devices by computing in ways that standard computers can hardly do.

Figure 13.1 Role of VLSI in IoT.

Current integrated circuit design methodologies (VLSI) enable the design of edge computing systems, which process data near its origin without full dependence on remote servers. Using this approach makes our systems run more quickly, keeps data safe, and makes it easier to transmit data without lagging. Edge computers merge AI calculators and memory stacks in one device to instantly process and choose data at the point where it's gathered. This chapter studies electrical circuits made with tiny elements that power efficient IoT software, expand performance, and keep data protected presented in Figure 13.1. It looks at three types of hardware accelerators (GPUs, TPUs, and neuromorphic processors) and also covers two new areas: 3D integrated circuits (which integrate things on multiple levels) and quantum computing. Using VLSI lets researchers and engineers improve both the way IoT systems perform and perform better while also being more reliable [4–7].

13.2 The Role of VLSI in IoT Data Processing

Using IoT technology, all industrial companies can now collaborate to collect data in real time and automate their processes [8]. The VLSI solution solves multiple challenges created by trillions of connected things that now generate significant amounts of data. This section explains what VLSI offers for IoT data processing: fast processing, power-efficient hardware, flexible capacity, instant feedback, and enhanced security capabilities [9].

13.2.1 High-Speed Data Processing

Real-time data processing helps industrial automation systems and smart city functions to work better with autonomous vehicles and healthcare monitoring. VLSI technology has many integrated transistors that enable ultra-fast computation without causing any delays and with high throughput. High-speed data processing in IoT devices means a faster response time, enhanced system efficiency, and reduced computational bottlenecks.

13.2.1.1 Parallel Processing and Multicore Architectures

VLSI technology provides high-performance parallel processing through multicore architectures, where several processing units execute complex tasks in parallel. Such parallelism increases speed in computation by enabling several cores to share computational work on data processing.

- **Pipeline Processing:** Pipeline processing is where a computer task is subdivided into small segments, which may differ at any stage during taking actions but not necessarily in the data stream.
- **Superscalar Execution:** Superscalar execution is capable of executing several instructions concurrently by different execution units in the processor, thus increasing processing efficiency and decreasing latency.
- **Heterogeneous Multicore Systems:** Heterogeneous multicore systems employ different types of processor cores that target different workloads efficiently, such as general-purpose CPUs for main workloads and specialized accelerators, e.g., GTPs and DSPs.
- **Task Scheduling and Load Balancing:** Modern VLSI-based IoT systems consist of dynamic scheduling algorithms that effectively allocate computing tasks among more than one core, solving bottlenecks and better system utilization.

VLSI technology allows these computing-intensive tasks in the IoT system using multiple parallel or multicore processing systems in real time for high-performance analytics, real-time monitoring, and smart decision-making [10].

13.2.1.2 Specialized Computing Accelerators

The progress with, on a big scale, some greater computations is being done by dedicated computing accelerators in the boards for its connection to application IoT. For these processors, they exert great effort in focusing on specific computation types to relieve general-purpose processors of their burden.

- **Digital Signal Processors (DSPs):** Real-time feeding of signals-processing speech, radar signals, noise filtering of high-fidelity audio signals, or normal bio-signal analysis. Fourier transformations, convolutions, or spectral analysis with embedded operation in IoT applications for healthcare monitoring, industrial process automation, and predictive maintenance. Hence, this allows quicker filtering, modulation, and compression through IoT technologies that have a considerable impact on data transmission and the respective storage relationship.

- **Graphical Processing Units (GPUs):** In that period—in the late 1980s—rendering graphics speed became an incredibly parallel engine for computing. The GPU has to provide similar acceleration in IoT applications, being used for image-based recognition, computer vision, or real-time video analytics. A housing for submatrix operations must be given on the GPU, along with deep learning inference on board for smart surveillance, autonomous vehicles, and augmented reality IoT applications.
- **Tensor Processing Units (TPUs):** The TPUs are specialized hardware dedicated to efficiently accelerate machine-learning and deep-learning applications. Dedicated hardware accelerators for computational tasks were developed, as well characterized by mathematical operations underlying artificial neural networks. TPU applications include real-time AI inference, pattern recognition, or automated decision-making in applications like smart traffic management in IoT, medical diagnostics, and AI predictive analytics.
- **Neuromorphic Processors:** The Spiking Neural Networks within neuromorphic processors—operating strikingly precisely like the working of the brain—use an event-based coding scheme to perform computations across biological neurons. Low-power cognitive processors such as anomaly detection, sensory data interpretation, or other various decision-making tasks in any IoT network. Such computing devices would undoubtedly prove their performance blows robotics, speech recognition, and completely autonomous systems in IoT applications.

13.2.2 Power Efficiency in IoT Devices

Power efficiency comes into view during the VLSI designs since that aspect introduces uniqueness to the devices—be they battery operated or powered through energy-harvesting from the natural resources. Performance enhancement got coupled with approaches tackling steps toward reducing energy consumption.

13.2.2.1 Low-Power Circuit Design Strategies

Low-power VLSI circuit design optimizes the power and time utility without incurring the logic penalization. Various approaches are used, such as:

- **Dynamic Voltage and Frequency Scaling (DVFS):** The operating speed of the processor is adjusted according to workloads, leading to clock frequency and voltage fitting within limits optimally for power. This helps to improve better energy efficiency during operation; this is the reason why IoT devices like environmental sensors and smart meters save a great deal of energy during their sleep periods.
- **Clock Gating and Power Gating:** The techniques shut off idle parts of the processor to save wasted power. Clock Gating itself does this by switching off clock signals to inactive portions of the circuit, thereby reducing dynamic power consumption. Power Gating itself switches off power for modules so that the module is rendered entirely inactive, thus taking charge for zero static power dissipation, which will be very critical in sections where certain applications must fall asleep and conserve energy.
- **Subthreshold Logic Circuits:** These circuits operate transistors in the subthreshold voltage region, and ultra-low operation makes them cost-effective IoT sensors and embedded devices—continuous monitoring with extremely low power. Subthreshold logic mainly caters to the biomedical implants, remote IoT sensors, and low-power edge computing nodes.

13.2.2.2 Energy-Efficient Transistors

The history of open systems has been compared to the advent of successive generations of the power-reduction technologies and transistors that are likely to provide a radical improvement in power efficiency for VLSI chips.

- **FinFET (Fin Field-Effect Transistor):** FinFET will provide better control over short-channel effects, producing low leakage currents and better power efficiency. Such designs are widely used in today's IoT processors and AI accelerators, thus continuing to save power while enabling unmatched computation speed.
- **Tunnel Field-Effect Transistors (TFETs):** TFETs involve research for very low voltages that could sufficiently diminish dynamic power dissipation, leading to longer battery operating time. Ever switching through the quantum tunneling mechanism, these transistors enable applications requiring ultra-low-power, thus making them ideal for wearable

health monitoring, environment sensing, and energy-autonomous IoT node.

- **Negative Capacitance Field-Effect Transistors (NCFETs):** Negative capacitance field-effect transistors exploit negative capacitance effect to lower the energy required for switching, thus being visible as energy-efficient devices for long-lasting IoT systems where battery replacement can be infrequent.
- **Carbon Nanotube Field-Effect Transistors (CNTFETs):** These transistors differ from standard silicon-based devices by utilizing materials based on carbon nanotubes, which offer much better effective power efficiency while providing much higher scalability. This has made these transistors a topic of research for almost a future ultra-low-power IoT application that expresses energy harvesting and sustainable computing as paramount.

13.2.3 Scalability and Integration

Growth in IoT devices requires responsiveness to process huge volumes of data. Greater requirements for IoT application in smart cities, healthcare, industrial automation, and transportation require integrated and strong solutions. VLSI technology is an essential building block by consolidating various systems' functions on a small chip compatible with scalable designs and having high performance, low latency, and power-efficient consumption. There is smooth exchange of information between these technologies, which also provide computation and processing capability without increasing the physical sizes of the systems.

13.2.3.1 System-on-Chip (SoC) and System-in-Package (SiP) Designs

Contemporary IoT systems need highly integrated strategies to lower power consumption while boosting computational abilities and optimizing communication effectiveness. Two key methodologies that support scalability and integration are System-on-Chip (SoC) designs and System-in-Package (SiP) configurations.

- **System-on-Chip (SoC):** In contrast to SoCs that integrate all functions in a single die configuration, SiPs package multiple integrated circuits into a single package form but not all necessarily integrated into a single part. This is to enhance modularity for customized configurations in a way that different ICs can be stacked or laid side by side in a single encapsulated part. SiPs are useful to IoT services with heavy processing requirements alongside enhanced communication abilities by eliminating delay due to interconnections while providing designers more flexibility in integrating more functions.

- **System-in-Package (SiP):** In contrast to SoCs that integrate all functions within one die design configuration, SiPs package a number of integrated circuits into a single package structure but not necessarily all integrated into a single component. This raises modularity to fit customized configurations so that different ICs may be stacked or arranged side by side in one encapsulated component. SiPs are useful to IoT services with high processing requirements paired with enhanced communication capabilities by eliminating delay introduced by interconnections while providing designers greater flexibility in integrating more functions.

13.2.3.2 3D Integrated Circuits (3D ICs)

All designs of integrated circuits are under the limitation of two-dimensional layouts, which are constrained in performance improvement *via* increased need for longer interconnects and a subsequent rise in power consumption. Stacking multiple layers of circuit elements in a three-dimensional IC configuration, 3D ICs promise a paradigm change in next-generation microprocessor design [11].

- **Vertical Stacking of Circuits:** In a 3D IC design, different silicon layers are stacked on each other, evolving a very dense architecture for the chip. This reduces the interconnect length and so increases speed because it increases signal propagation in a way that improves overall performance. It is particularly useful for certain applications such as AI-based IoT devices, rapid data processing, and more energy-efficient computing options.
- **Through-Silicon Vias (TSVs):** Combining all these semiconductor processes makes it extremely necessary to develop a high-speed transition between two silicon wafers. This requirement may be fulfilled by constructing

mechanical structures called TSVs. These are connections in three-dimensional ICs that allow input and output data from the different circuits in the upward or downward direction through the silicon layer. TSV lowers the power budget, latency, and complexity of interconnects in a circuit, thereby increasing the processing speed and improving the handling of thermal effects.

- **Heat Dissipation and Power Efficiency:** Heat management for increased power density is still one of the most important challenges in today's IoT networks. 3D ICs possess efficient heat dissipation capabilities with novel cooling schemes such as microfluidic cooling, on-chip heat spreaders, and thermal-dominated materials that can effectively dissipate heat away from critical components. With these technologies, increased performance can be achieved without thermal constraints, i.e., increased reliability and extended lifetime with constant working conditions in a harsh environment.

13.2.4 Security and Reliability in IoT Systems

The growing wave of cyber threats makes it impossible to protect IoT devices from attacks, data breaches, and unauthorized access. Given that IoT devices exist in open networks and varying environments, they are more susceptible to a good number of weaponry that would compromise their safety. Nonetheless, VLSI technology can enhance security and reliability, at least at the hardware level, through integration of security mechanisms, directly with chip architecture. Such hardware-based security features seek to secure IoT devices from cyber threats while still maintaining system integrity and reliability.

13.2.4.1 Secure Boot and Trusted Execution Environments (TEE)

Secure Boot is one of the iterations, through limited means of introducing an unconventional operating system, the characterizing process by which authorized and unapproved firmware or operating systems are not let into an IoT device. Cryptographic signature-based mechanisms employed by Secure Boot will prevent the trusted systems from starting due to tampering in the unknown code, thus guaranteeing the integrity of the systems. This is very important for critical applications such as smart healthcare, industrial automation, and connected vehicles, where the use of compromised firmware can cause disastrous effects.

- **How Secure Boot Works:** The boot loader uses digital signatures to authenticate every stage of the boot process. In case of a verification failure, the device does refuse to boot; and no unauthorized modification is allowed.
- **Benefits of Secure Boot for IoT:** Secure Boot mitigates against firmware-level attacks such as bootkits and rootkits and only allows software that the manufacturers have approved.

Trusted Execution Environments represent a secure space and isolation area within the processor factory working sensitive operations [12, 13]. TEEs protect from the control of malware running in the main environment sensitive computing, such as cryptographic key management, biometric authentication, and financial transactions.

- **TEE Advantages:**
    - It protects sensitive data against malware and unauthorized access.
    - It keeps encryption keys and security credentials private.
    - It countermeasures risks of cyber-attacks in IoT environments.

13.2.4.2 Hardware-Based Encryption

Encryption remains the only solution available to keep IoT communication secure, as it protects sensitive data not only when the data packets with the sensitive information are being sent out but also during storage. Performing hardware encryption and software encryption are considered different; while software encryption cannot be the ideal performance, hardware encryption is considered as the better option that offers speed. Thus, hardware encryption enables instant encryption of data with minimal processing overhead on the system. Architectural VLSI-based implementation of a security architecture integrates special-purpose cryptographic co-processors for real-time encryption and decryption operations.

- **AES (Advanced Encryption Standard):** With rapid, high-speed, and low-power encryption designed for securing communications over wireless channels, for example Bluetooth, Zigbee, and Wi-Fi, AES encryption has been predominant in such IoT security protocol applications.

- **RSA (Rivest-Shamir-Adleman):** Provides authentication and digital signing credentials in the IoT network and works in secure key exchanges.
- **Physically Unclonable Functions (PUFs):** PUFs, or Physically Unclonable Functions, produce unique, unclonable identifiers for a VLSI chip that in this manner function like a fingerprint, thwarting illegal duplication of the device and cloning. Thus, chips would never produce identical PUF outputs in response to the same inputs, making PUFs the most secure method for an authentication scheme for appliances.
- **Post-Quantum Cryptography (PQC):** In the face of quantum computing, the aim of researchers is to design post-quantum cryptography algorithms that can resist attack by quantum computers. Implementation of post-quantum cryptography on VLSI will go a long way in ensuring long-term security for IoT devices that are being used in critical infrastructures.

13.2.4.3 Fault-Tolerant Designs

Fault tolerance is an essential characteristic guaranteeing reliability and resilience to long-living IoT devices in rugged environments. VLSI-based fault tolerant architectures allow error-detection and correction schemes and redundancies in design, such that they promote system reliability.

- **Error Correction Code (ECC):** ECC aids IoT systems in detecting memory errors that are due to electromagnetic interference (EMI), due to radiation effects and hardware defects, and in cutting down the number of crashes, as well as data corruption.
- **Triple Modular Redundancy (TMR):** The functioning of this concept is based on redundancy and getting the correct outcome in real time from the majority votes. TMR is used in many areas, from aerospace to industrial automation and other sensitive IoT applications.
- **Watchdog Timers:** Watchdog timers ensure the systems are reset automatically if they have any critical fault so that IoT devices never become unresponsive, even in unknown failure conditions.

- **Self-Healing Circuits:** Most complex VLSI architectures integrate self-repair circuits, whereby defective portions of the chip are identified and isolated to prolong the lifespan of the device and bolster reliability in the field.

13.2.5 Real-Time Data Processing and Edge Computing

Edge computing makes real-time data processing the backbone of IoT systems to drive immediate action, enabling intelligent decision-making. Ultra-fast computations on data characterize autonomous vehicles, smart healthcare systems, and industrial robots. On this premise, the VLSI technology enables high-speed computation and low-latency architecture to facilitate AI-based edge computing for real-time data analytics.

13.2.5.1 Low-Latency Architectures

To reduce latency in real-time applications, VLSI chips are manufactured with fast memory hierarchies and high-speed interconnect. Low latency architecture defines the model of data processing and transmission such that the time spent on computation shall not exceed a few microseconds. Online trading, emergency response systems, and mission-critical IoT networks are some of the examples in which processing cannot tolerate even the slightest delay.

- **Fast Memory Layers:** Data that are accessed most often are kept by a VLSI chip in the cache memory, SRAM, and DRAM in some instances, thereby avoiding delay due to access from slower storage devices.
- **High-Speed Interconnects:** By means of advanced technologies, the latest communication is assured through Network-on-Chip (NoC) and Advanced Interconnect Technologies, which enable extremely high-speed communications between processor cores to memory units.
- **Edge-Based Data Prioritization:** Intelligent edge computing architectures give preference to time-sensitive data so that tasks with higher priority get processed first to provide faster response times.

- **Time-Sensitive Networking (TSN):** In industrial automation and automotive IoT, TSN protocols are deterministic, guaranteeing real-time data transmission with a guaranteed quality of service.

13.2.5.2 AI and Machine Learning on Edge Devices

VLSI-type AI accelerators bring advanced computing to the edge with IoT devices being able to run AI-driven workloads locally without offloading the computing to the cloud. Edge AI and ML models enable intelligent decision-making whenever and wherever it is needed, such as in autonomous drones, smart medical devices, and predictive maintenance systems.

- **AI-Specific Hardware Accelerators**
  - **Neural Processing Unit (NPU):** Circuits specifically made to speed up deep learning inference and, therefore, AI applications like speech recognition, picture identification, and natural language comprehension.
  - **Field Programmable Gate Array (FPGA):** Reconfigurable hardware that can be programmed for tasks relating to AI, thus giving a trade-off between flexibility and speed of execution.
  - **Application Specific Integrated Circuit (ASIC):** Chips that are specifically made for a certain AI task, significantly increasing productivity and lowering power use.
- **On-Device Learning and Adaptation**
  Continuous learning at the edge is made possible by VLSI-based AI processors, which enable real-time adaptation of IoT devices to changing circumstances. Real-time anomaly detection, industrial automation, and the creation of customized medical devices will all depend on this.
- **Energy-Efficient AI Processing**
  The VLSI chips' low-powered AI accelerators maximize calculations while using the least amount of energy possible, allowing the IoT devices to continue processing data. Approximate computing and quantized deep learning reduce computation while maintaining a respectable level of inference accuracy.

13.3 Hardware Accelerators for Real-Time Processing

13.3.1 Digital Signal Processors (DSPs)

These programmable processors are capable of carrying out complex signal processing tasks as well as simultaneous operations of filtering and compression for real-time application and feature extraction. The ubiquitous use of such processors is present in voice recognition and biomedical sensor systems as well as industrial automation, due to their ability to carry out important high-speed mathematical calculations. The introduction of Fourier transforms in conjunction with optimizations of the convolution operation and signal denoising.

13.3.2 Graphics Processing Units (GPUs)

GPUs speed up marketable IoT applications since these devices are great at carrying out parallel computations. Thus, they lend themselves to tasks such as intelligent solutions for image recognition and predictive analysis. These computing units propel deep learning models to efficiently handle large-scale data sets to benefit IoT decision systems. On edge AI devices, the GPU quickly processes sensor and video features in real-time because of the compute-intensive operations.

13.3.3 Tensor Processing Units (TPUs)

The Tensor Processing Unit serves as a custom-built acceleration platform to obtain the highest performance for deep learning algorithms utilized by IoT technologies. The processing units show special excellence in performing speech recognition functions along with natural language processing and real-time image analysis. Digital matrix computations being run on TPU give higher computational speed and energy savings because of optimized matrix operations and low memory latency management. When IoT devices are integrated with these specialized accelerators, the user gets accelerated AI inference along with lower power requirements.

13.3.4 Quantum Computing in IoT

Quantum computers are poised to revolutionize the processing of IoT networks by providing fast solutions for complex computational problems operated at record speed. Quantum algorithms offer optimum workflow implementations for IoT processes such as cryptography, optimization solutions, and big data processing. The initial research of implementing quantum computing in VLSI for the enhancement of IoT data processing is slowly gaining momentum; researchers predict this combination could offer large-scale efficient security in IoT analytics.

13.3.5 3D Integrated Circuits (3D ICs)

The vertical stacking of several circuit layers using 3D ICs improves both data transfer speed and energy efficiency in design. Product performance improves with shorter connection delays, pushing processing and memory units to closer integration points using this methodology. The implementation of 3D ICs allows IoT applications to develop compact, high-performance systems working more efficiently with lower power levels.

13.3.6 Challenges and Limitations in VLSI-Based IoT Systems

VLSI technology for IoT systems faces various issues that create barriers because of elevated manufacturing expenditures as well as thermal dissipation problems, together with complicated integration requirements and constrained IoT technology adaptation capabilities. Future scalability of IoT networks faces challenges that demand ongoing improvements in fabrication processes that must adapt to performance needs. The ongoing process of transistor miniaturization introduces reliability problems that can cause integrated circuits to fail and shortens their operational life expectancy. Sustainable fabrication techniques and affordable production methods present difficulties because developing scalable VLSI circuits for high performance requires substantial investment in semiconductor manufacturing technology development.

13.3.7 Advanced Security Techniques for VLSI-Based IoT Systems

VLSI-based IoT systems are facing major security vulnerabilities shown in Figure 13.2 from hardware trojans, together with side-channel attacks, as well as data breaches and malware infections. The use of hardware encryption along with trusted execution environments (TEE) paired with secure boot implementation together with fault-tolerant circuit designs improves device security. The implementation of physically unclonable functions (PUFs) generates unique device identifiers that stop device cloning and

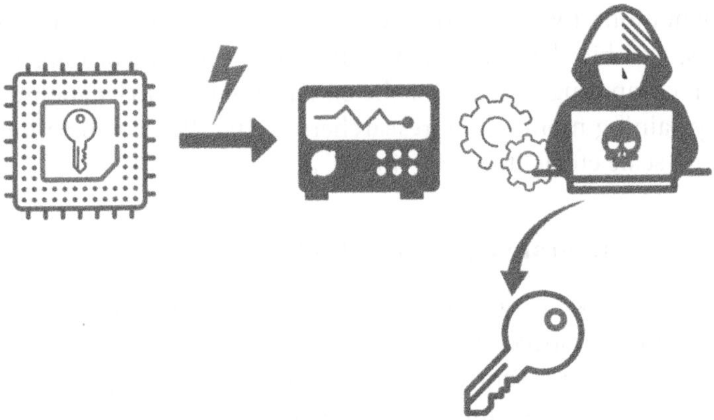

Figure 13.2 Security threats.

unauthorized user attempts. Modern quantum-resistant algorithms working together with autonomous anomaly detectors strengthen the cyber threat resilience of IoT VLSI systems.

13.3.8 Future Trends and Innovations in VLSI for IoT

Future VLSI developments for IoT include AI-enhanced VLSI chips coupled with chiplet architectures shown in Figure 13.3 and 3D IC stacking shown in Figure 13.4 and quantum computing shown in Figure 13.5 capabilities in addition to open-source hardware RISC-V and beyond. Memristor-based memory and photonic computing systems represent

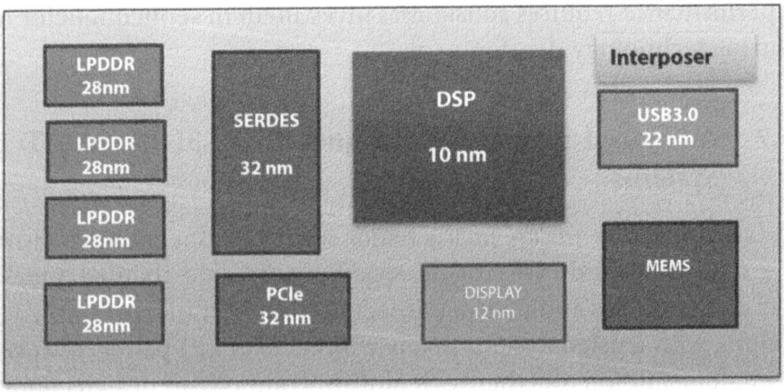

Figure 13.3 Chiplet architecture.

Figure 13.4 3D IC stacking.

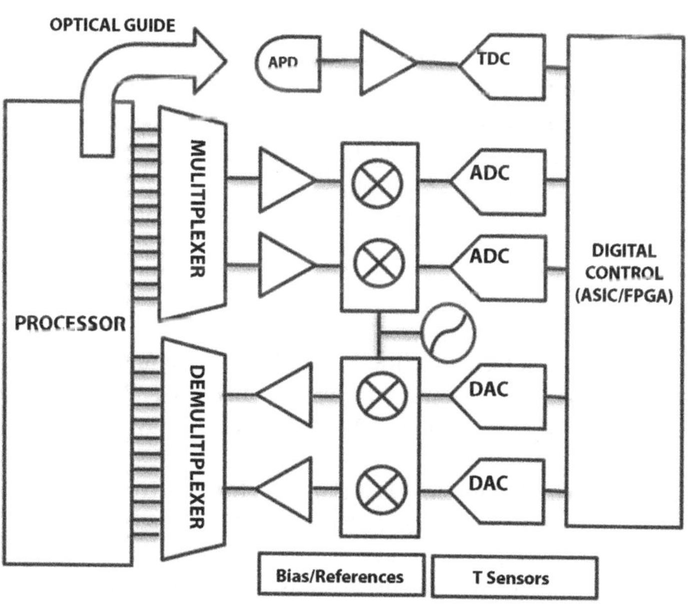

Figure 13.5 General quantum processor.

breakthrough technologies that advance both high-speed data transfer and power efficiency capabilities. Future VLSI systems will gain from extreme ultraviolet (EUV) lithography, which advances both transistor density and system performance.

13.9 Conclusion

High-speed processing, together with power efficiency, security, and scalability capabilities that VLSI technology delivers, are indispensable for real-time IoT networks. Modern IoT capabilities will gain additional capabilities through future innovations in machine-independent hardware, quantum computing, and three-dimensional integrated circuits. Exploring solutions to cost, heat dissipation, security flaws, and scalability barriers represents crucial requirements for advancing VLSI in IoT systems. Researchers and engineers can develop advanced IoT networks of tomorrow's connected environments by deploying novel, innovative approaches to make smarter IoT platforms that enhance efficiency and security. Future IoT success depends on secure VLSI frameworks combined with innovative computing approaches to achieve reliability with sustainability and efficiency outcomes.

AI disclosure statement: The author confirms that no artificial intelligence (AI)-generated content or AI-assisted writing tools were used in the preparation of this chapter (including summaries and the abstract). All text, analysis, and conclusions are the result of the author's original work.

References

1. Kabilamani, P., Efficient MRFFT Feedback-Commutator Technique for Handling Data Flow Control in Narrow Band-IoT Based LTE Systems. *IEEE Trans. Wirel. Commun.*, 22, 5, 1783–1795, May 2023. doi: 10.1109/TWC.2023.3156789.
2. Parimala Devi, M., Efficient Data Handling for Massive Internet of Medical Things. *IEEE Internet Things J.*, 9, 3, 456–472, March 2024. doi:10.1109/JIOT.2024.3279845.
3. Chible, H., Inexact Arithmetic Circuits for Energy-Efficient IoT Sensors Data Processing. *IEEE Trans. Circuits Syst. I Regul. Pap.*, 70, 1, 55–67, Jan. 2024. doi: 10.1109/TCSI.2024.3291873.

4. Chen, T.-F., A 3T1R Nonvolatile TCAM Using MLC ReRAM for Frequent-Off Instant-On Filters in IoT and Big-Data Processing. *IEEE J. Solid-State Circuits*, 58, 6, 934–946, June 2023. doi: 10.1109/JSSC.2023.3216721.
5. Joshi, R., Novel Bit-Sliced Near-Memory Computing Based VLSI Architecture for Fast Sobel Edge Detection in IoT Edge Devices. *IEEE Trans. Comput. Aided Des. Integr. Circuits Syst.*, 43, 2, 292–305, Feb. 2024. doi: 10.1109/TCAD.2024.3259763.
6. Kumar, M.R., Devi, B.R., Rangaswamy, K., Sangeetha, M., Kumar, K.V.R., IoT-Edge Computing for Efficient and Effective Information Processing on Industrial Automation. *IEEE Trans. Ind. Inf.*, 20, 1, 104–118, Jan. 2024. doi: 10.1109/TII.2024.3315123.
7. Gomes, M.M., da Rosa Righi, R., da Costa, C.A., Griebler, D., STEAM++: An Extensible End-to-End Framework for Developing IoT Data Processing Applications in the Fog. *IEEE Internet Comput.*, 28, 3, 36–47, May 2024. doi: 10.1109/MIC.2024.3378210.
8. Sun, B., Liu, D., Yu, L., Li, J., Liu, H., Zhang, W., Torng, T., MRAM Co-designed Processing-in-Memory CNN Accelerator for Mobile and IoT Applications. *IEEE Trans. Neural Netw. Learn. Syst.*, 35, 4, 678–692, April 2024. doi: 10.1109/TNNLS.2024.3385194.
9. Disabato, S., Roveri, M., Alippi, C., Distributed Deep Convolutional Neural Networks for the Internet-of-Things. *IEEE Trans. Artif. Intell.*, 2, 2, 167–180, June 2023. doi: 10.1109/TAI.2023.3297816.
10. Reddy, C.S. and Anand, K., Sensor Signal Processing using High-Level Synthesis and Internet of Things with a Layered Architecture. *IEEE Trans. Instrum. Meas.*, 73, 1, 281–295, Jan. 2024. doi: 10.1109/TIM.2024.3339417.
11. Mukherjee, A., Ghosh, S., Sen, S., VLSI Design of Energy Efficient Computational Centric Smart Objects for IoT, in: *Proceedings of the IEEE International Symposium on Smart Electronic Systems (ISES)*, Hyderabad, India, pp. 199–204, Dec. 2018, doi: 10.1109/iSES.2018.00052.
12. Al Faruque, M.A. and Vatanparvar, K., Energy Management-as-a-Service Over Fog Computing Platform. *IEEE Internet Things J.*, 3, 2, 161–169, April 2016. doi: 10.1109/JIOT.2015.2503267.
13. Yin, S., Ali, M.A., Yin, W., Prospects for Energy-Efficient Edge Computing with Integrated HfO_2-Based Ferroelectric Devices. *IEEE Trans. Electron Devices*, 66, 4, 2032–2037, April 2019. doi: 10.1109/TED.2019.2897767.

14

Using AMF Filtering Techniques, Compression of Clinical Images for Improvement in Compression Parameters

Suman Rani[1]* and Jaibir Singh[2]

[1]*Department of Electronics and Communication Engineering, Galgotias University, Greater Noida, Uttar Pradesh, India*
[2]*Department of Computer Science and Engineering, Galgotias University, Greater Noida, Uttar Pradesh, India*

Abstract

Image compression is deemed highly crucial in many different areas as it holds a significant role in numerous applications. Biomedical is a critical application and compression strategies are fundamentally categorized as lossy or lossless. The lossless method compresses the image without losing any information, but various information could be lost in the lossy approach. This study talks about various types of compression methods and focuses on brain snap shots for density strategy. Mutually lossy and lossless methods are employed by evaluating their pros and cons. Key quality procedures like compression ratio (CR) and peak signal-to-noise ratio are assessed in this research. The offered techniques like DCT, DFT, DWT, and Fractal are utilized, along with the introduction of new methods such as the fluctuation idea, BTC-SPIHT with Hybrid method incorporating flexible limits and the Quasi Fractal algorithm.

***Keywords*:** AMF, BTC-SPIHT, compression, compression ratio (CR), peak signal-to-noise ratio (PSNR)

**Corresponding author*: smn.bishnoi@gmail.com

Abhishek Kumar and Suman Lata Tripathi (eds.) Smart Chips for Smart Devices: VLSI Design for Next-Generation IoT Solutions, (269–278) © 2026 Scrivener Publishing LLC

14.1 Introduction

The research explains that lossy density is not always very advantageous for compressing an image. Compression is important in medical images because the processing time is long, so using compression algorithms can speed up the process. Different methods such as lossy and lossless are employed for compression, each with its own unique approach. The most challenging aspect of research is expanding the range of regulations. Several algorithms have progressed to achieve better results. The use of compression methods can lead to loss of important data, which is often utilized by researchers and professionals. Certain processes such as enhancement can also result in further degradation, prompting the need for more efficient image compression techniques. Using lossless compression is a superior choice for treatment. Several lossless methods rely mainly on linear calculation and outburst. Artistic medical images and better solution for storing large amounts of information and size is to reduce images. The utilization of adaptive threshold and Quasi Fractal algorithm in the density process, using the approaches of oscillation idea, BTC-SPIHT, and Hybrid method [1–5].

CR with PSNR were considered as important parameters in this study. Different methods including DCT (Discrete Cosine Transform), Discrete Fourier Transform (DFT), Discrete Wavelet Transform (DWT), and Fractal were utilized in the analysis. Fractal provides us with superior results. Through the utilization of Fractal image compression, achieved values of 3.51 for CR and 31.45 for PSNR. Additionally, the use of fractal resulted in a significantly large MSE value, suggesting that the high value of the figure is not suitable. To enhance CR and ensure a new approach, a cutting-edge methodology has be introduced. The fresh planned technique reveals that in Figure 14.1, there are changes in grey level intensities [6]; the changes are simply fluctuations in an image. This idea is used to detect differences in medical images, suitable oscillations are selected for image density by iterating the procedure we can extract the crucial component from the image. This is repeated until higher quality of essential part is obtained. The accurate superiority is obtained by isolating the crucial component. It provides improved compression levels and is simply a continuous signal for informational purposes. The improved results showed a Compression Ratio (CR) of 4.32 and PSNR of 32.56. MSE costs were also lowered compared to the current Fractal method, underscoring the importance of enhancing CR in the scientific field. Enhancement is applied before compressing images to improve their effects. In this case, a photo

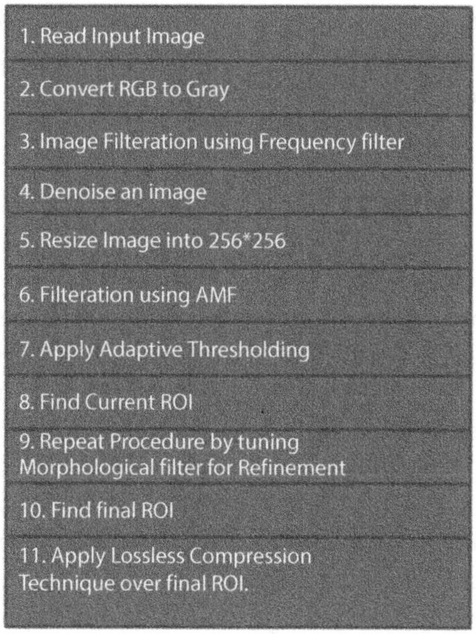

Figure 14.1 Algorithm.

enhancement technique is utilized. The CLAHE and DCS algorithms are utilized to improve a photograph. Better results like CR of 5.11 and PSNR of 32.77 are achieved by enhancing the image before compression [7–10].

These outcomes are superior to current methods. Additionally, the implementation of a good photo is preferable to using the Fractal and Oscillation method for compressing an image. The concept of oscillation is primarily utilized for lossy image compression methods. Hybrid strategies have been developed to enhance results, where two different algorithms are applied to ROI and Non-ROI. ROI Lossless and Non-ROI Lossy image compression methods were utilized, comparing various existing techniques and algorithms. A literature review was conducted, concluding that a new and improved algorithm is necessary for achieving complex values in high-quality parameter (CR, PSNR, MSE, and MSSIM). As a result, an innovative hybrid method was introduced. Initially, a hybrid algorithm utilizing BTC and SPIHT is utilized. BTC is a method of compressing data that results in some loss, while SPIHT is a method that compresses data without any loss. This hybrid position of rules is implemented post-enhancement. Implementing this fusion technique yielded enhanced results than attractive an image before compressing it. CR and PSNR have both increased by 5.65 and 33.01 attained correspondingly and also achieved superior image quality. A hybrid coding algorithm was developed to improve image quality for compression, leading to higher results in compression ratio (CR) of 24.98 and PSNR of 36.21.

14.2 Methodology

Two methods of image density are convenient: 1. Lossy, 2. Lossless. For these studies, important measures such as CR, PSNR are computed. Here, methods like DCT, DFT, DWT and Fractal are utilized. Fractal provides us with superior results. Through the utilization of Fractal image density, the principles obtained for CR and PSNR were 3.51 and 31.45, respectively. Furthermore, the utilization of fractal techniques revealed a significantly high MSE value, suggesting that the image quality is not adequate [11]. A new method has been introduced to improve CR and is exceptional. The suggested technique demonstrates that there are fluctuations in grey scale intensities within each image, which do not represent noise in the representation. This idea is utilized to find variation in medical descriptions, selecting proper oscillations for compressing images. By repeating the procedure, we are able to gain the important element of the representation. This continues until a higher quality is achieved. Essential importance is derived from receiving top-notch quality. It provides superior compression levels. It is simply a continuous signal for a specific purpose. Higher results, specifically CR and PSNR at 4.32 and 32.56, were achieved eventually, along with a decrease in MSE compared to the existing Fractal approach. Improving the CR is crucial and is a significant undertaking in the scientific realm. Enhancement is applied prior to image compression to improve results. An image enhancement technique is being utilized here. The CLAHE and DCS algorithms are employed to improve a photograph before compression, resulting in better outcomes with a PSNR of 32.77, surpassing current methods [12–15].

Furthermore, the proper picture quality achieved is superior to using the Fractal and Oscillation idea for compressing an image. The concept of oscillation is primarily utilized for lossy image compression methods. In order to improve results, hybrid strategies are developed. These strategies incorporate specialized algorithms for both ROI and non-ROI. ROI Lossless and non-ROI Lossy image compression techniques were utilized, comparing all current methods utilizing algorithms. Furthermore, after conducting a literature review, it was determined that updated algorithms may be necessary to enhance parameters like CR, MSSIM and PSNR, and MSE. As a result, innovative amalgam algorithms have been introduced.

Initially, a hybrid method incorporating both BTC and SPIHT is utilized. BTC is a method for compressing data with loss, while SPIHT is a process for compress data lacking any loss. This combination of rules is implemented post-enhancement and yielded enhanced grades than attractive a representation prior to compression. CR and PSNR excel, and also achieve greater image quality. An advanced mix coding algorithm was developed for image compression to enhance quality, resulting in improved CR and PSNR values of 24.98 and 36.21, respectively.

14.3 Quasi Fractal, Oscillation with Hybrid Coding: Morphological Filters with Lossless Fractal Image Compression (LFIC)

14.3.1 ROI, Non-ROI Detection

In initial stages, we have to distinguish the intellect element of the complete picture because the majority of the surroundings pixels are black, meaning they have zero depth. Therefore, excluding those pixels can also help decrease the number of bits needed for broadcast and can be used for cost-effective compression. Bounding box is a method utilized to part the foreground from the background. Similarly, a bounding container is worn to isolate the Region of Interest (ROI) generated by the turning point algorithm. ROI is chosen to demonstrate the application of bounding box in analysis. The selected portion of the image that captures the essence of the entire photograph is isolated as the Region of Interest and displayed prominently.

Methodology for ROI: RGB image is converted to grayscale, then downsized to 256x256 pixels. AMF is used for adaptive processing within the current ROI, followed by refinement and determination of the final ROI.

Flow of Hybrid Method:

1. AMF, adaptive threshold
2. Multilevel operations
3. Algorithm using ROI and non-ROI methods

14.3.2 Adaptive Threshold

Thresholding a grayscale representation in Figure 14.2 and Figure 14.3 with a fixed value is the most normally used operation in image processing. The threshold rate at every pixel position is determined based on adjacent pixel intensities. This is used to divide the unique representation into sub-descriptions and apply an international thresholding technique to each associate representation [16–18].

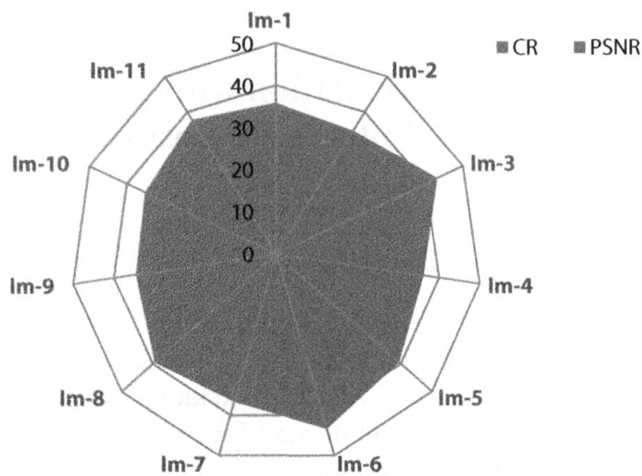

Figure 14.2 Brain images for CR and PSNR.

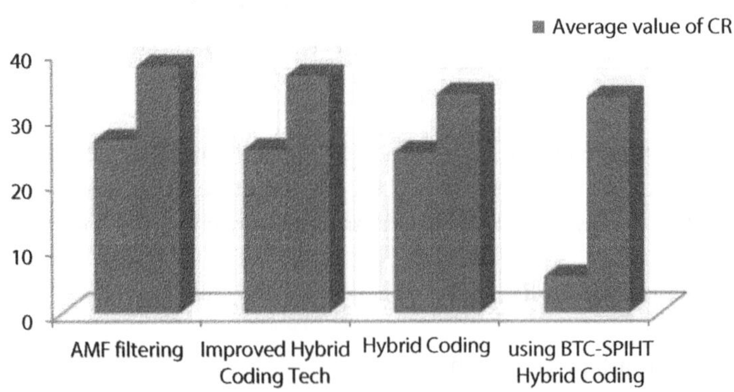

Figure 14.3 Statistical parameters CR and PSNR for various image compression techniques.

14.4 Result and Discussion

Improved Hybrid Method for different brain images from the hospitals. Table 14.1 shows the statistical parameters CR and PSNR for different brain images.

Table 14.1 Statistical parameters CR and PSNR.

Image/parameters	CR	PSNR
Im-1	26.28	35.82
Im-2	26.16	34.76
Im-3	28.07	43.09
Im-4	25.07	36.15
Im-5	27.14	39.37
Im-6	27.9	43.32
Im-7	25	36.63
Im-8	30.63	39.37
Im-9	25.46	34.6
Im-10	27.25	35.09
I-11	23.55	37.81
Average	26.49	37.82

Hybrid Method for CR and PSNR

Table 14.2 presents CR and PSNR for various hybrid techniques.

Table 14.2 CR and PSNR for hybrid techniques of statistical parameter.

Sr. no	Technique used for image compression	Parameters	
		Average value of CR	Average value of PSNR
1	Oscillation Concept & Quasi Fractal using AMF filtering	26.48	37.81
2	Improved Hybrid Coding using Oscillation Concept & Quasi Fractal	24.98	36.31
3	Hybrid Coding using Oscillation Concept & Quasi Fractal	24.62	33.52
4	Hybrid Coding using BTC-SPIHT	5.66	33.02

14.5 Conclusion

Following an assessment of the outcomes from current hybrid methods and the enhanced hybrid technique, it can be concluded that the advanced hybrid method algorithm has shown positive outcomes. Previous methods such as DFT, DWT and DCT, fractal produced density ratios ranging from 1 to 4. In contrast, the proposed approach has significantly amplified the compression ratio to a range of 1.01 to 26.48. Additionally, the PSNR values for existing algorithms range from 27.04 to 37.81, even as the improved hybrid method has notably enhanced image quality, indicating superior results.

AI disclosure statement: The author acknowledges the use of AI-based tools for minor language editing or grammatical correction. However, all intellectual contributions, analyses, and conclusions presented in this chapter are entirely the author's own.

References

1. Abo-Zahhad, M., Gharieb, R.R., Ahmed, S.M., Khaled, M., Brain Image Compression Techniques. *Int. J. Eng. Trends Technol. (IJETT)*, 19, 2, Jan 2015.
2. Vijayvargiya, G., A Survey: Various Techniques of Image Compression. *Int. J. Comput. Sci. Inf. Secur. (IJCSIS)*, 11, 10, October 2013.
3. Liang, S., Wang, G., Wang, S., Wang, Y., A New Method of Image Quality Assessment. *WSEAS Trans. Signal Process.*, 12, 2016.
4. Shaaban Ibraheem, M., Zahid Ahmed, S., Hachicha, K., Hochberg, S., Garda, P., Medical Image Compression with Clinical Diagnostic Quality Using Logarithmic DWT. *IEEE Xplore*, 2016.
5. Abo-Zahhad, M., Ragab Gharieb, R., Ahmed, S.M., Abd-Ellah, M.K., Huffman Image Compression Incorporating DPCM and DWT. *J. Signal Inf. Process.*, 6, 123–135, 2015.
6. Magar, S. and Sridharan, B., Comparative analysis of biomedical image compression using oscillation concept and existing method, in: *Lecture Notes in Computational Vision and Biomechanics, Book Series*, vol. 28, Springer.
7. Bhavani, S. and Thanushkodi, K., Neural Based Domain and Range Pool Partitioning Using Fractal Coding for Nearly Lossless medical Image Compress. *WSEAS Trans. Signal Process.*, 9, 1, January 2013.
8. Sharma, R., Kamargaonkar, C., Sharma, M., Hybrid Medical Image Compression: Survey. *Int. J. Adv. Res. Comput. Eng. Technol. (IJARCET)*, 5, 4, April 2016.

9. Alyaa, H., Medical Image Data Compression Using Hybrid Methods. *ARPN J. Eng. Appl. Sci.*, 13, 5, March 2018.
10. Magar, S. and Sridharan, B., Innovative approach to biomedical image compression using oscillation concept, in: *International Conference on Automotive Control and Dynamic Optimization Techniques (ICACDOT)*, IEEE Conference Publication, IEEE Xplore, pp. 124–128, 2016.
11. Markandeya, N. and Patil, S., Digital Image Compression Hybrid Technique Based on Block Truncation Coding and Discrete Cosine Transform. *Int. Conf. Trends Electron. Informatics ICEI*, 2017, 2017.
12. Singh, H. and Rana, S., Image Compression Hybrid using DCT, DWT, Huffman. *Int. J. Sci. Eng. Res. (IJSER)*, 8, August 2012.
13. Ghrare, S.E. and Khobaiz, A.R., Digital Image Compression Using Block Truncation Coding and Walsh Hadamard Transform Hybrid Technique, in: *IEEE 2014 International Conference on Computer, Communication, and Control Technology (I4CT 2014)*, September 2-4, 2014.
14. Soundarya, G., Comparison of Hybrid Codes for MRI Brain Image Compression. *Res. J. Appl. Sciences Eng. Technol.*, 4, 24, 2012.
15. Mathew, S. and Sebastian, S., Image Compression by using Morphological Operations and Edge-Based Segmentation Technique. *Int. J. Adv. Res. Comput. Commun. Eng.*, 4, 11. November 2015.
16. Kr. Saini, P., Brain Tumor Detection in Imaging Using MATLAB. *Int. Res. J. Eng. Technol. (IRJET)*, 2, 2, May 2015.
17. Cheng Yang, J. and Sun Park, D., Detecting Region of Interest (ROI) in Digital Mammogram by using Morphological Bandpass Filter. *IEEE*, 2004, 2004.
18. Kalaiselvi, T., Boopathiraja, S., Sriramakrishnan, P., ROI Based Hybrid Compression Techniques for Transferring MRI Brain Images. *Int. J. Comput. Sci. Trends Technol. (IJCST)*, 4, 4, Jul-Aug 2016.

15

Underwater Corrosion Detection: Utilizing Convolutional Neural Networks and Image Processing

Sai Srija Achukolu*, Bhushan Vasant Nanaware, Ravindra Singh, Vaibhav Singh, Aadarsh Kumar and Harpreet Singh Bedi

School of Electronics and Electrical Engineering, Lovely Professional University, Punjab, India

Abstract

Underwater structures, such as ship hulls, hydropower plants, turbines, and penstocks, among others, are susceptible to corrosion, a phenomenon that poses significant challenges in maintenance and safety. This is due to the fact that corrosion, especially when it occurs underwater, can compromise the structural integrity of these installations, potentially leading to catastrophic failures. Traditional methods of detecting corrosion, however, are often labor-intensive and require the use of costly, specialized equipment. Furthermore, these methods may not always be effective, especially in cases where the corrosion is not readily visible or accessible. This paper presents a novel and potentially transformative solution for detecting corrosion in underwater structures. This solution leverages the power of Convolutional Neural Networks (CNNs), a form of deep learning technology that has shown significant promise in image recognition tasks. Specifically, our approach involves training a CNN on images captured by the camera of an underwater Remotely Operated Vehicle (ROV) with 4 degrees of freedom (DOF). In this study, we have evaluated the effectiveness of our approach in detecting different types and degrees of corrosion, including pitting corrosion and galvanic corrosion, among others. The model we have developed is designed to accurately identify and classify different corrosion patterns, thus providing valuable information that can guide maintenance interventions. By doing so, our approach enables more efficient and timely maintenance, potentially saving significant costs and reducing safety risks.

**Corresponding author*: saisrijakrpu@gmail.com

Abhishek Kumar and Suman Lata Tripathi (eds.) Smart Chips for Smart Devices: VLSI Design for Next-Generation IoT Solutions, (279–290) © 2026 Scrivener Publishing LLC

***Keywords*:** Corrosion detection, underwater ROV, computer vision, deep learning, neural network

15.1 Introduction

Corrosion of underwater structures poses a significant challenge in various industries such as marine, offshore oil and gas, renewable energy, and infrastructure maintenance. The detrimental effects of corrosion not only lead to costly repairs but also jeopardize the safety and integrity of these vital structures [1]. Traditional methods of corrosion detection often involve manual inspections, which are not only labor-intensive but also limited in their ability to provide comprehensive and timely assessments of corrosion damage.

In recent years, the advancement of technology, particularly in the fields of computer vision and deep learning, has opened up new avenues for improving the efficiency and accuracy of underwater corrosion detection. One such promising technology is the utilization of Convolutional Neural Networks (CNNs), which have demonstrated remarkable capabilities in image recognition and classification tasks. This paper aims to address the challenges of underwater corrosion detection by proposing a novel approach that leverages CNNs trained on images captured by the camera of an underwater Remotely Operated Vehicle (ROV) with 4 degrees of freedom (DOF) [2, 3]. By harnessing the power of computer vision and deep learning, our proposed method offers a reliable and efficient solution for detecting, classifying, and quantifying corrosion in various underwater structures such as ship hulls, hydropower plants, turbines, penstocks, and more.

The primary focus of this research is to develop a CNN-based model capable of accurately identifying and classifying different corrosion patterns. This includes distinguishing between various types of corrosion such as pitting corrosion, galvanic corrosion, crevice corrosion, and others. By providing detailed insights into the extent and type of corrosion present, our model enables maintenance teams to make informed decisions and prioritize interventions based on the severity of the damage.

15.2 Designing of Underwater Remotely Operated Vehicle

15.2.1 Dimensions

The underwater Remotely Operated Vehicle (ROV) used in this study boasts a robust design that allows it to navigate challenging underwater environments. It has 4 degrees of freedom (DOF), which provide excellent maneuverability [4]. This includes the ability to move forward and backward, up and down, left and right, and to rotate about its vertical axis. This agility makes it an ideal tool for capturing images of various underwater structures from different angles and perspectives. The ROV's design is compact and aerodynamic, which is good for easy navigation under the water.

The ROV is completely 3D printed with PLA (Polylactic Acid) filament. The 3D printed body makes it lightweight and easily deployable. Figure 15.1 and Figure 15.2 present the ROV view from different views. The dimensions of the Remotely Operated Vehicle (ROV) are given in Table 15.1.

15.2.2 Computer-Aided Design

We designed our Remotely Operated Vehicle (ROV) using Fusion 360, a powerful 3D CAD, CAM, and CAE tool. The software enabled us to create a precise 3D model of the ROV, incorporating all crucial elements such as the camera, propulsion mechanisms, and the control system. The mechanical assembly of the ROV is shown in Figure 15.3.

In addition to the structural design of the ROV, careful attention was given to the modelling of the electronic enclosure. The electronic enclosure houses the essential components of the ROV, including control circuits, sensors, and communication devices. Given its vital role and the harsh underwater environment, creating a secure and waterproof electronic enclosure was of paramount importance. The electronic enclosure was also designed with a focus on waterproofing, as any water ingress could lead to catastrophic failure of the system. This necessitated the use of durable, corrosion-resistant materials and secure sealing mechanisms.

The final design of the electronic enclosure in Figure 15.4 was a compact, thermally efficient, and waterproof unit that could safely house all the electronic components of the ROV. The enclosure was designed to be easily

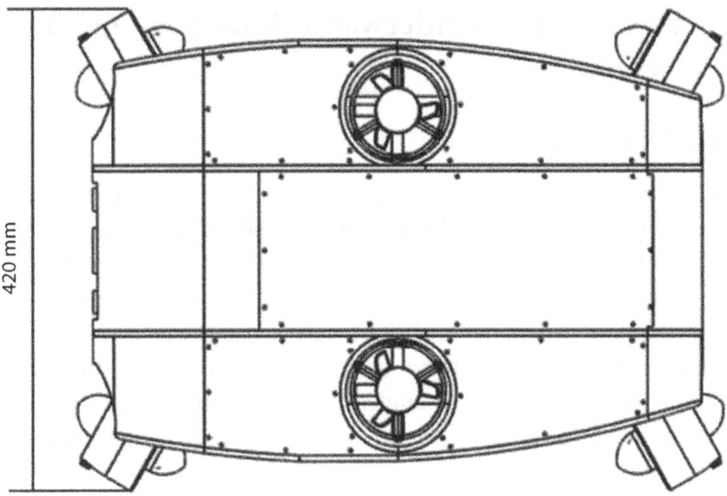

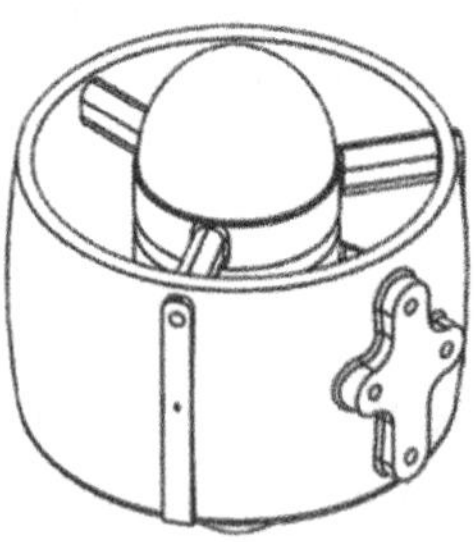

Length	110 mm
Diameter	90 mm
Material	Polylactic Acid
Weight	200 g

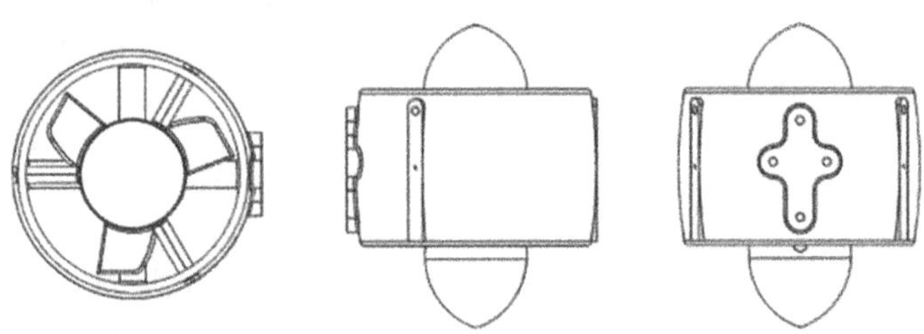

Figure 15.1 Top view and Thruster specifications.

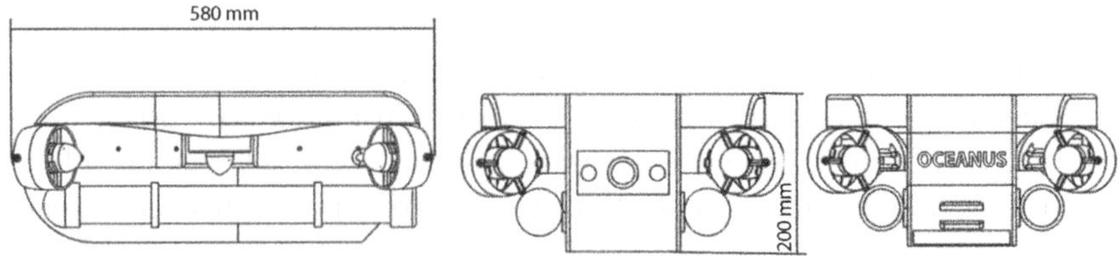

Figure 15.2 Side view, front view, rear view.

Table 15.1 Dimensions of ROV.

Parameter	Measure
Length	580mm
Width	420mm
Height	200mm

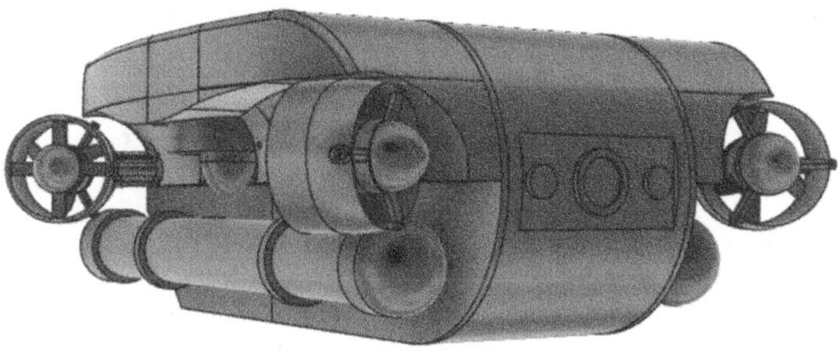

Figure 15.3 Computer-aided design of remotely operated vehicle in fusion 360.

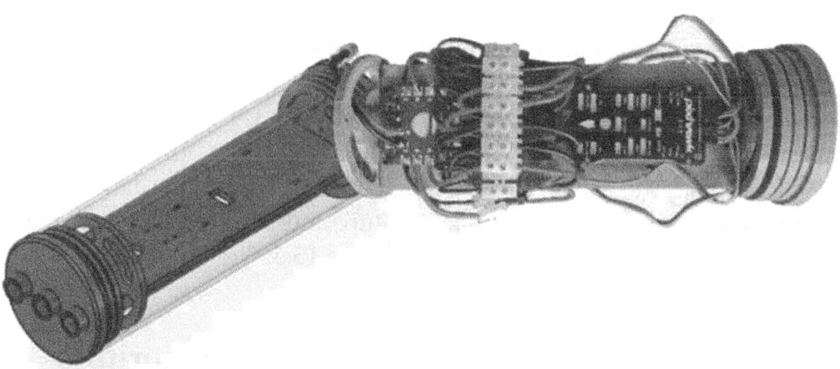

Figure 15.4 Electronic enclosure.

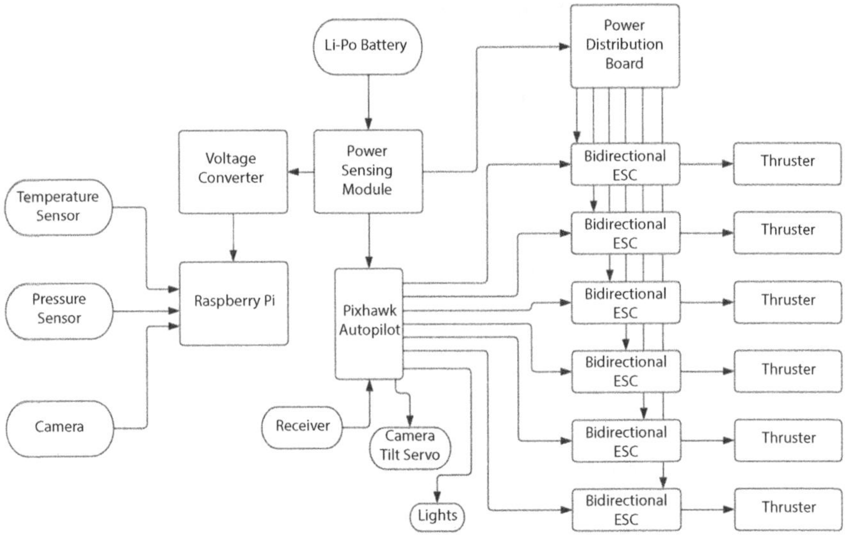

Figure 15.5 Block diagram of connections of electronic components of ROV.

accessible for routine maintenance and potential upgrades, ensuring the longevity and adaptability of the ROV system.

The ROV is controlled using the Pixhawk Autopilot Controller onboard. It includes MPU6000: Main gyro and accelerometer, ST Micro 16-bit gyroscope: Gyroscope, ST Micro 14-bit accelerometer/compass (magnetometer): Accelerometer and compass. Two 10000mAh 4S Li-Po batteries are used to power the circuitry of the ROV, one to power the thrusters and the other to power other electronics [5–7]. EMAX 2213-935KV BLDC motors are used in the 6 thrusters which are controlled using 30A bidirectional ESCs. Raspberry Pi Model 3 B is used to control and store the data received by the camera and sensors like temperature and pressure sensor. A block diagram of the electronics is shown in Figure 15.5.

15.3 Designing of Corrosion Detection Algorithm

15.3.1 Workflow and Algorithm

The flowchart in Figure 15.6 presents the corrosion detection using the ROV. It begins with the ROV being deployed into the water and navigating to the underwater structure. The onboard camera then captures images

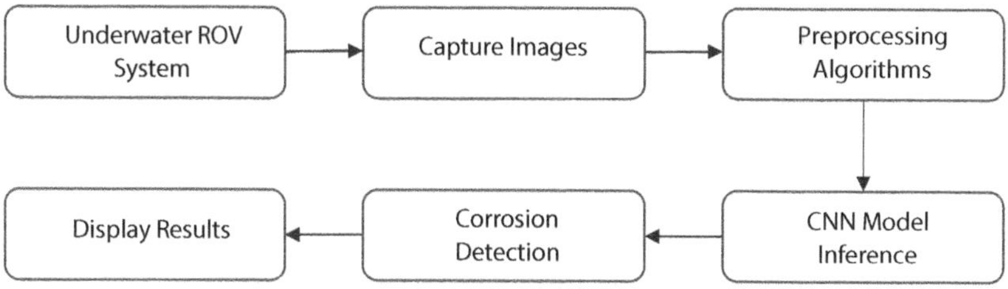

Figure 15.6 Workflow diagram of corrosion detection.

from various angles. These images are fed into the Convolutional Neural Network (CNN) for processing. The CNN identifies and classifies corrosion patterns, distinguishing between different types of corrosion. This information is then used by maintenance teams to prioritize interventions and repairs.

The process begins with data collection, involving the capture of images showcasing both corroded and non-corroded underwater structures using the ROV's camera. A diverse dataset is meticulously constructed, encompassing varying degrees of corrosion and a spectrum of structure types. Following this, data preprocessing steps ensue, where images are normalized and resized to a uniform format suitable for the Convolutional Neural Network (CNN) model [8]. Augmentation techniques are then employed to enrich the dataset, enhancing its variability and bolstering the model's ability to generalize. Moving on to the CNN model design phase, a pre-trained architecture is chosen as the foundation. This model undergoes fine-tuning specifically tailored for corrosion detection *via* transfer learning methods. Subsequently, the dataset is split into training and validation sets for model training. The CNN model is trained with a focus on optimizing for high accuracy in corrosion detection, with hyperparameters adjusted through validation set performance. Post-training, the model undergoes rigorous testing on a separate dataset featuring unseen images of both corroded and non-corroded structures. Performance metrics such as accuracy, precision, recall, and F1-score are meticulously measured to gauge the model's effectiveness presented in Figure 15.7. Finally, for real-world application, the trained CNN model is deployed onto the ROV's onboard system. This implementation is meticulously tailored to ensure seamless compatibility with the ROV's camera board and processing capabilities, enabling real-time corrosion detection during underwater missions.

```
Test Accuracy: 0.9433962106704712
2/2 [==============================] - 1s 223ms/step
Confusion Matrix:
 [[12  0]
 [ 3 38]]

Classification Report:
              precision    recall  f1-score   support

           0       0.80      1.00      0.89        12
           1       1.00      0.93      0.96        41

    accuracy                           0.94        53
   macro avg       0.90      0.96      0.93        53
weighted avg       0.95      0.94      0.95        53
```

Figure 15.7 Accuracy, confusion matrix and classification report.

15.3.2 Corrosion Image Classification

The CNN algorithm, which excels at image recognition tasks, analyzes the images. It seeks to identify and classify different corrosion patterns based on the training it received. The pattern recognition capabilities of CNNs enable them to distinguish between different types of corrosion, such as pitting, galvanic, and crevice corrosion, among others shown in Figure 15.8.

Once the CNN has processed the images, it outputs data detailing the type, location, and extent of any corrosion it has detected. This data is then used by maintenance teams to make informed decisions about necessary interventions and repairs. By utilizing the CNN algorithm, the corrosion detection process becomes much more efficient, accurate, and safe compared to traditional manual inspections.

The dataset used in our study consisted of a large number of corrosion images, collected from various underwater structures such as ship hulls, hydropower plants, and turbines. The dataset was diverse, encompassing different types of corrosion such as pitting, galvanic, and crevice corrosion, among others. Prior to feeding these images into our Convolutional Neural Network (CNN) model, they were pre-processed to ensure optimal results. This pre-processing step involved resizing the images, enhancing the contrast for better visibility of corrosion patterns, and normalizing the pixel values. The rich and diverse dataset of corrosion images played a crucial role in training our CNN model to accurately identify and classify different corrosion patterns.

Figure 15.8 Dataset of corrosion images.

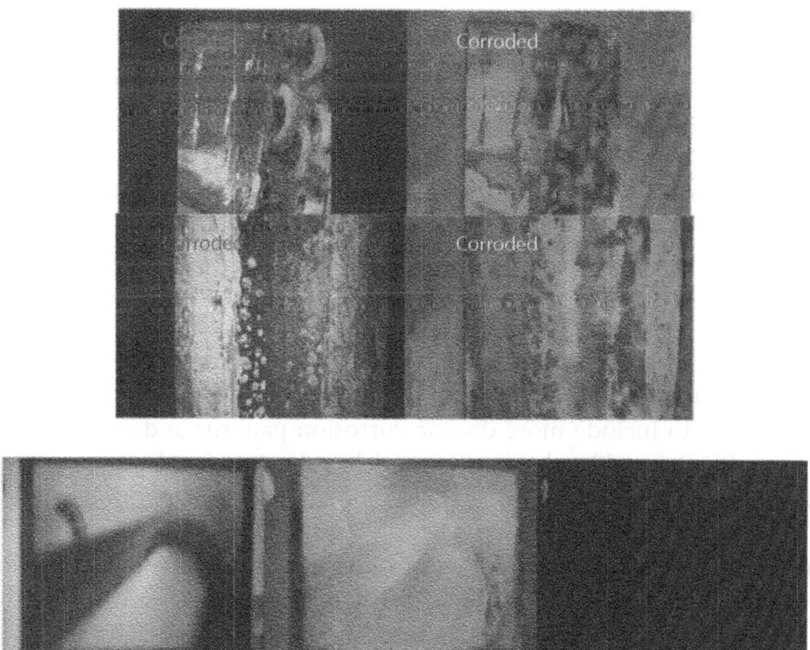

Figure 15.9 Results of corrosion detection using CNN classification model.

15.4 Experimental Results

The experimental results demonstrated in Figure 15.9 show the effectiveness of our CNN-based approach for underwater corrosion detection. With an accuracy of 94.3% and an overall F1-score of 95%, the model showcases robust performance in identifying and classifying various corrosion patterns. This level of accuracy is crucial for maintenance teams, as it provides reliable insights into the condition of underwater structures, enabling timely interventions to prevent catastrophic failures.

The high precision and recall values further highlight the model's ability to minimize false positives and false negatives. This ensures that maintenance efforts are targeted towards areas with genuine corrosion issues, optimizing resource allocation and minimizing downtime for underwater installations. The confusion matrix reveals that the model performed exceptionally well across all types of corrosion, with minimal misclassifications. This indicates that the CNN model was successful in learning distinctive features of each corrosion pattern, enabling accurate classification.

15.5 Conclusion

In conclusion, our proposed CNN-based approach offers a transformative solution for underwater corrosion detection, leveraging the power of computer vision and deep learning. By deploying this model on a Remotely Operated Vehicle (ROV) with 4 degrees of freedom, maintenance teams can efficiently and accurately assess the integrity of underwater structures. This not only reduces the risks associated with corrosion-related failures but also leads to substantial cost savings through targeted and timely maintenance interventions. Further research avenues could include real-time implementation of the model on the ROV, allowing for on-the-fly corrosion detection during underwater missions. Additionally, the expansion of the dataset to include more diverse corrosion patterns and environmental conditions could enhance the model's robustness and generalization capabilities.

AI disclosure statement: The author confirms that no artificial intelligence (AI)-generated content or AI-assisted writing tools were used in the preparation of this chapter (including summaries and the abstract). All text, analysis, and conclusions are the result of the author's original work.

References

1. Aguirre-Castro, O.A., Inzunza-González, E., García-Guerrero, E.E., Tlelo-Cuautle, E., López-Bonilla, O.R., Olguín-Tiznado, J.E., Cárdenas-Valdez, J.R., Design and Construction of an ROV for Underwater Exploration. *Sensors*, *19*, 24, 5387, 2019.
2. Azis, F.A., Aras, M.S.M., Rashid, M.Z.A., Othman, M.N., Abdullah, S.S., Problem identification for underwater remotely operated vehicle (ROV): A case study. *Procedia Eng.*, *41*, 554–560, 2012.
3. Bogue, R., Underwater robots: a review of technologies and applications. *Ind. Robot Int. J.*, *42*, 3, 186–191, 2015.
4. Tehrani, N.H., Heidari, M., Zakeri, Y., Ghaisari, J., Development, depth control and stability analysis of an underwater Remotely Operated Vehicle (ROV), in: *IEEE ICCA 2010*, 2010, June, IEEE, pp. 814–819.
5. Xie, L. and Yuille, A., Genetic CNN, in: *Proceedings of the IEEE International Conference on Computer Vision*, pp. 1379–1388, 2017.
6. Alzubaidi, L., Zhang, J., Humaidi, A.J., Al-Dujaili, A., Duan, Y., Al-Shamma, O., Farhan, L., Review of deep learning: concepts, CNN architectures, challenges, applications, future directions. *J. Big Data*, *8*, 1–74, 2021.
7. Hu, J.Y., Zhang, S.S., Chen, E., Li, W.G., A review on corrosion detection and protection of existing reinforced concrete (RC) structures. *Constr. Build. Mater.*, *325*, 126718, 2022.
8. Atha, D.J. and Jahanshahi, M.R., Evaluation of deep learning approaches based on convolutional neural networks for corrosion detection. *Struct. Health Monit.*, *17*, 5, 1110–1128, 2018.

16

Metamaterial Integrated Wearable Ultrawide Bandwidth Antenna with Low Specific Absorption Rate

S. Prasad Jones Christydass[1]*, S. Suresh Kumar[2] and Yerumbu Nandakishora[3]

[1]Department of ECE, SRM Institute of Science and Technology, Kattankulathur Campus, Chennai, Tamil Nadu, India
[2]Department of Computational Intelligence, SRM Institute of Science and Technology, Kattankulathur Campus, Chennai, Tamil Nadu, India
[3]Department of Mechanical Engineering, Presidency University, Innovation & Translational Research Hub, Presidency University, Bangalore, Karnataka, India

Abstract

This article presents a wearable antenna capable of functioning at ultra-wideband frequencies. The characteristics of denim jean fabrics include a dielectric constant of 1.7 and a loss tangent of 0.02, both of which are utilised in antenna manufacturing. The antenna's overall dimensions are $45 \times 44.92 \times 0.5$ mm^3. It utilises a coplanar waveguide feeding method for signal transmission. To get a total bandwidth of 6767 MHz, it functions throughout a broad frequency spectrum ranging from 4.29 to 11.09 GHz. The antenna generates an omni-directional radiation pattern with a gain exceeding 2.4 dBi throughout the entire operational bandwidth. It is an optimal selection due to its simple design, ease of fabrication, consistent radiation properties, and dependable performance for gain and directivity in C and X band wireless applications.

***Keywords*:** Textile, ultrawide band, coplanar waveguide, jean

**Corresponding author*: prasadjc@srmist.edu.in

Abhishek Kumar and Suman Lata Tripathi (eds.) Smart Chips for Smart Devices: VLSI Design for Next-Generation IoT Solutions, (291–306) © 2026 Scrivener Publishing LLC

16.1 Introduction

The tremendous advancements that have been made in wearable technology over the course of the past two decades have had a significant impact on a sizeable number of different industries, including the military, healthcare, and medical applications, amongst others [1]. Wearable technology of the modern era is distinguished by important characteristics such as a lightweight design, low cost, and versatility [2–5]. Real-time monitoring, emergency warnings, and position detection are just some of the applications that can be made possible by these devices, which are characterised by their mobility, which is a crucial feature. Due to the fact that these devices are essentially mobile, antennas represent an essential component in the process of transferring information to remote monitoring centres [6–9]. Both a low return loss and a decent gain are required for an antenna to be deemed efficient. Only then can it be considered efficient. Because of their consistent radiation patterns, low profile, and ease of manufacture, microstrip patch antennas [10, 11] are chosen over conventional forms among the available options. This is among the reasons why they are preferred. Because of this, they are suitable for the wireless standards that are now being utilised [12, 13]. Wearable antennas are often constructed using a radiating element that is positioned on a textile substrate. This is the case in the majority of cases. These antennas perform extraordinarily well because of the fact that they are located so close to the human body, which is naturally lossy. The fact that they are particularly favourable is one of the reasons why they are so. Not only are habiliment antennas lightweight and stretchy, but they are also inexpensive, easy to produce, and may be easily incorporated into clothing. In addition, they are lightweight and stretchy. Because of the increasing demand for wearable devices that are able to operate across several frequency bands and are small in size, antennas that are able to handle numerous frequency bands while yet preserving their compactness have become an absolute necessity [14]. These requirements can be satisfied by using several antennas, but doing so will result in an increase in the size of the device [15]. Developing antennas with multiband or larger impedance bandwidth characteristics is one way to overcome this limitation. This can be accomplished through the use of techniques such as defective ground structures, slots in the patch or ground, parasitic patches, fractal designs, and metamaterials [16–19]. The substrate material is an important component in applications that include wearable technology. Properties such as dielectric constant, adhesive strength, and mechanical durability are frequently taken into consideration when selecting an appropriate textile material [20–26]. Examples of such materials include cotton, felt, and denim. Miniaturisation can be accomplished with substrates that have a high dielectric constant [27–30], although use of such substrates may result in spurious radiation and decreased flexibility. Consequently, wearable antennas have become important for applications that include Wireless Body Area Networks (WBAN) [31], Body-Centric Electromagnetic Telecommunications (BCEC) [32], and on-body and off-body networks [33]. There are a great number of distinct antenna designs that have been given in the published works for consideration. Planar Upturned F Antennas [34], Substrate Incorporated Waveguides [35], Microstrip Patch A [36], Electronic Band Gap [37], and metamaterials are some of the designs that have been developed. On the other hand, a sizeable portion of these systems are hampered by irregular radiation patterns, low gain, narrow bandwidth, and single-band operation. These are all problems that they face. This article presents an easy printed antenna in the shape of a half moon that is inspired by metamaterial engineering. The antenna is presented in this article. Specifically designed and built out of denim material, the antenna has a compact footprint that measures 45 millimetres by 44.92 millimetres. It was conceived and made utilising denim material. This gadget is able to function across a wide bandwidth that spans from 4.29 GHz all the way up to 11.09 GHz. A significant increase in the antenna's bandwidth can be achieved through the usage of a metamaterial structure, which does not result in the antenna in any way being significantly larger. Following the discussion of the data in Section 16.3, which is followed by the presentation of the conclusion in Section 16.4, Section 16.2 discusses the technique that will be used for the projected antenna.

16.2 Design of the Half-Moon-Shaped Metamaterial-Inspired Wearable Printed Antenna

The proposed antenna is made up of a Half-Moon-Shaped patch that is combined with a complementary split ring resonator. This patch serves as the radiating part of the antenna. A coplanar waveguide (CPW) line is employed in order to get it as a result of this usage. Denim is a type of textile material that is utilised in the design and construction of the antenna. It possesses a dielectric constant of 1.7 and a loss tangent of 0.02, and it is utilised in both of these aspects. In this particular instance, the denim substrate that was utilised had a thickness of 0.5 millimetres. As a result of the ground plane being incorporated into the construction, there is a defect in the ground plane. Wg and Lg are the abbreviations that are used to refer to the width and length of the ground plane, respectively. Lf is the abbreviation for the length of the feed line, and Wf is the abbreviation for the width of the feed line. The feed line is the component that is responsible for supplying radiation to the antenna. For the denim substrate, which is the base material, the complete measurements are expressed as W for width and L for length. This is because the denim substrate acts as the basis material. There are two unique stages of development that are involved in the process of generating the antenna, as depicted in Figure 16.1. Figure 16.2 provides a visual representation of the parameters, as well as the geometry of the final antenna design. The millimetre values of the parameters are listed in Table 16.1. The simulation and design optimisation process was carried out in its entirety with the assistance of CST EM Studio.

In order to create the proposed antenna, it is necessary to migrate from a basic half-moon-shaped patch to a more advanced configuration that incorporates an element that is inspired by hexagonal metamaterials. Not only does this metamaterial design contribute significantly to the enhancement of the antenna's radiation performance and bandwidth, but it also helps to maintain the antenna's tiny size. It is now extremely appropriate for applications like wireless local area networks (WLAN), satellite communication, and ultra-wideband (UWB) systems as a result of these advances. The early design, which consisted of nothing more than the patch in the shape of a half moon, was constrained by its narrower bandwidth and moderate efficiency standards. The developed design, on the other hand, makes use of the one-of-a-kind electromagnetic features of metamaterials in order to obtain superior operational characteristics. In addition, the utilisation of denim as the substrate guarantees flexibility, durability, and

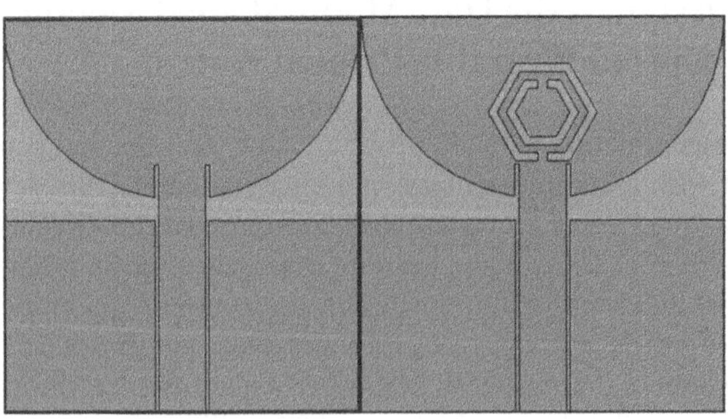

Figure 16.1 Stages of half-moon-shaped metamaterial-inspired wearable printed antenna.

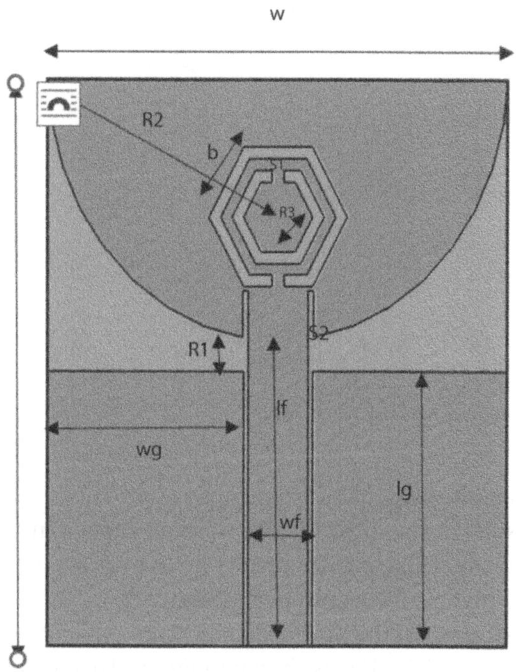

Figure 16.2 Geometry parameters of half-moon-shaped metamaterial-inspired wearable printed antenna.

Table 16.1 Parameters of half-moon-shaped metamaterial-inspired wearable printed antenna (values in mm).

S2	**wg**	**b**	**S1**	**wf**	**lf**	**l**	**w**
0.50	45	12.6	0.75	2	11.5	30	24
R1	R2	R3	t	h	x	y	lg
4.5	10	3.8	0.0035	0.5	0.25	1	13.5

comfort, which makes the antenna suitable for applications that include wearable technology.

Figure 16.3 depicts the S11 plot for Antenna A, which has a straightforward patch design in the shape of a half moon. The frequency range that this antenna covers is from 3.2 to 8.27 gigahertz, and it demonstrates wideband resonance when it is used. Through the incorporation of metamaterial components into the design, the wideband frequency range is

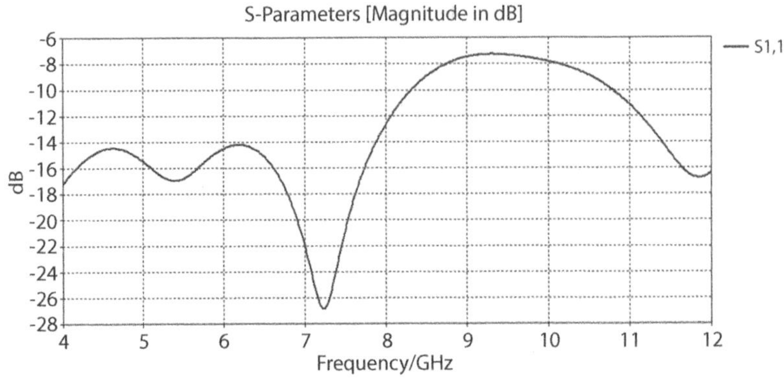

Figure 16.3 Antenna A S11 performance.

turned into an ultrawideband frequency range. The incorporation of metamaterials results in the introduction of more capacitance, which shifts to a higher frequency. This, in turn, results in an improvement in the antenna's bandwidth as well as its overall performance.

A demonstration of the proposed antenna's capacity to create an ultrawide bandwidth of 6767 MHz is shown in Figure 16.4, which displays the S11 plot of the antenna. The fractal structure that was included into the design of the antenna is responsible for the wide bandwidth that it possesses. The impedance is efficiently matched across the whole resonant range using an inset feed in conjunction with a decreased ground structure. This is made possible by the fact that the ground structure is reduced. The voltage standing wave ratio (VSWR) plot is shown in Figure 16.5, which demonstrates

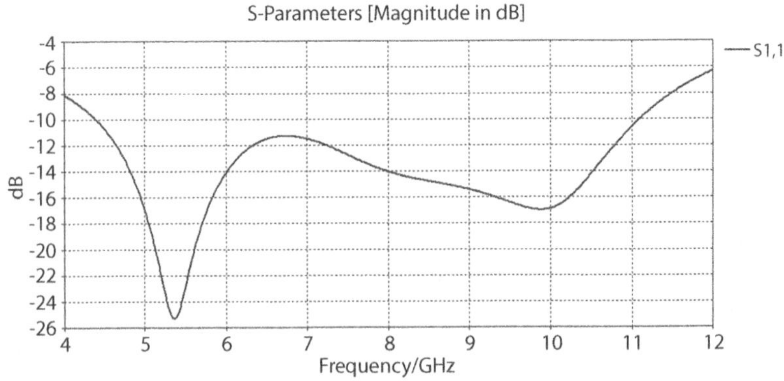

Figure 16.4 Return loss plot of proposed half-moon-shaped metamaterial-inspired wearable printed antenna.

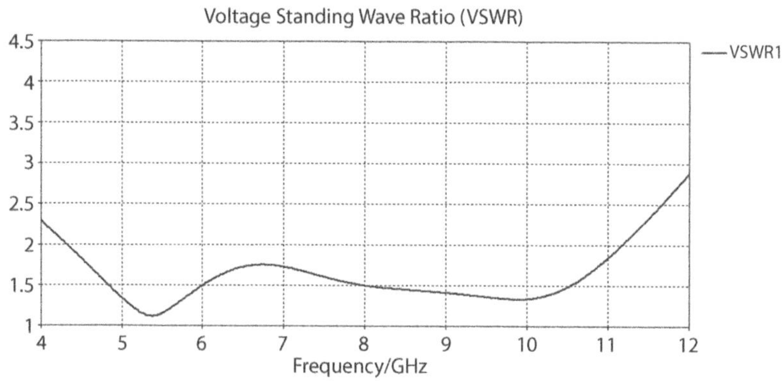

Figure 16.5 VSWR of proposed half-moon-shaped metamaterial-inspired wearable printed antenna.

that the VSWR percentage is constantly kept below 2. Exhibiting its efficiency and adaptability for ultrawideband applications, the antenna operates between 4.29 and 11.09 GHz, demonstrating its overall functionality.

16.3 Result and Discussion

Through the use of parametric analysis in CST EM Studio, the important parameters of the antenna were analysed in order to ascertain the final optimised values for construction. The hexagonal side length (s), the ground length (lg), and the inner radiating slot radius (R3) were found to be the three most important characteristics that directly influence the performance of the antenna. The ground length of the antenna was gradually changed in increments of one millimetre, ranging from 9.65 millimetres to 11.65 millimetres, and the effect that this process had on return loss was explored. This can be seen in Figure 16.6, which shows the results of the investigation. When the ground length is 10.56 millimetres, it is evident that the suggested antenna achieves optimal impedance matching and displays an enhanced impedance bandwidth. This is evident when the findings shown in Figure 16.6 are taken into consideration. Consequently, the antenna is suitable for the applications that were intended to be carried out with it. The inner radius of the hexagonal metamaterial slot begins at 4.6 millimetres and gradually increases to 6.6 millimetres. This steady increase in inner radius is characterised by a progressive increase of one mm. Figure 16.7 provides a visual representation of the impact that varying slot radii have on the frequency at which the device operates. The figure makes it

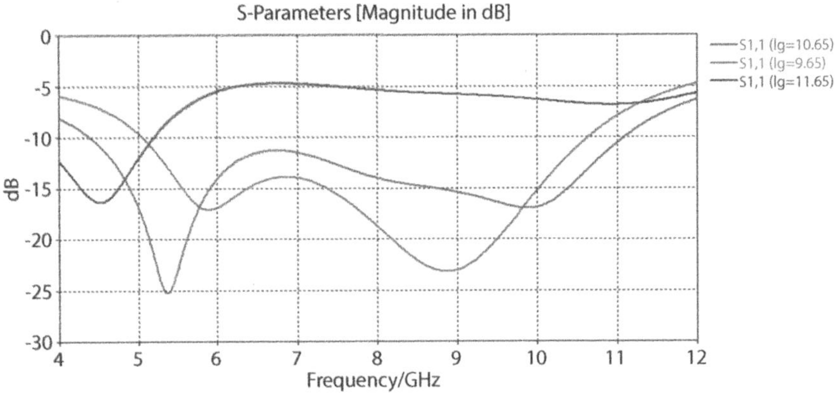

Figure 16.6 Effect of round length in return loss.

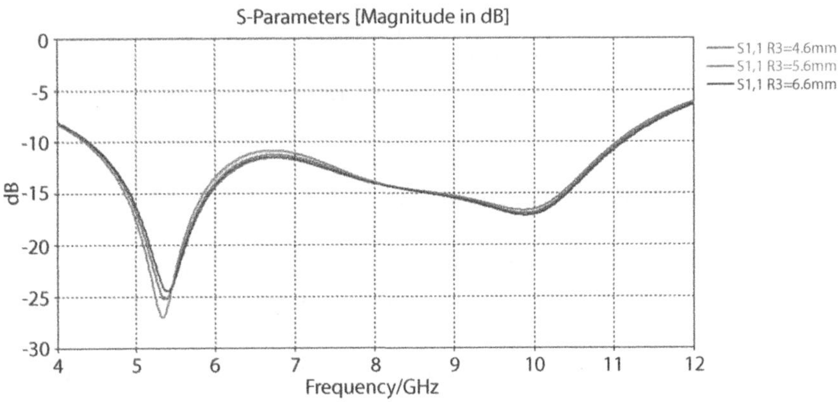

Figure 16.7 Effect of R3 in return loss.

plainly evident that the slot radius of 5.6 mm provides the best possible impedance matching as compared to other variations. After careful consideration, this particular value has been chosen to represent the culmination of the manufacturing process.

Immediately after that, a parametric analysis is performed on the length of the hexagonal side that is most external to the structure. Figure 16.8 is shown in order to illustrate the influence that the side length has on the S11 parameter. In order to alter the length of the side, increments of 0.6 millimetres are used, and the range of adjustment is from 12 millimetres to 13.2 millimetres. It has been found, on the basis of the findings of the research, that a side length of 12.6 millimetres results in an extremely wide

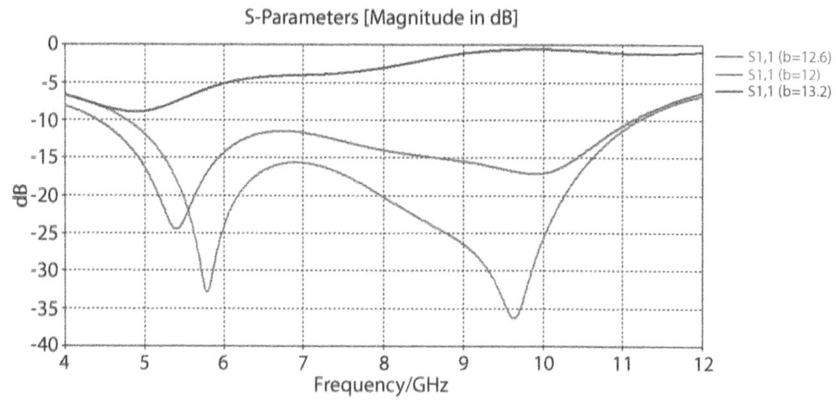

Figure 16.8 Effect of hexagonal side length in return loss.

bandwidth for the structure that has been suggested. As a consequence of this, this strategy is the most appropriate choice for the construction that will be completed.

A graphical representation of the surface current circulation of the proposed textile antenna is shown in Figure 16.9 at two different resonant frequencies. At a frequency of 5.56 gigahertz, the most significant current is concentrated at the lengthwise boundaries of the hexagonal slot that is situated on the exterior. When, on the other hand, the frequency is 9.95 GHz, the majority of the accumulation occurs at the intersections of the metamaterial structure. Because the current is distributed uniformly throughout

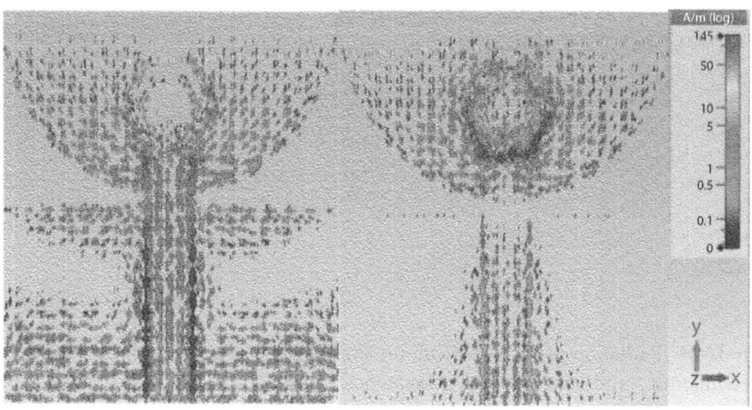

Figure 16.9 Half-moon-shaped metamaterial-inspired wearable printed antenna – surface current at different resonating frquencies (in GHz).

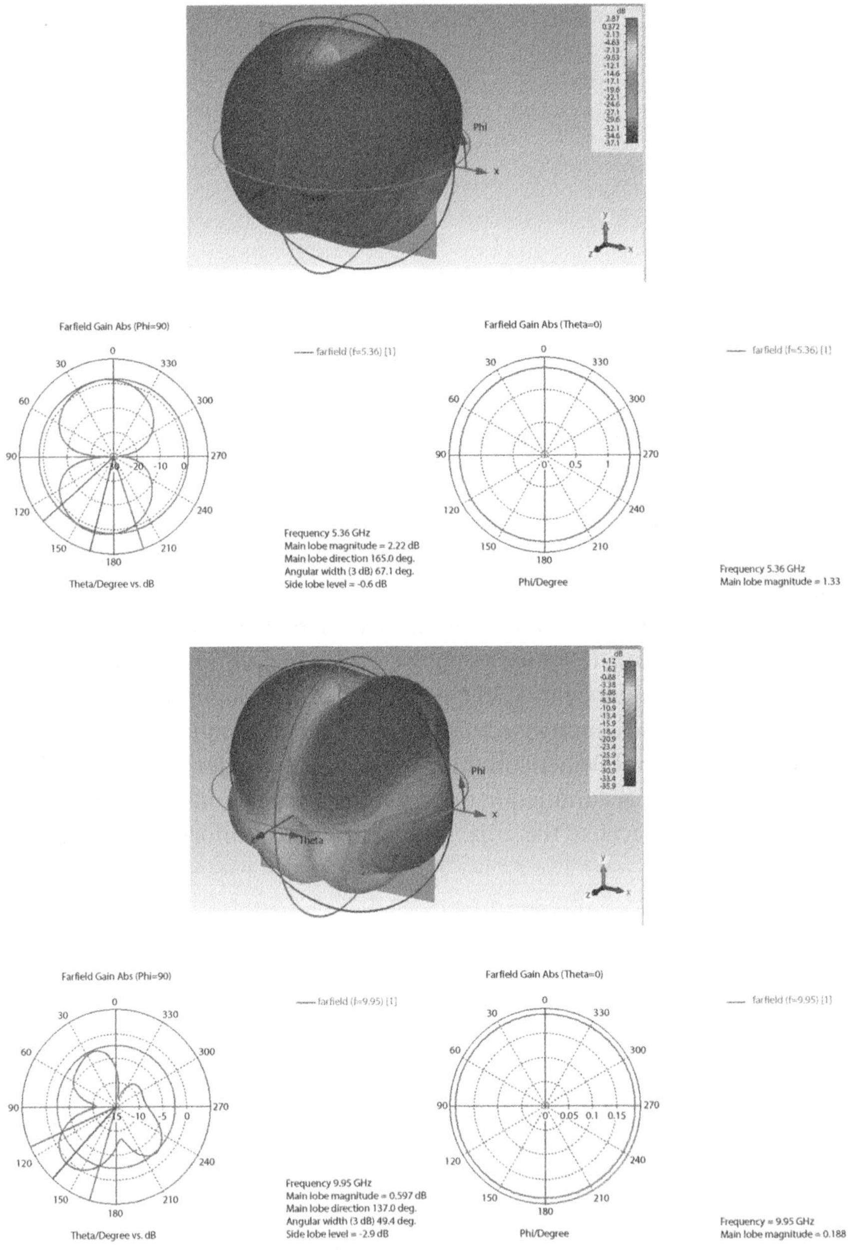

Figure 16.10 Radiation pattern - half-moon-shaped metamaterial-inspired wearable printed antenna – different resonating frequencies (in GHz).

the radiating element across the entirety of the resonating band, an omnidirectional radiation pattern is produced as a result of this distribution.

In Figure 16.10, the three-dimensional radiation pattern is depicted for frequencies of 5.56 and 9.95, respectively. Within both frequency ranges, the antenna displays a radiation pattern that is omnidirectional in nature. This pattern is shown by the antenna. As opposed to the H-plane radiation pattern, which is distinguished by its omnidirectional nature, the E-plane

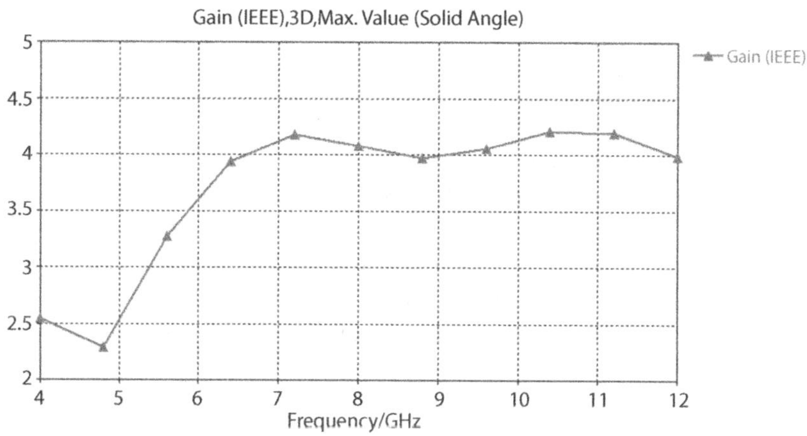

Figure 16.11 Half moon-shaped metamaterial-inspired wearable printed antenna – gain vs. frequency plot.

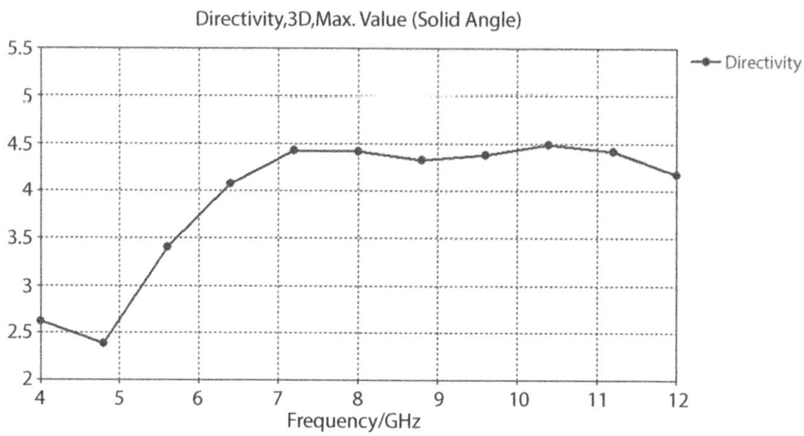

Figure 16.12 Half-moon-shaped metamaterial-inspired wearable printed antenna – directivity vs. frequency plot.

radiation pattern is formed like a figure eight. This is in contrast to the H-plane radiation pattern.

Figure 16.11 is a representation of the antenna gain. The graphic makes it abundantly evident that the antenna is capable of achieving a gain that is larger than 2.4 dBi across the entirety of the resonant band. This figure was created to illustrate this capability. In Figure 16.12, the directivity of the antenna is depicted. It is also demonstrated that the antenna's directivity remains consistently high over the whole frequency range in which it operates.

16.4 SAR Analysis

Figure 16.13 depicts the proposed antenna in conjunction with the fragmented human phantom model that is located within the CST simulation environment. According to what was mentioned earlier, the proximity of the antenna to the skin, which is a lossy medium, has a considerable impact on the frequency response of the antenna. The IEEE/IEC 62704-1 standard, which was implemented on an average basis across 10 grammes, was utilised in order to evaluate the specific absorption rate (SAR) of biological tissues. The surface area ratio (SAR) was calculated by employing the following formula: SAR = $\sigma|E|2/\rho$, where σ constitutes the conductivity

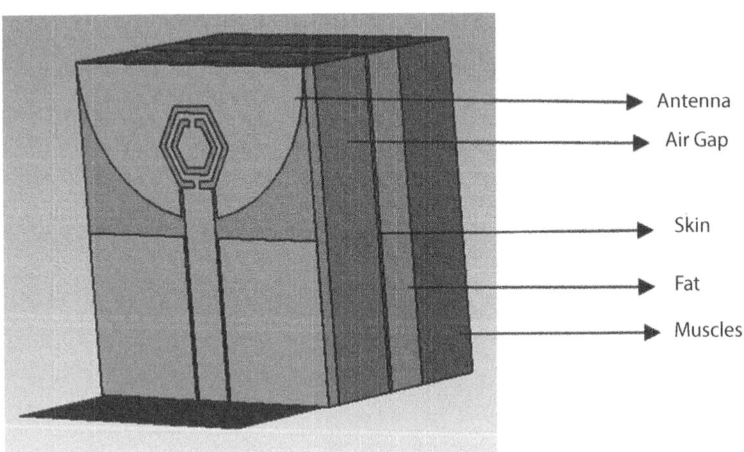

Figure 16.13 Half-moon-shaped metamaterial-inspired wearable printed antenna – SAR analysis setup.

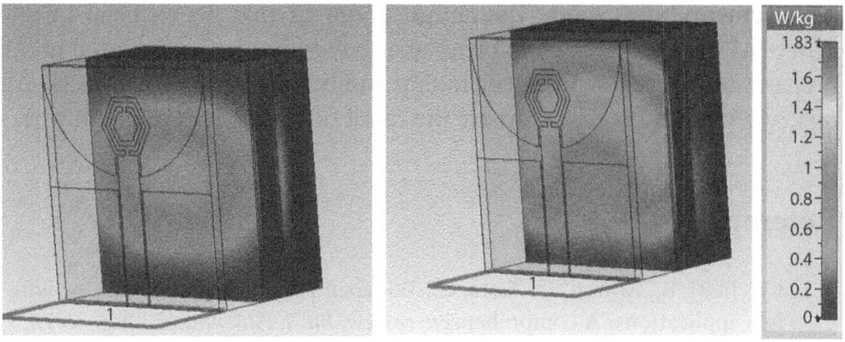

Figure 16.14 Half-moon-shaped metamaterial-inspired wearable printed antenna - SAR value at different resonant frequencies (in GHz).

of the tissue, E signifies the magnitude of the electric field, and ρ represents the mass density of the tissue. The peak SAR values that were determined stayed lower than 2 W/kg, which is an indication that they complied with the safety guidelines.

The results of the simulated SAR are shown in Figure 16.14, and they demonstrate that the SAR values at all resonant frequencies remain below the limit of 2 W/kg. Taking into consideration these observations, it appears that the proposed antenna is suitable for use in wearable applications.

16.5 Conclusion

In this article, an easy hexagonal fractal antenna is presented. This antenna is designed to be utilised in applications that involve the S, C, and X bands. It is possible for the antenna to achieve a wide bandwidth of 6767 MHz by making use of the fractal design feature. In order to be of aid to the design process, the results of the parametric analysis were presented to the audience. Critical parameter values were discovered while the procedure was being carried out. It is observed that the antenna has a directivity of 4.5 dBi. The manufacture of the antenna, which operates within a frequency band that extends from 4.29 to 11.09 GHz, is accomplished using denim serving as the substrate. With a maximum gain of 4.25 dBi, it displays a radiation pattern that is uniform and stable, which makes it appropriate for applications in the C band as well as the X band.

AI disclosure statement: The author confirms that no artificial intelligence (AI)-generated content or AI-assisted writing tools were used in the preparation of this chapter (including summaries and the abstract). All text, analysis, and conclusions are the result of the author's original work.

References

1. Cicchetti, R., Miozzi, E., Testa, O., Wideband and UWB antennas for wireless applications: A comprehensive review. *Int. J. Antennas Propag., 2017*, 1, 2390808, 2017.
2. Poffelie, L.A.Y., Soh, P.J., Yan, S., Vandenbosch, G.A., A high-fidelity all-textile UWB antenna with low back radiation for off-body WBAN applications. *IEEE Trans. Antennas Propag., 64*, 2, 757–760, 2015.
3. Ashyap, A.Y., Abidin, Z.Z., Dahlan, S.H., Majid, H.A., Waddah, A.M.A., Kamarudin, M.R., Noras, J.M., Inverted E-shaped wearable textile antenna for medical applications. *IEEE Access, 6*, 35214–35222, 2018.
4. Ashyap, A.Y., Abidin, Z.Z., Dahlan, S.H., Majid, H.A., Shah, S.M., Kamarudin, M.R., Alomainy, A., Compact and low-profile textile EBG-based antenna for wearable medical applications. *IEEE Antennas Wirel. Propag. Lett., 16*, 2550–2553, 2017.
5. Soh, P.J., Van den Bergh, B., Xu, H., Aliakbarian, H., Farsi, S., Samal, P., Nauwelaers, B.K., A smart wearable textile array system for biomedical telemetry applications. *IEEE Trans. Microwave Theory Tech., 61*, 5, 2253–2261, 2013.
6. Crespo-Bardera, E., Sanchez-Fernandez, M., Garcia-Armada, A., Martin, A.G., Duran, A.F., Analysis of a LTE-based textile massive MIMO proposal for public safety networks, in: *2017 IEEE 86th Vehicular Technology Conference (VTC-Fall)*, 2017, September, IEEE, pp. 1–5.
7. Mohd Hussin, E.F.N., Soh, P.J., Abdul Malek, M.F., Jamlos, M., Fauzi, M., Lago, H., Rahim, H., A wearable textile dipole for search and rescue application, 2016.
8. Kaivanto, E.K., Berg, M., Salonen, E., De Maagt, P., Wearable circularly polarized antenna for personal satellite communication and navigation. *IEEE Trans. Antennas Propag., 59*, 12, 4490–4496, 2011.
9. Shanthi, S., Jayasankar, T., Christydass, P.J., Venkatesh, P.M., Wearable textile antenna for GPS application. *Int. J. Sci. Technol. Res., 8*, 11, 2019.
10. Ramya, N., Sujatha, M., Jayasankar, T., Christydass, P.J., Metamaterial inspired circular antenna with DGS for tetra band application. *Int. J. Control Autom., 13*, 2, 877–882, 2020.
11. Baudha, S. and Kumar, V.D., Miniaturized dual broadband printed slot antenna with parasitic slot and patch. *Microwave Opt. Technol. Lett., 56*, 10, 2260–2265, 2014.

12. Chaurasia, P., Kanaujia, B.K., Dwari, S., Khandelwal, M.K., Design and analysis of seven-bands-slot-antenna with small frequency ratio for different wireless applications. *AEU Int. J. Electron. Commun.*, *99*, 100–109, 2019.
13. Liu, G., Fang, M., Zhi, R., Bai, J., Zeng, Z., Compact CPW-fed multiband antenna for TD-LTE/WLAN/WiMAX applications. *Prog. Electromagn. Res. Lett.*, *65*, 9–14, 2017.
14. Chougule, J.S. and Wali, D.U.V., Design of Flexible Microstrip Antenna for wearable application. *Int. J. Res. Emerging Sci. Technol.*, *2*, 6, 144–147, 2015.
15. Sidhu, E., Mittal, D., Saini, S.S., Singh, C., Kaur, R., Flexible microstrip patch antenna designs for Bluetooth, IMT, WLAN and WiMAX applications, in: *2017 Progress in Electromagnetics Research Symposium - Spring (PIERS)*, 2017, May, IEEE, pp. 953–957.
16. Christydass, S.P.J. and Gunavathi, N., Octa-Band Metamaterial Inspired Multiband Monopole Antenna for Wireless Application. *Prog. Electromagn. Res. C*, *113*, 2021.
17. Wa'il, A., Shaaban, R.M., Tahir, A., Design, simulation and measurement of triple band annular ring microstrip antenna based on shape of crescent moon. *AEU Int. J. Electron. Commun.*, *117*, 153133, 2020.
18. Shan, K., Ruan, C.L., Peng, L., Design of a novel planar ultrawideband antenna with 3.5 and 5.5 GHz dual band-notched characteristics. *Microwave Opt. Technol. Lett.*, *53*, 2, 370–375, 2011.
19. Sam, P.J.C. and Gunavathi, N., A tri-band monopole antenna loaded with circular electric–inductive–capacitive metamaterial resonator for wireless application. *Appl. Phys. A*, *126*, 1–11, 2020.
20. Iqbal, A., Bouazizi, A., Saraereh, O.A., Basir, A., Gangwar, R.K., Design of multiple band, meandered strips connected patch antenna. *Prog. Electromagn. Res. Lett.*, *79*, 51–57, 2018.
21. Basar, M.R., Hossain, M.A., Hoque, M.R.U., Morshed, K.M., Compact dual l-slit slotted antenna for spacecraft, WLAN, Wi-Fi, and bluetooth application, in: *2012 7th International Conference on Electrical and Computer Engineering*, 2012, December, IEEE, pp. 706–708.
22. Sarkar, S., Three Dual-Band and Dual-Linearly Polarized Antenna Configurations for UHF-RFID and WLAN Applications. *IEEE J. Radio Freq. Identif.*, 2024.
23. Aldhaibani, J.A., Mohammed, M.Q., Mahmood, A.A., Development of wearable textile patch antenna 2.43 GHz for biomedical applications. *Int. J. Adv. Technol. Eng. Explor.*, *11*, 111, 177, 2024.
24. Rais, N.H.M., Soh, P.J., Malek, F., Ahmad, S., Hashim, N.B.M., Hall, P.S., A review of wearable antenna, in: *2009 Loughborough Antennas & Propagation Conference*, 2009, November, IEEE, pp. 225–228.
25. Corchia, L., De Benedetto, E., Monti, G., Cataldo, A., Tarricone, L., Wearable antennas for applications in remote assistance to elderly people, in: *2017 IEEE International Workshop on Measurement and Networking (M&N)*, 2017, September, IEEE, pp. 1–6.

26. Gao, Z., Zhou, Y., Zhang, J., Foroughi, J., Peng, S., Baughman, R.H., Wang, C.H., Advanced Energy Harvesters and Energy Storage for Powering Wearable and Implantable Medical Devices. *Adv. Mater.*, *36*, 42, 2404492, 2024.
27. Priya, A., Kumar, A., Chauhan, B., A review of textile and cloth fabric wearable antennas. *Int. J. Comput. Appl.*, *116*, 17, 2015.
28. Nadh, B.P., Madhav, B.T., Kumar, M.S., Rao, M.V., Anilkumar, T., Asymmetric ground structured circularly polarized antenna for ISM and WLAN band applications. *Prog. Electromagn. Res. M*, *76*, 167–175, 2018.
29. Krishna, R.R. and Kumar, R., A dual-polarized square-ring slot antenna for UWB, imaging, and radar applications. *IEEE Antennas Wirel. Propag. Lett.*, *15*, 195–198, 2015.
30. Hong, Y., Tak, J., Choi, J., An all-textile SIW cavity-backed circular ring-slot antenna for WBAN applications. *IEEE Antennas Wirel. Propag. Lett.*, *15*, 1995–1999, 2016.
31. Yan, S., Soh, P.J., Vandenbosch, G.A., Wearable dual-band magneto-electric dipole antenna for WBAN/WLAN applications. *IEEE Trans. Antennas Propag.*, *63*, 9, 4165–4169, 2015.
32. Simorangkir, R.B., Yang, Y., Matekovits, L., Esselle, K.P., Dual-band dual-mode textile antenna on PDMS substrate for body-centric communications. *IEEE Antennas Wirel. Propag. Lett.*, *16*, 677–680, 2016.
33. Yan, S., Volskiy, V., Vandenbosch, G.A., Compact dual-band textile PIFA for 433-MHz/2.4-GHz ISM bands. *IEEE Antennas Wirel. Propag. Lett.*, *16*, 2436–2439, 2017.
34. Moro, R., Agneessens, S., Rogier, H., Bozzi, M., Circularly-polarised cavity-backed wearable antenna in SIW technology. *IET Microwaves Antennas Propag.*, *12*, 1, 127–131, 2018.
35. Yahya, R., Kamarudin, M.R., Seman, N., Effect of rainwater and seawater on the permittivity of denim jean substrate and performance of UWB eye-shaped antenna. *IEEE Antennas Wirel. Propag. Lett.*, *13*, 806–809, 2014.
36. Lin, X., Seet, B.C., Joseph, F., Li, E., Flexible fractal electromagnetic bandgap for millimeter-wave wearable antennas. *IEEE Antennas Wirel. Propag. Lett.*, *17*, 7, 1281–1285, 2018.
37. Tak, J. and Choi, J., A wearable metamaterial microwave absorber. *IEEE Antennas Wirel. Propag. Lett.*, *16*, 784–787, 2016.

17

Industry 5.0: Opportunities, Challenges, and Potential Directions for Future Research

Pankaj Kumar Keshri

Computer Science Engineering, Lovely Professional University, Phagwara, Punjab, India

Abstract
This chapter describes Industry 5.0 and the challenges, opportunities and potential areas for future research that it presents. Industry 5.0 is bringing about a revolution in technology innovation and human-machine cooperation for the foundation of future growth. By using customized products, the industrial revolution is increasing client happiness. Industry 5.0 is necessary for today's business with the paid technology advancements to provide factories with economic growth and competitive advantages.

***Keywords*:** Industry 5.0, cloud computing, 6G, big data analytics, IoT, AI, ML

17.1 Introduction

Industry 4.0 is now on edge; after one decade of success in addressing the industry's deficiencies, the time has come for Industry 5.0. Industry 4.0 is limited since smart factories are raising corporate productivity. Industry 5.0, which emphasizes the integration of cutting-edge technologies with human-centric approaches, marks a revolutionary transformation in the manufacturing environment [1–3]. This new paradigm aims to improve human-machine collaboration while balancing automation and artificial intelligence with ecological practices. Industries must manage worker

Email: pankaj.16384@lpu.co.in

Abhishek Kumar and Suman Lata Tripathi (eds.) Smart Chips for Smart Devices: VLSI Design for Next-Generation IoT Solutions, (307–324) © 2026 Scrivener Publishing LLC

adaptation, ethical issues, and technology integration while they take advantage of the opportunities brought forth by greater customization, efficiency, and resilience. To solve these issues and find creative solutions that maximize Industry 5.0's potential while maintaining environmental sustainability and social responsibility, future research areas will be essential [4–5].

17.1.1 Industry 5.0 Technologies

Cloud computing, blockchain, big data analytics, the Internet of Things, and 6G networks are some of the enabling technologies for Industry 5.0 as represented in Figure 17.1.

17.1.2 Cloud Computing

Industry 5.0, which emphasizes the collaboration of humans and cutting-edge technologies to develop more customized and effective production processes, heavily relies on cloud computing. The following is how Industry 5.0 technologies are supported by cloud computing:

(a) **Data Integration and Management:** Cloud computing offers the framework for integrating and managing the

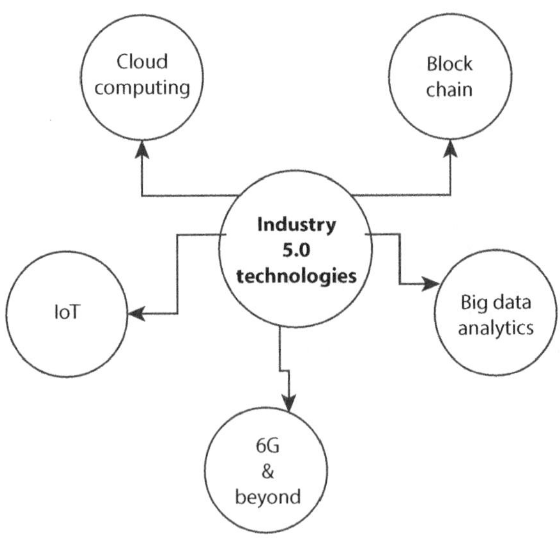

Figure 17.1 Industry 5.0 technologies.

massive volume of data produced by Industry 5.0 technologies, including robotics, sensors, and Internet of Things devices. Making educated decisions and streamlining procedures depend on the ability to store data centrally and interpret it in real time.

(b) **Scalability:** The cloud provides adaptive resources that can be changed in response to demand. For Industry 5.0 applications, where workloads may fluctuate owing to shifting production requirements or adjustments in data volumes, this flexibility is essential.

(c) **Collaboration and Connectivity:** Cloud platforms provide smooth communication and cooperation across various teams and departments as well as between humans and robots. A common feature of Industry 5.0 is the combination of AI- and human-driven operations; cloud computing facilitates data exchange and easy communication between these entities.

(d) **Advanced Analytics and AI:** Access to strong analytics tools and AI services that can analyze large, complicated data sets and produce insights is made possible by cloud computing. In Industry 5.0 scenarios, these tools are critical for quality control, predictive maintenance, and optimization.

(e) **Cost-effectiveness:** Businesses can lower the capital costs related to maintaining on-premises technology and software by utilizing cloud infrastructure. Thanks to this cost-effectiveness, businesses can invest more in innovation and cutting-edge Industry 5.0 technologies.

(f) **Security and Compliance:** To safeguard sensitive data and guarantee compliance with industry laws, cloud providers provide strong security features and compliance solutions. In Industry 5.0, where data security and privacy are critical, this is essential.

(g) **Flexibility and Innovation:** New technologies may be quickly tested and implemented thanks to the cloud. Without having to make large upfront investments, businesses can quickly embrace and test new Industry 5.0 developments, creating an environment that is more inventive and nimble.

17.1.3 Blockchain

Industry 5.0 can be greatly improved by blockchain technology since it offers efficient, transparent, and safe methods of managing transactions and data [6–10]. How blockchain helps Industry 5.0 technology is as follows:

(a) **Enhanced Security:** Data is protected against alteration and illegal access thanks to the blockchain's decentralized and cryptographic structure. Blockchain contributes to data integrity and fraud prevention in Industry 5.0, where sensitive data from IoT devices, manufacturing processes, and AI systems are regularly exchanged.

(b) **Transparency and Traceability:** Blockchain technology offers an unchangeable ledger that records each transaction and modification and prevents them from being changed in the past. Ensuring compliance with norms and standards and tracking the lifespan of products from raw ingredients to completed items depend heavily on transparency.

(c) **Supply Chain Management:** By offering a single, shared version of the truth, blockchain technology can enhance supply chain visibility. This aids in monitoring the source of resources, confirming the legitimacy of goods, and cutting down on fraud and inefficiencies in the supply chain.

(d) **Data Integrity:** Blockchain guarantees the accuracy and immutability of data entered into the system. Maintaining the quality and dependability of data is essential for AI-driven decision-making, predictive maintenance, and sophisticated manufacturing processes.

(e) **Decentralization:** Blockchain runs on a distributed network of nodes, as opposed to conventional centralized databases. By ensuring that data and transactions are resistant to intrusions and system failures, this decentralization lowers the possibility of single points of failure.

(f) **Authentication and Identity Management:** Blockchain can improve security and expedite procedures for confirming an individual's identity. Blockchain technology can effectively handle digital identities and verify users and devices in Industry 5.0, where various stakeholders are involved.

17.1.4 Big Data Analytics

Big data analytics is essential to Industry 5.0 because it uses vast amounts of data to improve operational effectiveness, inform decisions, and develop more customized and sophisticated production processes. Big data analytics helps Industry 5.0 technologies in the following ways:

(a) **Predictive Maintenance:** By analyzing enormous volumes of data from machines and Internet of Things sensors, big data analytics can foresee equipment faults before they happen. Predictive maintenance helps minimize downtime, optimize maintenance schedules, and prolong the life of machinery by evaluating previous performance and real-time data.

(b) **Quality Control:** Big data analytics can be used to find trends and abnormalities that impact the quality of products by analyzing data gathered from different manufacturing stages. This enables modifications and enhancements in real time, guaranteeing that goods fulfill strict requirements and cutting down on waste.

(c) **Process Optimization:** By offering insights into manufacturing processes, big data analytics helps manufacturers find inefficiencies and streamline operations. Businesses may increase productivity, optimize processes, and save energy costs by evaluating data from sensors, supply networks, and production lines.

(d) **Customization and Personalization:** Industry 5.0 places a strong emphasis on producing individualized goods. By analyzing consumer preferences, habits, and feedback, big data analytics enables businesses to create personalized goods that live up to individual needs and expectations.

(e) **Supply Chain Management:** Better demand forecasting, inventory control, and logistical optimization are all made possible by analyzing data from all points of the supply chain. Real-time tracking and improved supplier collaboration are made possible by big data analytics, which lowers interruptions and boosts productivity.

17.1.5 6G and Beyond

Industry 5.0 is expected to undergo significant changes due to the upgraded connectivity, ultra-low latency, and improved data capacity offered by 6G and beyond [10–13]. These next-generation wireless technologies help Industry 5.0 in the following ways:

(a) **Ultra-Low Latency**
- 6G Capabilities: Offers incredibly low latency, maybe in the microsecond range.
- Encouragement of Industry 5.0: permits real-time communication between humans, machines, and sensors. This is critical for applications like synchronized human-machine cooperation, real-time quality control, and autonomous robotics that need instant response.

(b) **High-Speed Data Transfer**
- 6G Capabilities: Provides bandwidth and data speeds that are much higher than those of preceding generations.
- In favor of Industry 5.0: Allows for the quick transport of enormous amounts of data, which is necessary for applications like edge computing's complicated AI models, real-time data analytics, and high-definition video surveillance.

(c) **Massive Device Connectivity**
- 6G capabilities: A large number of connected devices can be supported per square kilometer.
- Support for Industry 5.0: Guarantees the smooth integration of a wide range of sensors, IoT devices, and intelligent machinery in production settings. Across the whole industrial ecosystem, complete data gathering, monitoring, and control are made possible by this interconnectedness.

(d) **Advanced Edge Computing**
- Using edge computing, 6G capabilities process data closer to the source.
- Encouragement of Industry 5.0: permits real-time data analysis at the edge, therefore lowering latency and bandwidth consumption. Applications like autonomous decision-making, predictive maintenance, and real-time process optimization depend on this.

(e) **Enhanced Security**
- 6G capabilities: Contains cutting-edge security mechanisms and features.
- Support for Industry 5.0: Uses strong encryption and security methods to shield confidential industrial data from online attacks. By doing this, devices can communicate securely while preserving their operational integrity and intellectual property.

(f) **Energy Efficiency**
- 6G Capabilities: Focuses on network operations and energy-efficient technology.
- Support for Industry 5.0: Lowers operating expenses and promotes sustainable production practices by reducing the energy consumption of communication networks.

(g) **Global and Reliable Coverage**
- 6G capabilities: Provides better coverage and connectivity in hard-to-reach or difficult-to-reach areas.
- Encouragement of Industry 5.0: Enables seamless operations and global data integration by guaranteeing consistent and dependable data communication across geographically scattered and remote industrial locations.

17.1.6 Internet of Things (IoT)

Industry 5.0 relies heavily on the Internet of Things (IoT) to link disparate systems and devices to facilitate smarter, more effective, and human-centered manufacturing processes. IoT helps Industry 5.0 technologies in the following ways:

(a) **Real-Time Data Collection**
- IoT Capabilities: Real-time data from machines, manufacturing lines, and environments is collected by sensors and devices.
- Encouragement of Industry 5.0: Gives constant insight into manufacturing processes, allowing for prompt modifications and guaranteeing that operations are optimized by the most recent data.

(b) **Predictive Maintenance**
- IoT Capabilities: IoT sensors keep an eye on the functionality and health of equipment.

- Encouragement of Industry 5.0: Uses data analysis to forecast possible equipment malfunctions before they happen. This minimizes maintenance expenses, increases the lifespan of machinery, and decreases unscheduled downtime.

(c) **Process Optimization**
- IoT Capabilities: Devices and sensors monitor a range of parameters about energy consumption, production efficiency, and quality.
- Encouragement of Industry 5.0: Makes use of data to find inefficiencies, streamline operations, and improve production as a whole. Better resource usage, decreased waste, and more productivity result from this.

(d) **Enhanced Quality Control**
- IoT Capabilities: Throughout the production process, monitors and logs quality metrics.
- Encouragement of Industry 5.0: Gives instant access to information on the quality of products, allowing for quick correction and guaranteeing that goods live up to expectations.

(e) **Personalization and Customization**
- IoT Capabilities: Compiles information on manufacturing needs and consumer preferences.
- Support for Industry 5.0: Makes it possible to create personalized goods using data collected in real time and input from customers. More adaptable production procedures that can meet specific requirements are made possible by this.

(f) **Supply Chain Management**
- IoT Capabilities: Keeps tabs on supply chain, logistics, and inventory activities.
- Encouragement of Industry 5.0: Improves inventory control, demand forecasting, and supply chain visibility and coordination while lowering disruptions.

(g) **Improved Human-Machine Interaction**
- IoT Capabilities: Uses real-time data and user-friendly interfaces to link human operators with equipment and systems.
- Encouragement of Industry 5.0: Improves decision-making, makes it easier for humans and machines to

work together, and makes sure that operators can communicate with and operate systems more effectively.

(h) Remote Monitoring and Control

- IoT capabilities: Makes it possible to access devices and systems remotely.
- Support for Industry 5.0: Enables remote manufacturing process monitoring and control, which is especially helpful for managing operations across several locations and addressing problems without the need for on-site presence.

(i) Integration with AI and Machine Learning

- IoT Capabilities: Offers data that machine learning and artificial intelligence algorithms can use.
- Encouragement of Industry 5.0: provides real-time data to algorithms, enhancing AI-driven insights and automation and resulting in more informed decisions, predictive analytics, and streamlined procedures.

17.2 Challenges of Industry 5.0

Industry 5.0, which blends cutting-edge technologies and prioritizes human-centric methods, presents several challenges in addition to the

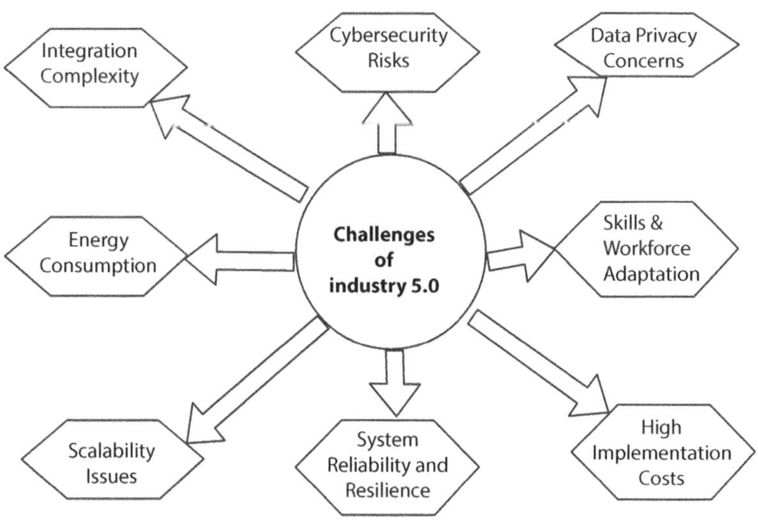

Figure 17.2 Industry 5.0 challenges.

potential for considerable breakthroughs in manufacturing and industrial processes. The following are some of the main challenges that Industry 5.0 faces as reflected in Figure 17.2.

17.2.1 Integration Complexity

- Challenge: Including cutting-edge technology like robotics, IoT, and AI in current systems can be expensive and time-consuming.
- Impact: Companies would have trouble attaining seamless interoperability and might have to make significant infrastructure upgrades.

17.2.2 Cybersecurity Risks

- Challenge: Systems are more vulnerable to cybersecurity threats because of Industry 5.0's enhanced connection and data interchange.
- Impact: Strong security measures and ongoing attention are necessary due to the increased risk of cyberattacks, data breaches, and industrial espionage.

17.2.3 Data Privacy Concerns

- Challenge: The collection and processing of massive volumes of data raises questions about data privacy and regulatory compliance.
- Impact: By navigating the complex data protection standards, organizations must ensure that sensitive and personal data is handled securely and morally.

17.2.4 Skills and Workforce Adaptation

- Challenge: The workforce needs new skills and abilities as a result of the shift to more advanced technologies.
- Impact: Businesses may need to close a skills gap by retraining employees, and hiring people with experience in developing technology, and other means.

17.2.5 High Implementation Costs

- Challenge: Investing a substantial amount of capital in new machinery, systems, and technologies can be necessary to implement Industry 5.0 technology.
- Impact: Small and medium-sized businesses (SMEs) can find it difficult to afford the cost of implementing these cutting-edge solutions.

17.2.6 System Reliability and Resilience

- Challenge: Keeping intricate, interconnected systems dependable and robust in a variety of scenarios is a challenge.
- Impact: To avoid having a significant negative influence on operations, systems must be built to elegantly accept errors and interruptions.

17.2.7 Scalability Issues

- Challenge: It can be difficult to scale Industry 5.0 solutions across many activities and geographical areas.
- Impact: In order to meet the diverse demands of various industries and business sizes, solutions must be flexible and scalable.

17.2.8 Energy Consumption

- Challenge: Enhanced connectivity and cutting-edge technologies may result in higher energy usage.
- Impact: To promote sustainability goals, businesses must take the environment into account and look for energy-efficient solutions.

17.3 Industry 5.0 Creative Applications

17.3.1 Intelligent Healthcare

ML models are being used by doctors these days to aid in patient diagnosis. This aids in increasing the precision of disease diagnosis, saving the patient a great deal of time and money. But given the circumstances, this is insufficient. The world urgently needs technology that can guarantee individualized patient requirements, such as tracking blood pressure, blood sugar levels, etc., and providing patients with individualized treatment with medical help. This can be made feasible *via* Industry 5.0. Smart watches, intelligent sensors, and other wearable technology can continuously record a patient's medical history in real time, and this information can be saved on the cloud. The patients' medical conditions can subsequently be diagnosed using ML algorithms. These smart gadgets can interact with one another. Should a medical professional be called upon, these devices can relay the patient's current condition and notify the physician to attend to the patient. By using cobots, doctors can perform surgery on patients with the assistance of robots that can converse with one another. These are just a few instances of how Industry 5.0 could transform the healthcare sector [14–16]. This revolution facilitates the production of implants, customized gadgets, etc. By Industry 5.0, everyday tasks such as doctor check-ups can be handled by cobots. Doctors can focus on higher-level work in this way.

17.3.2 Cloud Manufacturing

A unique approach to transforming the conventional manufacturing paradigm into an advanced manufacturing process is cloud manufacturing, which incorporates the newest technologies including virtualization, cloud and EC, IoT, and service-oriented technologies. Multinational stakeholders will work together to run an economical and effective manufacturing process in a cloud manufacturing setup. Reliability, excellent quality, cost-effectiveness, and on-demand capabilities are what set cloud manufacturing apart. Additionally, it benefits the environment because cloud manufacturing can do away with the need for raw materials to be sent across great distances for the manufacturing process. The next generation of cloud manufacturing systems, known as Industry 5.0, is anticipated to support increasingly complex and varied needs in the contexts of engineering, production, and logistics. The development of EC features, 5G-based telecommunication networks, and AI/ML technologies opens up new ways to greatly increase the capabilities of cloud manufacturing systems in the future [17–18].

17.3.3 Supply Chain Management

Disruptive technologies that facilitate Industry 5.0, such as cobots, ML, IoT, EC (edge computing), 5G, and beyond, DT (digital twins), and others, when combined with human intelligence and creativity, can assist companies in fulfilling demand and producing individualized and customized goods more quickly.

17.3.4 Manufacturing/Production

Industry 5.0 is the next stage of industrial development, emphasizing human-machine cooperation to create more efficient, sustainable, and customized manufacturing processes. In contrast to Industry 4.0, which placed a strong emphasis on automation and data interchange *via* IoT and AI technologies, Industry 5.0 restores human-centric innovation to the forefront. In the manufacturing sector, this entails fusing human ingenuity and problem-solving abilities with sophisticated robotics and artificial intelligence. As a result, production becomes more customized, with humans supervising and optimizing processes to ensure flexibility and adaptability while machines take care of repetitive chores. Furthermore, Industry 5.0 emphasizes social responsibility and sustainability heavily, pushing manufacturers to embrace eco-friendly procedures and make constructive contributions to society [19–22].

17.3.5 Smart Education

Industry 5.0, which combines cutting-edge technology with a human-centered approach to learning, has the potential to completely transform intelligent education. In contrast to other industrial revolutions that prioritized technology alone, Industry 5.0 highlights how artificial intelligence (AI) and human intelligence can work together to improve personalized learning. This applies to AI-driven teaching technologies in smart education that adjust to each student's unique learning preferences and requirements, delivering personalized lesson plans and immediate feedback. While AI takes care of administrative duties and data analysis, instructors can employ these tools to concentrate more on emotional intelligence, creativity, and critical thinking. Industry 5.0 also promotes sustainability and ethical technology use by encouraging the integration of social and ethical issues into schooling. Industry 5.0 seeks to build a more flexible, inclusive, and engaging learning environment that equips students for the challenges of the future by fusing technical innovation with human understanding.

17.3.6 Disaster Management

By combining cutting-edge technology with human knowledge to improve preparedness, response, and recovery activities, Industry 5.0 revolutionizes disaster management. These days, the emphasis is on developing intelligent systems that collaborate with people rather than just automating activities. Real-time data analysis and early warning systems are made possible by the increased accuracy with which robots and AI-driven tools can predict and monitor crises. Autonomous vehicles and drones can help with search and rescue efforts by entering places that might be too hazardous for people. Furthermore, Industry 5.0 highlights the significance of sustainability and social responsibility, promoting the creation of environmentally friendly solutions and resilient infrastructure that lessen the effects of disasters on communities. Industry 5.0 improves disaster management techniques by fusing technical innovations with human intuition and creativity. This makes disaster management tactics more effective, flexible, and adaptable to changing circumstances.

17.4 Potential Directions for Future Research

Prospective areas for further research on Industry 5.0 could include extending its impact across multiple sectors and expanding our understanding in several areas, as described below.

17.4.1 Human-Machine Collaboration

Looking at the seamless integration of robotics, advanced AI, and human intelligence across several industries. This involves researching the most effective methods of combining mechanical efficiency and human creativity and problem-solving to maximize results.

17.4.2 Ethical and Social Implications

Investigating the socioeconomic effects and moral dilemmas raised by Industry 5.0 technology. Research may concentrate on how Industry 5.0 may affect employment, privacy, and data security, as well as how to make sure that the advantages are shared fairly throughout society.

17.4.3 Sustainability and Green Technologies

Analyzing the potential contributions of Industry 5.0 to environmental sustainability. This entails creating environmentally friendly production methods, cutting waste, and investigating how circular economy ideas apply to smart manufacturing and other industries.

17.4.4 Personalization and Customization

Looking into how Industry 5.0 technology can make it possible for goods and services to be hyper-personalized. This study can concentrate on sectors where customized solutions can significantly enhance user experiences and results, such as healthcare, education, and retail.

17.4.5 Resilience and Adaptation

Investigating how Industry 5.0 may improve supply chains, healthcare, and disaster management, among other vital systems, in terms of resilience. The goal of research can be to create flexible systems that require little human intervention to respond to unforeseen obstacles and interruptions.

17.4.6 Interdisciplinary Applications

Increasing the amount of research being done on the potential applications of Industry 5.0 in smart cities, agriculture, and energy, among other domains. Recognizing how various industries may use human-machine cooperation to solve difficult issues and improve innovation.

17.4.7 Skill Development and Workforce Transition

To ensure that workers can adapt to new technology and prosper in a changing labor market, research can examine educational initiatives, training programs, and legislative interventions.

17.5 Conclusion

Industry 5.0, which combines human-centric approaches with cutting-edge technology like AI, robotics, and IoT, offers much potential. It improves sustainability, encourages creativity, and customizes output. Nonetheless, obstacles encompass moral dilemmas, changes in the workforce, and the requirement for strong cybersecurity. Attention must also be paid to the challenge of merging human and machine collaboration. Subsequent investigations ought to concentrate on tackling these obstacles, examining the possibilities of Industry 5.0 in other domains, and formulating structures that guarantee fair and enduring advantages for industries and the community as a whole. Industry 5.0 can realize its full potential by tackling these issues.

AI disclosure statement: The author acknowledges the use of AI-based tools for minor language editing or grammatical correction. However, all intellectual contributions, analyses, and conclusions presented in this chapter are entirely the author's own.

References

1. Varela-Vaca, Á., *et al.*, Towards the Realization of Industry 5.0: Human-Centric AI for Sustainable Manufacturing. *IEEE Access*, 11, 36578–36593, 2023, doi: 10.1109/ACCESS.2023.3251269.
2. Zhang, Y. and Li, W., Industry 5.0: Evolution, Impacts, and Future Prospects. *IEEE Trans. Eng. Manage.*, 70, 3, 1016–1029, 2023, doi: 10.1109/TEM.2023.3249854.
3. John, S. and Mehta, A., Sustainable Innovation in Industry 5.0: Integration of Digital Twins and Circular Economy. *IEEE Trans. Ind. Inf.*, 19, 2, 854–863, 2023, doi: 10.1109/TII.2023.3268975.
4. Ahmed, A. and Chen, Z., The Role of Artificial Intelligence in Industry 5.0: Opportunities and Challenges. *IEEE Trans. SMC Syst.*, 53, 4, 732–740, 2023, doi: 10.1109/TSMC.2023.3276871.
5. Sharma, R. and Gupta, S., Human-Centric Smart Factories: Bridging the Gap Between Industry 4.0 and Industry 5.0. *IEEE IoT J.*, 10, 7, 4876–4885, 2023, doi: 10.1109/JIOT.2023.3279856.
6. Na, X. and Wang, Y., Steps Toward Industry 5.0: Building '6S' Parallel Industries With Cyber-Physical-Social Intelligence. *IEEE/CAA J. Autom. Sin.*, 10, 6, 2390–2402, 2023, doi: 10.1109/JAS.2023.3267854.
7. Wang, F.-Y. and Yang, J., Industry 5.0: From Cyber-Physical Systems to Cyber-Physical-Social Systems. *IEEE Intell. Syst.*, 38, 2, 32–45, 2023, doi: 10.1109/MIS.2023.3247520.
8. Kaur, J. and Singh, S., Industry 5.0: Leveraging Human-Machine Collaboration for Enhanced Productivity. *IEEE Trans. Autom. Sci. Eng.*, 20, 4, 1515–1525, 2023, doi: 10.1109/TASE.2023.3281178.

9. Chen, H. and Liu, Y., Exploring Industry 5.0's Potential in Autonomous Manufacturing Systems. *IEEE Rob. Autom. Lett.*, 8, 1, 65–73, 2023, doi: 10.1109/LRA.2023.3245189.
10. Pérez, J. and Garcia, R., Human-in-the-Loop Automation: A Pathway to Industry 5.0. *IEEE Trans. Autom. Sci. Eng.*, 20, 5, 1725–1737, 2023, doi: 10.1109/TASE.2023.3290456.
11. Morales, R. and Chen, X., Challenges and Opportunities in the Transition to Industry 5.0: A Case Study in the Automotive Sector. *IEEE Access*, 11, 12345–12359, 2023, doi: 10.1109/ACCESS.2023.3280912.
12. Gupta, P. and Srivastava, R., AI-Driven Supply Chains: Industry 5.0 Perspectives. *IEEE Trans. Eng. Manage.*, 70, 4, 890–899, 2023, doi: 10.1109/TEM.2023.3257894.
13. Shukla, P. and Patel, R., Towards Green and Human-Centric Manufacturing: The Role of Industry 5.0. *IEEE Trans. Ind. Electron.*, 71, 3, 1456–1464, 2023, doi: 10.1109/TIE.2023.3245782.
14. Wang, L. and Zhang, Y., Leveraging Edge Computing in Industry 5.0 for Real-Time Decision Making. *IEEE Internet Things J.*, 10, 6, 4876–4888, 2023, doi: 10.1109/JIOT.2023.3284785.
15. Kumar, A. and Verma, P., Cybersecurity Challenges in Industry 5.0: A Review. *IEEE Secur. Privacy*, 21, 2, 30–38, 2023, doi: 10.1109/MSEC.2023.3287456.
16. Ali, M. and Khan, N., Industry 5.0 and the Circular Economy: A Path towards Sustainable Development. *IEEE Access*, 11, 22356–22368, 2023, doi: 10.1109/ACCESS.2023.3296712.
17. Peterson, D. and Moore, A., AI-Augmented Robotics in Industry 5.0: Enhancing Human-Robot Collaboration. *IEEE Trans. Rob.*, 39, 1, 175–188, 2023, doi: 10.1109/TRO.2023.3289125.
18. Roberts, J. and Hall, K., Human-Centric AI in Industry 5.0: From Concept to Implementation. *IEEE Intell. Syst.*, 38, 3, 56–66, 2023, doi: 10.1109/MIS.2023.3257810.
19. Rossi, M. and Bianchi, P., Integrating Digital Twins with Industry 5.0: A New Paradigm in Manufacturing. *IEEE Trans. Syst. Man Cybern. Syst.*, 53, 5, 1840–1850, 2023, doi: 10.1109/TSMC.2023.3279104.
20. Lee, C. and Kim, H., Industry 5.0: Synergies Between Human and Artificial Intelligence. *IEEE Trans. Neural Netw. Learn. Syst.*, 34, 3, 960–970, 2023, doi: 10.1109/TNNLS.2023.3285714.
21. Tripathi, S.L., Ganguli, S., Kumar, A., Magradze, T., *Intelligent Green Technologies for Sustainable Smart Cities*, Wiley-Scrivener Publishing LLC, USA, 2022, doi: 10.1002/9781119816096 https://onlinelibrary.wiley.com/doi/book/10.1002/9781119816096.
22. Thillaiarasu, N., Tripathi, S.L., Dhinakaran, V. (Eds.), *Artificial Intelligence for Internet of Things: Design Principle, Modernization, and Techniques*, 1st ed., CRC Press, Boca Raton, 2022, https://doi.org/10.1201/9781003335801.

18

Design and Simulation of an Audio Watch to Address the Problem Faced by Elderly People

Shahid Shabir*, **Krishan Arora and Anshul Mahajan**

School of Electronics and Electrical Engineering, Lovely Professional University, Phagwara, Punjab, India

Abstract

The research work "Design and simulation of an audio watch to eliminate the problem faced by elderly people" aims to provide a solution for people especially those who are elderly and are dependent on others to get access to the current time of the day. I came to know about this when I conducted different surveys around me. I conducted an offline survey, in which at least more than half of the people said they have people at home, or someone they know, who are experiencing this problem because of old age, blindness etc. Also, I conducted an online survey. I made Google forms and circulated them among my family, university groups and neighbourhood. Then, I asked for their suggestions, and after considering these suggestions, I decided to come up with a solution to make an audio watch which will help them in knowing the time. So far, I was able to do this in the English language. Now, I have accomplished the objective to create it in our own native language. Overall, the research work demonstrates the effectiveness of knowing the time through just the click of a button in the local Kashmiri language. The project can prove to be beneficial for those who don't understand any language other than Kashmiri.

***Keywords*:** Time, microprocessor, microcontroller, RTC, library, embedded system

**Corresponding author*: shahidshabir01@gmail.com

Abhishek Kumar and Suman Lata Tripathi (eds.) Smart Chips for Smart Devices: VLSI Design for Next-Generation IoT Solutions, (325–340) © 2026 Scrivener Publishing LLC

18.1 Introduction

A lot of elderly people in Kashmir are not able to determine the time on their own; they are dependent on others to know the time because they are either blind or don't understand any language other than Kashmiri. To overcome this problem, I am designing a watch, which, on just pressing a button will convey the time in the Kashmiri language. It will be an Arduino-based work. I will prepare a complete set of voice files in the Kashmiri language for every different hour of the day. Each set of voice files will be uniquely designated with some special code. After preparing a complete set of voice files, I will program a watch. A button will be provided on the watch. On pressing the button the clock's current time will be the input for the microprocessor and that current time will sync with the already stored voice files and it will play the time using a speaker.

- **Overview of embedded systems**

"Embedded" refers to something that is affixed to another object, as the word implies. One method of conceptualizing an embedded system is a piece of computer hardware with software integrated into it. An embedded system can function independently or as a part of a larger system. A microcontroller or microprocessor-based system built to carry out a certain function is called an embedded system. For example, an embedded device that can only detect smoke is a fire alarm. Figure 18.1 displays the general embedded system's block diagram.

By following a strategy to limit latencies, the Real-Time Operating System (RTOS) provides a means of enabling the processor to carry out a process in accordance with a timetable. The system's operation is defined by the RTOS. It establishes the guidelines when the application software is running. RTOS might not be present in a small-scale embedded system. Thus, an embedded system is a software-driven, microcontroller-based, dependable, real-time control system [2].

18.2 Characteristics of an Embedded System

Single-functioned: An embedded system typically carries out a certain task and repeats it. For instance: A pager always does what it's supposed to do.

- **Strictly limited:** Design metrics are restricted in all computing systems, but they can be particularly severe in embedded systems. The cost, size, power, and performance of an implementation are all measured by design metrics. It needs to be small enough to fit on a single chip, fast enough to process data instantly, and sufficiently power-efficient to prolong battery life.
- **Reactive and real time:** A lot of embedded systems have to compute specific outcomes instantly and continuously respond to changes in their surroundings. Take the example of a car cruise controller, which continuously scans and responds to brake and speed sensors. It must repeatedly calculate acceleration or deacceleration in a certain amount of time; if the computation is delayed, control over the automobile may be lost.

  It needs to be based on microprocessors or microcontrollers.
- **Memory:** Since its program typically embeds in ROM, it needs a memory. The computer doesn't require any secondary memory.

Connectivity: It has to have linked peripherals in order to connect input and output devices.

We can see this in Figure 18.1 [3].

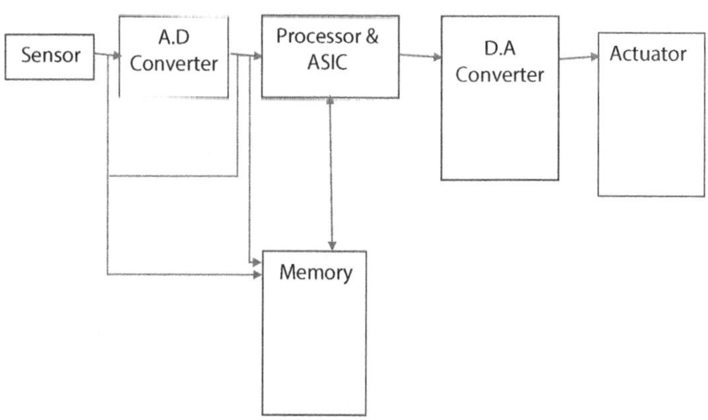

Figure 18.1 Block diagram of embedded systems.

- **Sensor** – When the physical quantity is measured, an electrical signal is generated that can be read by an observer or any electronic equipment, including an A2D converter. A sensor stores the measured quantity in the memory.
- **A-D Converter** – An analog-to-digital converter converts the analog signal from the sensor into a digital signal.
- **Processor and ASICs** – Processors process the data and store it in memory for measurement purposes.
- **D-A Converter** – The digital data that the processor receives is converted into analog data by a digital-to-analog converter.
- **Actuator** – An actuator saves the permitted output after comparing the D-A converter's output with the real (anticipated) output that was recorded in it.

▪ **Processors in embedded systems**

The processor is the main part of an embedded system. It is the essential part that receives inputs and processes data, and outputs the results. Designing embedded systems requires an understanding of both microprocessors and microcontrollers.

To retrieve instructions from memory, the CU has a fetch unit. Data conversion and data transfer directives are implemented by circuits throughout the European Union.

The Arithmetic and Logical Unit (ALU) and circuits that carry out directives for program control operations, such as interrupts or jumps to a different sequence of instructions, are included in the EU. The instructions are executed in the same order that they are acquired from memory by a processor that performs fetch cycles.

The following categories of processors are possible:

Microprocessor General Purpose Processor (GPP) Application Specific System Processor (ASSP), Digital signal processors, media processors, microcontroller embedded processors, application-specific instruction processors (ASIPs), GPP cores, or ASIP cores on either an Application-Specific Integrated.

A single VLSI chip with a CPU is called a microprocessor. Additionally, it might feature extra units that provide faster instruction processing, for coaches, an arithmetic unit for floating point processing, and pipelining units. Earlier generation microprocessors used a clock frequency of roughly 1 MHz to regulate the fetch-and-execute cycle. These days, processors operate at a frequency of 2GHz. Figure 18.2 displays the general block diagram.

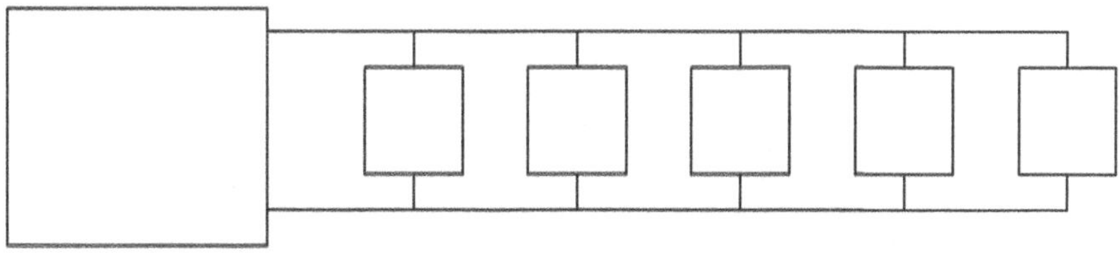

Figure 18.2 Block diagram of microprocessor [4].

Table 18.1 Microcontroller model.

RAM	CPU	ROM
TIM	I/O PORT	Serial COM Port

Often referred to as a microcomputer, a microcontroller is a single-chip VLSI device that has several on-chip functional units and improved input/output capacity despite having limited computing power. For real-time control applications like on-chip program memory and devices, microcontrollers are essential components of embedded systems. Table 18.1 illustrates the microcontroller model.

Arduino is an open-source tool for building computers that have greater physical sensing and control than desktop computers [1]. It is an open-source physical computing platform built on a simple microcontroller board and a programming environment for making software for the board. C or C++ is the programming language used to write the software. An implementation of wiring, the Arduino development board is a physical computing platform that is similar to the Arduino setup built around the Atmel microprocessor. This makes it simple and popular to use in a variety of do-it-yourself projects. The AVR processor only has an 8-bit core and runs at 16 MHz, with a very little amount of accessible memory consisting of 2 kilobytes of random access memory and 32 kilobytes of storage. Figure 18.3 displays the Arduino.

Many systems that were once built using a collection of intricate electronic circuits are now being replaced by embedded systems. A microcontroller is typically the central component of an embedded system. Arduino is one type of microcontroller. It is a prototype platform that is open source and used to detect and manage physical objects.

Figure 18.3 Arduino [5].

Arduino uses regular C/C++ (although simplified) as its programming language.
An Arduino sketch must have the setup () and loop () routines.
Configuration (): When your program starts, the function is invoked. It can be used to pin modes, initialize variables, begin using libraries, and more. The setup function will only run once, after each reset or power-up of the Arduino board.
Loop (): The loop () function performs exactly what its name implies, looping successively to enable your program to modify and react, after a setup () function that sets and initializes the initial variables has been created.

The Arduino board can be actively controlled with it.
Function declarations are the only context in which the void keyword is used. It signifies that the function should not return any data to the function that invoked it. Libraries are C or C++ files that provide your designs functionality (e.g., the ability to control an LED matrix, read an encoder, etc.). Additionally, they provide easier methods of controlling the hardware for which they are designed and expand the capabilities of the Arduino IDE. These allow you to store a value, and in order to instruct the computer on what type of value to anticipate, variables must be specified.

- **Introduction to Mechatronics**

A comprehensive technical method for creating electromechanical machinery and equipment is called mechatronics design. Equipment, including mechanical, electrical, electronic, and computer systems, is typically constructed as a collection of distinct systems. These systems are frequently composed of generic parts that work with any kind of machine, even ones that are different from one another. Systems are not viewed as distinct entities in mechatronics design; rather, they are integrated into the machine as a whole, with each component made to function best with every other component in that particular machine. Mechatronics design can thus create highly effective machinery to carry out almost any duty as efficiently as possible. Because each industry has various equipment requirements, mechatronics design is based on a wide range of concepts that differ from one another. Nonetheless, a prevalent notion in mechatronics design is the idea that each component of a machine should be designed with the machine's overall goal in mind, rather than just its own. In this manner, the performance of one component is not sacrificed in order to make up for that of another. For instance, every part of a machine made to operate in or near water would be made to be waterproof. When designing new equipment from the standpoint of mechatronics, engineers' education is also a significant consideration. The majority of engineers focus on just one field of study or competence, like electrical engineering. An engineer must be knowledgeable about every facet of the machine, from the computers and programming that will be used to operate it to the makeup of the materials that will be utilized in its construction, in order to pursue a true mechatronics design. Engineers that are knowledgeable in every facet of a machine's design and functioning can provide a variety of fresh approaches to long-standing issues, such as chemical fixes for computer issues or vice versa – this becomes accessible through the design of mechatronics. Many equipment manufacturers are looking for multi-disciplinary engineers with real-world experience in a variety of sectors because of the challenges associated with properly engineering machines and equipment under the mechatronics design approach. There aren't many engineers like that. As a result, new disciplines in engineering and education have emerged. In order to teach and train new kinds of design engineers, numerous colleges and universities are increasingly collaborating with different companies. Instead of earning degrees in a single traditional area, these engineers' degrees demonstrate their proficiency in all fields needed by a particular industry. Concurrent engineering ideas are supported by mechatronic design. Coordinating the relevant information and expertise across several specialist groups is essential when building a mechatronic device. Concurrent engineering is a design methodology wherein a product's manufacturing and design are uniquely integrated. There are no longer any conventional barriers between design and manufacture. There has been some impact from the knowledge that a large portion of the high production costs are established during the product design phase. Even throughout the design phase, quality, dependability, sales, life cycle performance, market analysis, and customer perception are all taken into consideration. Planning and product design happen at the same time.

With a focus on group decision-making, the organization's overall concurrent engineering mindset is ideal for team-oriented project management. The effective implementation of concurrent engineering requires coordinating adequate information exchange and resolving organizational barriers to cross-functional cooperation. Concurrent engineering has reduced the traditional divisions between design and manufacturing, but it has also made information transfer more difficult due to the lack of a common interface language. In a complex system, integrating sensors and control systems lowers capital costs, preserves a high level of flexibility, and increases machine utilization. The organization must begin with a long-term plan that can account for the evolving needs of process, data functions, control, and integration tools to create a concurrent engineering system that is mechatronic and capable of achieving these objectives. In addition to producing high-quality products, the mechatronic design process, sometimes referred to as life cycle design, focuses on maintaining them. The following describes a number of significant life cycle elements.

- Delivery: Cost, time, and mode
- Reliability: Materials, tolerances, and failure rate. Modular design ensures maintainability.
- Serviceability: modular design, prognostics, and onboard diagnostics.
- Upgradeability: The ability to use existing designs in the future.
- Disposability: Hazardous material recycling and disposal.

Three stages make up the mechatronic design process: modeling and simulation, prototyping, and deployment. All modeling should have a modular structure, regardless of whether it is based on more intricate physics or first principles (basic equations). A first principle model is a simple model that illustrates some of the fundamental behavior of a subsystem. An expansion of the basic principle model, a detailed model offers greater functionality and precision than the first-level model. Complex models can be created by connecting the modules, or blocks, together. Each block is a subsystem that corresponds to certain physically or functionally realizable operations, with input/output limited to input signals, parameters, and output signals. Although this restriction may not always be feasible or desirable, its implementation will result in modular subsystem blocks that are simple to maintain, can be used independently, can be swapped out for one another (detailed blocks can be swapped out for first principle blocks, and vice versa), and can be reused in different applications. The life cycle functions start after deployment, and data from these functions ought to be made accessible for product upgrades. Mechatronical systems are ideal for applications requiring reconfigurations due to their modularity. These products can be redesigned by changing different subsystem modules during the design phase or as they are being manufactured. Because information and other computer-based tasks are necessary for many of the steps in mechatronic design processes it is crucial to have an effective computer-aided prototyping environment, as well as fusion, management, and design testing. Orthopedic and endoscopic surgery. • Vehicles that are magnetically levitated. • Robotics for space applications and nuclear inspection.

18.3 Online Survey

I conducted an online survey by creating a Google form and then shared it among our family, neighbourhood and university circle. I was amazed as well as shocked to see how many people are dependent on others for knowing the real time. So, I turned that data into the pie charts shown in Figure 18.4 and Figure 18.5.

In applications where we need to keep track of the current time, such as data recording, clock timers, and alarms, real-time clocks (RTCs) like the DS3231 RTC are necessary. A timer that can record longer time periods, such as minutes or days, is integrated into the majority of microcontrollers, including the Arduino. Nevertheless, this timer only records the amount of time since the microcontroller was last powered on, thus it instantly resets to zero anytime the power is cut off. On the other hand, clocks are merely timepieces that, thanks to power backups, maintain the time and date even when the microcontroller's power source is cut off. An overview of the DS3231 RTC hardware is depicted in Figure 18.6.

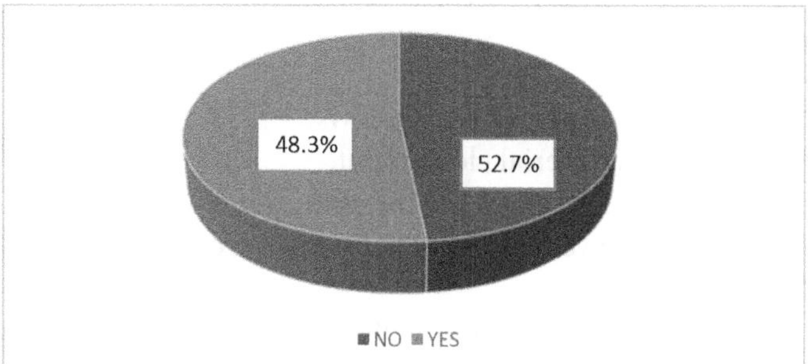

Figure 18.4 Results on the basis of online survey.

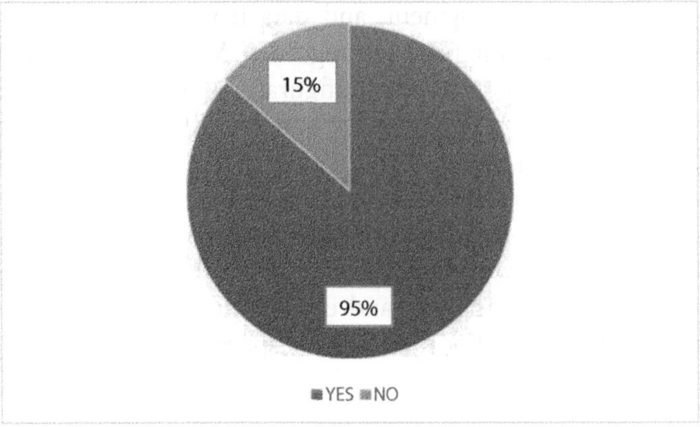

Figure 18.5 Results on the basis of offline survey.

Figure 18.6 RTC interfacing.

There are I/O pins on both sides of the DS3231 RTC module. Depending on the application, you can use either side of these pins, which have a similar arrangement.

The DS3231RTC RTC's pinout is as follows:
SCL: The I2C interface's serial clock input.
The I2C serial interface's serial data input/output is known as SDA.
VCC: The module's power supply, which ranges from 3.3V to 5V.
Ground pin, or GND.
When the primary power source is switched off, the backup 3V power supply, or BAT, is used to keep track of the time.

Square-wave frequencies of 1 Hz, 4 Hz, 8 Hz, or 32 Hz can be output using SQW.
32K: Provides a precise, temperature-compensated reference clock.

The DS3231 chip, a cheap and incredibly accurate RTC chip, is the main part of this RTC module. All timekeeping tasks are handled by it, and it is simple to interface with any microcontroller using the I2C standard. In Figure 18.7, the real-time clock, or RTC, is displayed. A 32kHz Temperature Compensated Crystal Oscillator (TCXO) integrated within the chip powers the DS3231 RTC. This Crystal Oscillator is more accurate in keeping time than conventional RTC modules because it is extremely resistant to external temperature fluctuations, which frequently alter oscillation frequency. Adjacent to the embedded crystal is a temperature sensor that adds or subtracts clock ticks to compensate for frequency variations and maintain accurate timekeeping.

The 32-byte 24C32 EEPROM chip on board can be utilized for storing settings or anything else. The 24C32 EEPROM shares an I2C bus with the DS3231 and communicates *via* an I2C interface. The three solder jumpers at the back, A0, A1, and A2, make it simple to adjust the EEPROM's I2C address. The address is hardcoded into each of these. The address is set if a jumper is soldered shorted.

A 20mm 3V lithium coin cell (any CR2032 battery would work fine) is stored in a battery backup at the bottom of the board to provide precise timekeeping in the event that the device's main power supply is cut off. Because your MCU can still keep track of time, you don't have to worry about power interruptions.

The Arduino Uno board will be connected to the DS3231 RTC, which communicates *via* I2C. GND to Arduino ground VCC to Arduino 5V SCL

Figure 18.7 RTC (real-time clock) [6].

to Pin A5 SDA to Pin A4. We must first update the DS3231 RTC module's time and date settings before we can use it. The configured time and date can be shown on the serial display by attaching the RTC module to an Arduino board and utilizing pre-made libraries. Because it uses your computer's time to automatically set the current date and time, the RTClib.h library is simple to use.

Step 1: Identify these components in Figure 18.8 to Figure 18.11.
Step 2: I connected these components as shown in Figure 18.11.

Figure 18.8 Amplifier pam8403.

Figure 18.9 Loudspeaker.

Figure 18.10 Arduino UNO.

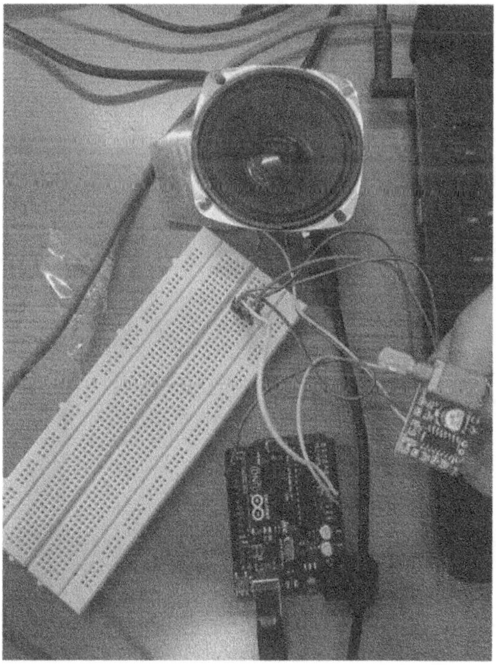

Figure 18.11 Interfacing of speaker with Arduino.

Step 3: I created a code out of an audio recording. To do this, follow the instructions below.

a) I obtained the Audacity program. This is a free and open-source digital audio editor and recording program as shown in Figure 18.12.
b) Now, launch this software and connect a microphone to the computer. Record your voice after that.
c) Next, convert the soundtrack to a 16-bit, single-track format.
d) After that, export this clip to any directory on your computer as a WAV file.

Step 4: Paste the sound clip code to the sample array.
Step 5: Choose board and port now. Upload this code to the Arduino board after that.

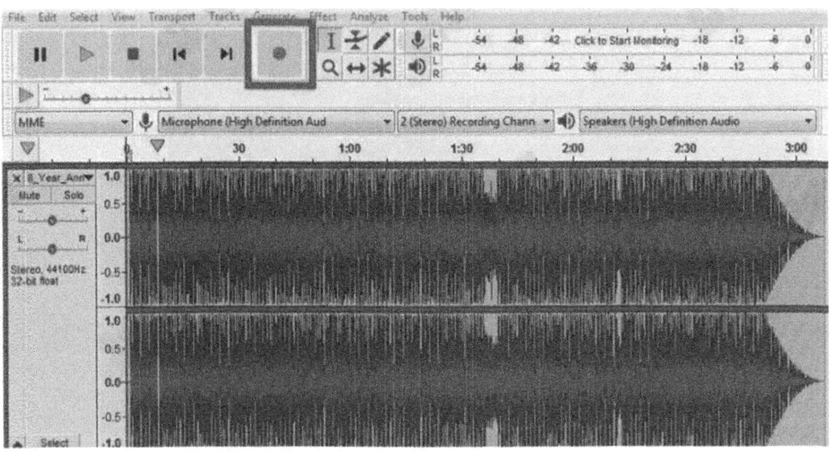

Figure 18.12 Audacity software.

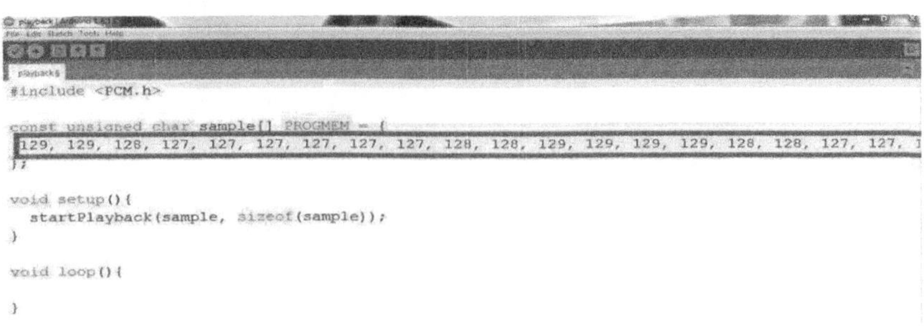

```
#include <PCM.h>

const unsigned char sample[] PROGMEM = {
129, 129, 128, 127, 127, 127, 127, 127, 127, 128, 128, 129, 129, 129, 129, 128, 128, 127, 127,
};

void setup(){
  startPlayback(sample, sizeof(sample));
}

void loop(){

}
```

Figure 18.13 Code.

Now, I'll use the talkie library to output a voice. Let's use this library's voice for that.

Use the circuit design and components mentioned above for this as well.

Step 1: Code is written as Figure 18.13.
Step 2: After selecting the board and port, upload this code to the Arduino board as in Figure 18.14.
Step 3: Figure 18.15 shows the begin of sumulation (run).

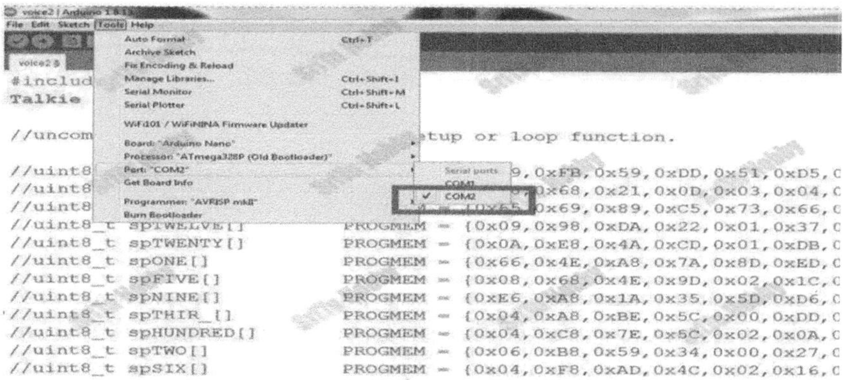

Figure 18.14 Selection of port.

Step 4: A interfaced diagram of the working system architectur is presented in Figure 18.16.

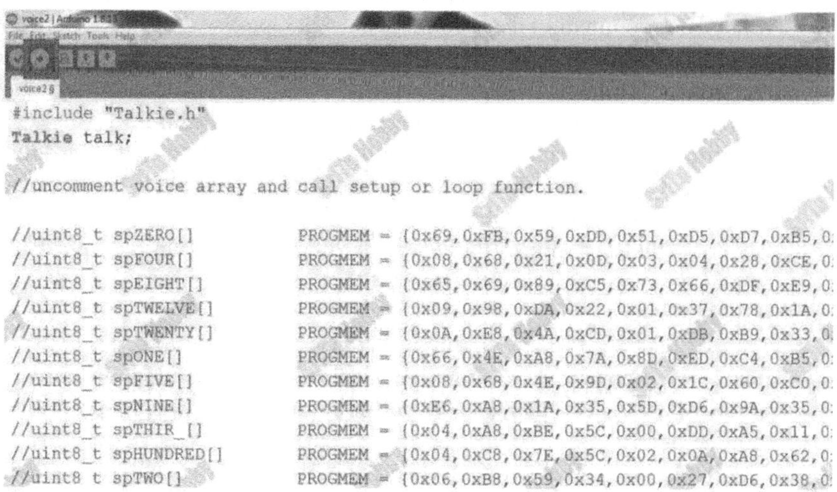

Figure 18.15 Run the code.

The schematic diagram is given below:

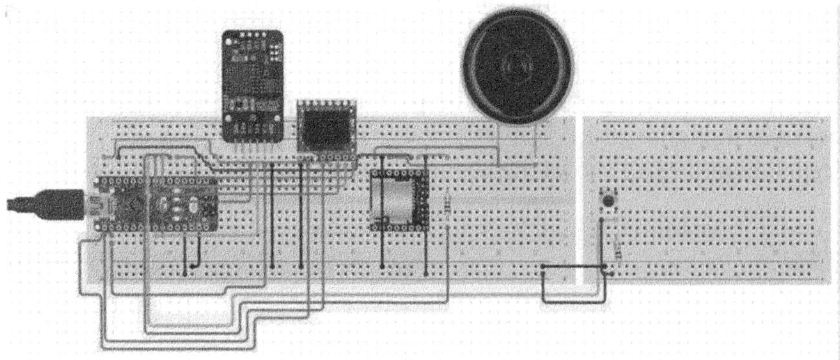

Figure 18.16 Schematic of the circuit.

AI disclosure statement: The author acknowledges the use of AI-based tools for minor language editing or grammatical correction. However, all intellectual contributions, analyses, and conclusions presented in this chapter are entirely the author's own.

References

1. Arduino Uno - R3, Wikipedia, [Online]. Available: https://en.wikipedia.org/wiki/Arduino#/media/File:Arduino_Uno_-_R3.jpg.
2. DS3231 AT24C32 IIC Precision RTC Real-Time Clock Module, Kuongshun Electronics [Online]. Available: https://www.kuongshun-ks.com/uno/uno-board-shield/ds3231-at24c32-iic-precision-rtc-real-time.html.
3. Arduino, Arduino Nano, Arduino [Online]. Available: https://store.arduino.cc/products/arduino-nano?srsltid=AfmBOorp-Sr2-jFiOiWJfJFbbkwguIsy-royJhp97bKateR7O8FSQhilv.
4. Generic, DFPlayer Mini MP3 Player Module, Amazon [Online]. Available: https://www.amazon.in/Generic-DFPlayer-Player-Module-Arduino/dp/B0130LMJ4K.
5. Type-C TP4056 1S 3.7V Lithium Battery Charging Protection Module Board, MyInnovation, [Online]. Available: https://www.myinnovation.in/product-page/type-c-tp4056-1s-3-7volt-lithium-battery-charging-protection-module-board.
6. Mazdi, M., The 8051 Microcontrollers & Embedded Systems, 1 January 2007.

Index

Also of Interest

By the same editors

Waste Management and Utilization for a Sustainable Environment, edited by Amit Srivastava and Suman Lata Tripathi, ISBN: 9781394292974. Waste Management and Utilization for a Sustainable Environment is an essential guide to understanding how cutting-edge technologies and innovative engineering solutions can transform waste management into a cornerstone of a sustainable future.

Energy-Efficient Communication Networks, edited by Shakti Raj Chopra, Krishan Arora, Suman Lata Tripathi, and Vikram Kumar, ISBN: 9781394271658. *Energy-Efficient Communication Networks* is essential for anyone looking to understand and implement cutting-edge energy optimization strategies for communication systems, ensuring they meet growing energy demands while seamlessly integrating renewable energy sources and enhancing battery life in embedded applications.

Smart Textiles and Wearables for Health and Fitness, edited by Jyotirmoy Pathak, Abhishek Kumar, Suman Lata Tripathi, and Balwinder Raj, ISBN: 9781394302949. *Smart Textiles and Wearables for Health: The Flexible Electronics Revolution* provides an in-depth exploration of how innovative technologies and materials are reshaping healthcare, making it an essential resource for anyone looking to understand the transformative power of smart textiles and wearables in patient monitoring, diagnosis, and rehabilitation.

Development of 6G Networks and Technology, edited by Suman Lata Tripathi, Mufti Mahmood, C. Narmadha, and S. Albert Alexander, ISBN: 9781394230655. *Development of 6G Networks and Technology* provides an in-depth exploration of the potential impact of 6G networks on various industries, including healthcare, agriculture, transport, and national security, making it an essential resource for researchers, scholars, and students working in the field of wireless networks and high-speed data processing systems.

Integrated Devices for Artificial Intelligence and VLSI: VLSI Design, Simulation and Applications, edited by Balwinder Raj, Suman Lata Tripathi, Tarun Chaudhary, K. Srinivasa Rao, and Mandeep Singh, ISBN: 9781394204359. With its in-depth exploration of the close connection between microelectronics, AI, and VLSI technology, this book offers valuable insights into the cutting-edge techniques and tools used in VLSI design automation, making it an essential resource for anyone seeking to stay ahead in the rapidly evolving field of VLSI design.

Decentralized Systems And Distributed Computing, edited by Sandhya Avasthi, Suman Lata Tripathi, Namrata Dhanda, and Satya Bhushan Verma, ISBN: 978139420436. This book provides a comprehensive exploration of next-generation internet, distributed systems, and distributed computing, offering valuable insights into their impact on society and the future of technology.

Electric Vehicle Design: Design, Simulation and Applications, edited by Krishan Arora, Suman Lata Tripathi, and Himashu Sharma, ISBN: 9781394204373. This book will serve as a definitive guide to conceptual and practical knowledge about the design of hybrid electrical vehicles (HEV), battery electrical vehicles (BEV), fuel cell electrical vehicles (FCEV), plug-in hybrid electrical vehicles (PHEV), and efficient EV charging techniques with advanced tools and methodologies for students, engineers, and academics alike.

Industrial Control Systems, edited by Vipin Chandra Pal, Suman Lata Tripathi, and Souvik Ganguli, ISBN: 9781119829256. This volume serves as a comprehensive guide in the journey of industrial control systems with a multidisciplinary approach to the key engineering problems in the 21st century.

Modern Automotive Electrical Systems: Theory and Applications, edited by Pedram Asef, Sanjeevikumar Padmanaban, and Andrew Lapthorn, ISBN: 1119801047. Presenting the concepts and advances of modern automotive electrical systems, this volume, written and edited by a global team of experts, also goes into the practical applications for the engineer, student, and other industry professionals.

Nanodevices For Integrated Circuit Design, edited by Suman Lata Tripathi, Abhishek Kumar, K. Srinivasa Rao, and Prasantha R. Mudimela, ISBN: 9781394185788. Written and edited by a team of experts in the field, this important new volume broadly covers the design of nano-devices and their integrated applications in digital and analog integrated circuits (IC) design.

Explainable Machine Learning Models and Architectures: Real-Time System Implementation, edited by Suman Lata Tripathi and Mufti Mahmud, ISBN: 9781394185849. This cutting-edge new volume covers the hardware architecture implementation, the software implementation approach, and the efficient hardware of machine learning applications.

Machine Learning Techniques for VLSI Chip Design, edited by Abhishek Kumar, Suman Lata Tripathi, and K. Srinivasa Rao, ISBN: 9781119910398. This cutting-edge new volume covers the hardware architecture implementation, the software implementation approach, and the efficient hardware of machine learning applications with FPGA or CMOS circuits, and many other aspects and applications of machine learning techniques for VLSI chip design.

Intelligent Green Technologies for Smart Cities, edited by Suman Lata Tripathi, Souvik Ganguli, Abhishek Kumar, and Tengiz Magradze, ISBN: 9781119816065. Presenting the concepts and fundamentals of smart cities and developing "green" technologies, this volume, written and edited by a global team of experts, also goes into the practical applications that can be utilized across multiple disciplines and industries, for both the engineer and the student.

Design and Development of Efficient Energy Systems, edited by Suman Lata Tripathi, Dushyant Kumar Singh, Sanjeevikumar Padmanaban, and P. Raja, ISBN 9781119761631. Covering the concepts and fundamentals of efficient energy systems, this volume, written and edited by a global team of experts, also goes into the practical applications that can be utilized across multiple industries, for both the engineer and the student.

Electrical and Electronic Devices, Circuits, and Materials: Technical Challenges and Solutions, edited by Suman Lata Tripathi, Parvej Ahmad Alvi, and Umashankar Subramaniam, ISBN 9781119750369. Covering every aspect of the design and improvement needed for solid-state electronic devices and circuit and their reliability issues, this new volume also includes overall system design for all kinds of analog and digital applications and developments in power systems.

Green Energy: Solar Energy, Photovoltaics, and Smart Cities, edited by Suman Lata Tripathi and Sanjeevikumar Padmanaban, ISBN 9781119760764. Covering the concepts and fundamentals of green energy, this volume, written and edited by a global team of experts, also goes into the practical applications that can be utilized across multiple industries, for both the engineer and the student.

Check out these other related titles from Scrivener Publishing

Hydrogen Electrical Vehicles, edited by Mehmet Sankir and Nurdan Demirci Sankir, ISBN: 9781394166381. Hydrogen electrical vehicles are an essential component of the "Green New Deal" and this book covers cutting-edge technologies designed for fuel-cell-powered cars.

Pavement Asset Management, by Ralph Haas, W.R. Hudson, and Lynne Cowe Falls, ISBN: 9781119038702. Whether you are new to managing pavement assets or a seasoned practitioner, this book provides the most up-to-date information in the field on the history, technology, practice and look ahead.

Printed and bound by CPI Group (UK) Ltd, Croydon, CR0 4YY
17/09/2026
14964264-0003